Zero Trust Architecture

Identity, Microsegmentation, and Continuous Verification for Modern Hybrid Enterprises

Second Edition

Jordan O'Neal

Cybersoft Publishing LLC

Fort Washington, MD 20744

Drclaude.net

Second Edition, March 2026

Table of Contents

Author's Note

Zero trust is one of the most invoked phrases in enterprise security today, and one of the most poorly understood. Vendor marketing has wrapped it in product bundles and platform branding, making the underlying architectural logic nearly invisible. Meanwhile, the real pressure building inside hybrid enterprises is not coming from marketing decks. It is coming from auditors, federal regulators, breach post-mortems, and the growing recognition that perimeter-based security has been operating on borrowed time for a decade.

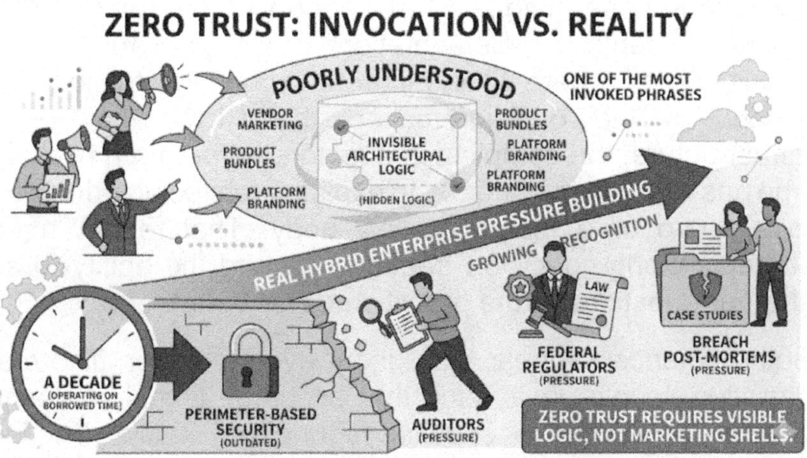

This book exists because that gap needs to be closed. The gap between what zero trust is sold as and what zero trust architecture actually requires is significant, and the cost of misunderstanding it compounds every year. Lateral movement is now the dominant breach pattern. An attacker who gains a foothold inside a trusted network can traverse enormous distances before any control triggers. The organizations suffering the worst outcomes are not necessarily the ones with the weakest perimeters; they are the ones that never built meaningful internal controls because they trusted the perimeter to do all the work.

Federal mandates have accelerated the urgency. Executive Order 14028, OMB M-22-09, FISMA modernization guidance, and CMMC have placed zero-trust architecture on the compliance agenda for any organization that touches federal contracts or federal data. Commercial frameworks, including HIPAA, PCI-DSS 4.0, SOX, and SOC 2, have followed closely behind, each tightening expectations around identity, access, and continuous verification. The organizations reading this book are operating inside that pressure, not outside it.

I wrote this book for two kinds of readers who are usually in the same room but rarely read the same material. The first is the practitioner: the security architect, the network engineer, the identity team lead, the person responsible for actually building and operating zero trust controls. That reader needs technical depth, implementation guidance, and honest discussion of where theory meets production friction. The second is the decision-maker: the CISO, the VP of Infrastructure, the compliance director, the program manager who has to allocate budget and explain architectural decisions to a board or a federal review body. That reader needs strategic clarity, decision frameworks, and the ability to ask sharper questions of the practitioners on their team.

Both audiences appear throughout every chapter. The main narrative speaks in a voice that assumes technical fluency without assuming a specific technology stack. Sections marked "Geeks Only" go deeper on protocols, configurations, and implementation specifics for readers who want that level of detail. "Manager's Decision Guide" sections at the end of each chapter translate the architectural concepts into the criteria and trade-offs that non-technical leaders need to make sound decisions. Neither track talks down to its audience. Engineers do not need acronym definitions; managers do not need condescension.

The protagonist throughout the book is Marcus Reyes, a Security Architect at Meridian Logistics Group, a hybrid enterprise with three on-premises data centers, two cloud tenants, and roughly 4,000 employees supported by a rotating contractor ecosystem. Marcus is not a superhero. He inherits an environment built by people who made reasonable decisions for their time, and he has to move it toward a zero-trust architecture while keeping the business running. The scenarios he navigates are composites drawn from real hybrid enterprise deployments. They are designed to surface the actual decision points and failure modes, not the ideal-condition case studies that populate vendor whitepapers.

The three pillars organizing the book are identity, microsegmentation, and continuous verification. Every access decision begins with verified identity. Microsegmentation limits the blast radius when something goes wrong, accepting that containment is more reliable than prevention. Continuous verification treats trust as a variable that must be re-earned with each transaction, rather than as a binary state granted at login. These pillars do not compete; they reinforce each other, and the later chapters show how they compound when implemented together.

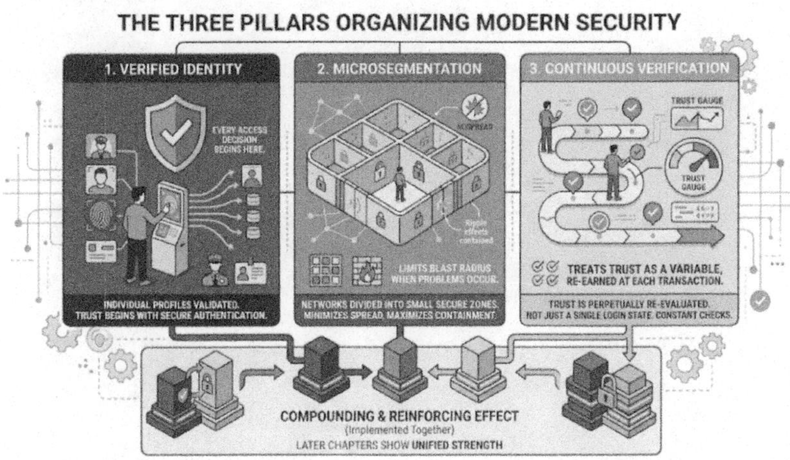

The chapters are designed to be read for the full arc, but each stands on its own as a targeted reference. If you are in the middle of an identity deployment and you need to work through the trade-offs right now, start there. If you are preparing for a federal compliance review, the chapters covering EO 14028 alignment and NIST 800-207 implementation will give you what you need without requiring you to read everything before them. The book will be waiting for the rest of the journey whenever you are ready.

1 Introduction

Marcus Reyes did not expect a Tuesday morning in February to reframe his understanding of enterprise security. The incident his team was reviewing had started the previous Friday: a contractor account with valid credentials, a flat-trusted interior, and 12 hours of undetected lateral movement across Meridian Logistics Group's hybrid environment. By the time the anomaly detection platform flagged the activity, the attacker had touched three on-premises data centers and one cloud tenant. No perimeter alarm had fired. None was designed to. The perimeter had done exactly what it was built to do: keep outsiders out, and it had done nothing about the insider moving freely across a trusted interior that no one had ever designed to be untrusted. That Tuesday morning, Marcus began working backward from the incident to the architecture, and the architecture kept pointing to the same answer: once implicit trust was granted, it became the attack surface.

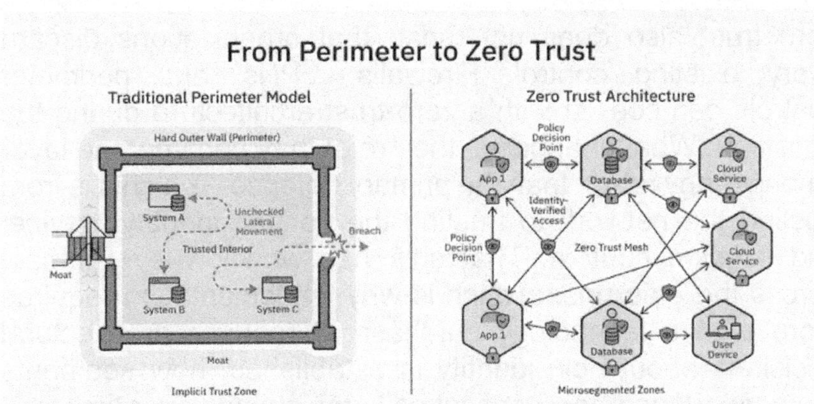

From Perimeter to Zero Trust

1.1 What Zero Trust Actually Is

Zero trust is an architecture, not a product. That distinction matters more than any other sentence in this book. No vendor ships a box that makes an enterprise zero trust. No platform migration completes the journey on its own. Zero trust is a set of architectural principles that govern how access decisions are made, enforced, and continuously re-evaluated across an enterprise environment. The principles originate in NIST Special Publication 800-207, which defines zero trust as an evolving set of cybersecurity paradigms that move defenses from static, network-based perimeters to focus on users, assets, and resources.

The most common misreading of zero trust is treating it as a synonym for "never trust." That framing is too blunt to be useful. The actual principle is "never trust implicitly, always verify explicitly." The difference is significant. Access is still granted, but only based on verified identity, device health, request context, and policy, rather than on network location. A user inside the corporate firewall gets no more inherent trust than a user connecting from a coffee shop. A contractor who has passed identity verification gets exactly the level of access the policy permits and no more. The architecture assumes that any segment of the network may already be compromised, and it designs access controls accordingly.

Zero trust also does not mean that organizations discard every existing control. Firewalls, VPNs, and perimeter controls can coexist with a zero-trust architecture during the transition. What changes is their role: they become one layer among many rather than the primary defense. The shift is from trusting the network to trusting the verified identity, device, and session context. That shift has workflow-level impact across the enterprise, which is why implementation requires more than a technology purchase. It requires architectural decisions about how identity is established, how segments are defined, and how verification is maintained over time.

1.2 The Three Pillars

Identity is the first pillar and the foundation of every zero-trust access decision. Every access request begins with the question of who or what is making the request, and the answer must be verifiable rather than assumed. For Marcus at

Meridian, the contractor account that enabled the February breach had valid credentials but had never been subjected to the kind of continuous authentication and behavioral analysis that would have flagged the anomalous access pattern. Identity in a zero-trust architecture means more than a username and password. It means multi-factor authentication, device enrollment, session risk scoring, and the ability to revoke or step up authentication at any point in a session when context changes.

Microsegmentation is the second pillar, and it directly addresses the problem Marcus encountered: once inside, movement should be hard. Microsegmentation divides the network interior into discrete zones with enforced boundaries, so that a credential compromise in one zone does not automatically grant access to adjacent zones. The attacker who breached Meridian moved through twelve hours of flat network space because the interior trusted itself. Microsegmentation is not primarily about prevention; it is about containment. The working assumption is that breaches will occur, and the architectural goal is to limit the blast radius to the smallest possible scope. A well-segmented environment turns a potential catastrophe into a contained incident.

Continuous verification is the third pillar, transforming trust from a binary state into a variable that is re-evaluated throughout each session. Traditional access models grant trust at login and hold it until logout. Zero-trust architecture re-evaluates trust at every significant transaction, checking whether device health has changed, whether the request aligns with the user's behavioral profile, and whether the accessed data is sensitive. This ongoing evaluation is what NIST 800-207 calls the policy decision point function, and it is the mechanism that makes zero trust adaptive rather than static. For an organization like Meridian, with its hybrid mix of on-premises and cloud assets, continuous verification is what closes the gap between the authentication moment and the access event.

The Three Pillars Working Together

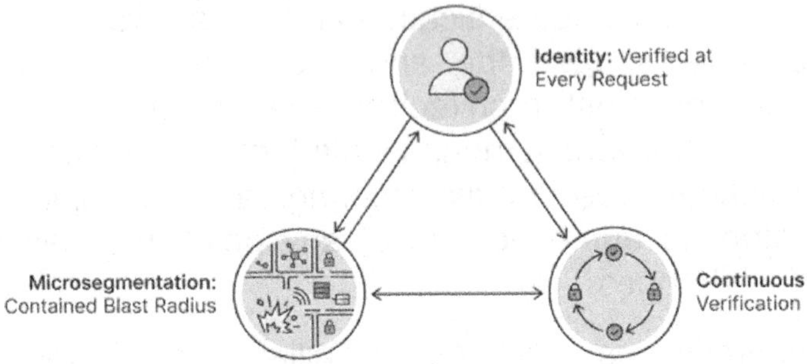

The Three Pillars Working Together

1.3 Who This Book Is For

This book is written for practitioners and decision-makers responsible for actually moving a real organization toward a zero-trust architecture. It is not a vendor-neutral overview for an audience with no context. It assumes that readers already operate inside complex environments with legacy infrastructure, budget constraints, compliance deadlines, and executive stakeholders who want clear answers to unclear questions. The gap it addresses is not a knowledge gap about what zero trust is in theory. It is a practical gap between the architectural principles and the decisions that have to be made in real infrastructure, with real constraints and under real timelines.

The practitioner audience includes security architects like Marcus, network engineers making segmentation decisions, identity and access management teams configuring authentication policies, and DevSecOps professionals embedding security controls into deployment pipelines. These readers need depth of implementation, and they will find it in the main narrative and in the "Geeks Only" sections that appear in each chapter. Those sections go into protocol-level detail, configuration patterns, and the specific trade-offs that arise when ideal architecture meets production constraints. They cover topics including SAML and OIDC federation across hybrid boundaries, software-defined perimeter implementations, microsegmentation tagging strategies for containerized workloads, and the behavioral analytics that power continuous verification at scale. The practitioner does not need to be told that security matters; they need guidance on where the difficult decisions actually live.

The decision-maker audience includes CISOs, VPs of Infrastructure, compliance directors, program managers, and federal-adjacent contractors who need to understand what zero trust architecture requires in terms of investment, organizational change, and governance. These readers will find the main narrative accessible and the "Manager's Decision Guide" sections at the end of each chapter directly applicable. Those sections translate architectural concepts into decision criteria: what to ask, what to watch for, and how to evaluate whether an implementation is on the right track. Mission-aligned decision-making requires understanding the architectural logic, not just the product catalog. A leader who can read a vendor proposal and identify which zero trust principles it actually addresses and which ones it leaves unresolved is a far more effective steward of security investment than one who relies entirely on vendor-provided evaluation criteria.

Federal-adjacent organizations face a particular kind of dual pressure: the technical requirements of NIST 800-207 and EO 14028 on one side, and the operational reality of running a hybrid enterprise on the other. This book addresses both sides of that pressure throughout every chapter, rather than compartmentalizing federal guidance into a separate chapter that most readers will skip. The federal mandate requirements and the commercial compliance framework requirements are treated as context throughout, referenced where they are directly relevant to the architectural decisions at hand, rather than isolated in a compliance appendix that loses its connection to the technical reasoning.

Readers who sit between these two audiences, such as security managers who came up through engineering or architects who have taken on program management responsibilities, will find that the dual-track structure maps well to their dual roles. The main narrative serves as the connective tissue between the technical details and the organizational decision. The Geeks Only sections and the Manager's Decision Guide sections are complements, not substitutes, and readers with breadth across both dimensions are encouraged to use both.

1.4 How to Read This Book

The chapters build on each other in the order presented, moving from the foundational trust crisis through identity architecture, microsegmentation design, continuous verification implementation, federal compliance alignment, and the operational patterns that sustain a zero trust program over time. Reading in order gives the fullest picture of how the three pillars interact and compound. The arc of Marcus's work at Meridian runs through every chapter, providing a consistent reference environment for the technical and organizational decisions addressed. Meridian's hybrid environment, with its three on-premises data centers, two cloud tenants, and a contractor ecosystem, is designed to reflect the infrastructure reality most enterprise security programs actually face, not the clean, greenfield deployment that vendor architecture diagrams often assume.

At the same time, each chapter is designed to function as a targeted reference. If your organization is mid-deployment on an identity initiative, the identity chapters will give you what you need without requiring that you revisit the foundational material. If you are preparing for a federal compliance review under OMB M-22-09, the chapters on mandate alignment are self-contained and can be used directly. The pilot-to-production pathway runs through the book as a structural thread, meaning that each chapter addresses not just the design but the operational transition from controlled pilot to enterprise-wide enforcement, so a reader who picks up a chapter in the middle will find both the architectural logic and the implementation guidance in the same place. Each pilot-to-production section includes the specific success criteria that should gate a transition from limited scope to broader rollout, because the most common implementation failure mode is moving to production scale before the pilot has generated reliable data.

Each chapter ends with a set of key takeaways that distill the main architectural decisions and trade-offs covered. These are useful as a review for readers working through the book in sequence, and as an orientation for readers coming to a chapter directly. They are written to be shared with colleagues who have not read the full chapter, so that a practitioner who has worked through the technical content can hand a colleague the takeaways and have a productive conversation about next steps without requiring that colleague to read the full chapter first.

The "Geeks Only" sections are clearly marked and can be skipped by readers whose responsibilities are organizational rather than technical. The "Manager's Decision Guide" sections are equally self-contained and are worth reading even for practitioners, because they surface the questions that executives and auditors will ask and that practitioners need to be ready to answer. A practitioner who understands the decision framework a CISO uses can frame their technical recommendations more effectively. A manager who understands what a Geeks Only section covers can ask better questions when reviewing an implementation proposal, even without reading the technical details directly.

1.5 What You Will Walk Away With

By the end of this book, you will have a working architectural framework for zero trust that is grounded in NIST 800-207, applicable across hybrid enterprise environments, and aligned with the federal mandates and commercial compliance frameworks that govern most of the organizations likely to be reading it. The framework is responsible by design, meaning it builds security controls into architectural decisions rather than layering them on afterward. It provides the operational clarity to distinguish between implementations that satisfy audit requirements and implementations that actually reduce risk. And it connects the three pillars into a coherent program rather than a collection of point solutions, so that the identity work, the segmentation work, and the continuous verification work reinforce each other rather than operating in parallel silos.

You will also have a concrete reference environment at Meridian Logistics Group that grounds every architectural concept in the hybrid infrastructure reality most enterprises actually operate in. The decisions Marcus makes, the trade-offs his team navigates, the organizational friction he encounters when asking established teams to change long-standing practices: these are not pedagogical constructs. They are the actual texture of zero trust implementation, and walking through them with a consistent protagonist makes it possible to see how decisions in one domain, such as identity federation, create both opportunities and constraints in adjacent domains, such as microsegmentation policy enforcement. The book aims to leave you with the ability to reason through that texture in your own environment, not just to apply a template.

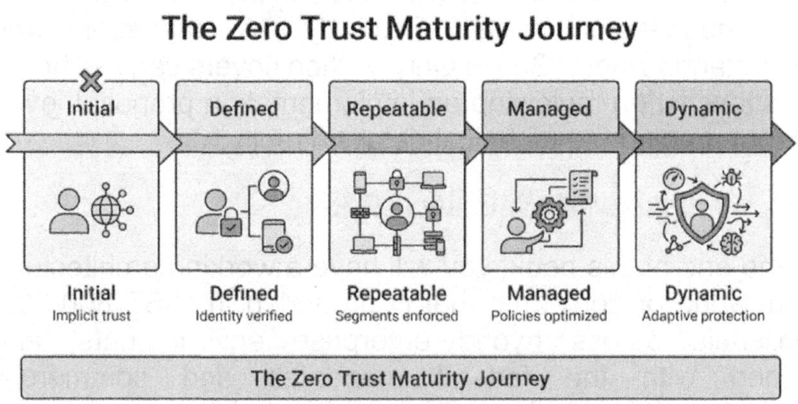

Chapter 1 begins where Marcus began: at the incident review that forced a reckoning with the limits of perimeter-based security. The trust crisis he encountered at Meridian Logistics Group is not unique to Meridian. It is the defining architectural problem of the hybrid enterprise era, and the work of resolving it starts with understanding exactly how implicit trust became an attack surface in the first place.

2 The Trust Crisis in Hybrid Enterprises

Marcus Reyes arrived at the Monday morning incident review with a cup of coffee and a reasonable expectation of a quiet week. He had been in the Security Architect role at Meridian Logistics Group for only four months, having been recently promoted from Senior Network Engineer after his predecessor's departure. He was still building his mental map of the organization's hybrid environment. Meridian was the kind of company security professionals called "sprawling": three on-premises data centers, two cloud tenants in separate regions, a remote workforce of roughly four thousand employees, and a contractor ecosystem of several hundred firms whose personnel came and went with project cycles. The Monday review was supposed to cover routine patch compliance metrics and a minor result from a phishing simulation. What it actually uncovered changed Marcus's professional understanding of the threat landscape permanently.

The incident had been flagged late Friday by an automated anomaly alert from the network traffic analysis platform. A contractor account associated with a facilities management firm had been pulling database queries at two in the morning, targeting records that had nothing to do with facilities management: employee payroll data, client shipping contracts, and a directory of partner access credentials stored in a legacy file share that had not been migrated to the new secrets management system. The account was valid. The VPN session originated from a registered contractor device. The credentials had passed multi-factor authentication the first time they were used, forty-seven days earlier, and had never been challenged again. In those forty-seven days, the session token had been quietly reused across seventeen distinct network segments, moving through the flat internal architecture like water through a cracked foundation, finding no walls because no walls had been built.

Marcus spent three hours in that Monday meeting reconstructing the timeline on a whiteboard. The contractor whose credentials were used had left the project thirty-one days prior. Their access had been deprovisioned from the contractor portal, but not from the VPN gateway, which ran on a separate identity store that synchronized weekly rather than in real time. The gap between those two systems, a gap nobody had formally owned or documented, had been wide enough for an attacker to park inside the network for over a month. Marcus looked at the network diagram pinned to the conference room wall and, with uncomfortable clarity, understood that there was no diagram in the building showing the real trust boundaries. The diagram showed topology. It said nothing about which devices trusted which other devices, which accounts had access to which resources, or what would happen if any single credential anywhere in the chain was silently handed to an adversary.

This moment, an incident review that began as a routine Monday exercise and ended with a fundamental architectural reckoning, is where every serious zero-trust conversation begins. A missing firewall rule or a failed antivirus scan did not cause the breach at Meridian. It was caused by a structural assumption baked so deeply into the architecture that nobody had thought to question it: the assumption that a device on the internal network, with valid credentials, was trustworthy by definition. That assumption is the trust crisis this book addresses. Marcus's whiteboard session is a scene that plays out, with different names and different organizations, across federal agencies, commercial enterprises, healthcare systems, and manufacturing operations every week. The details change. The structural failure is always the same.

2.1 Why It Matters

The trust crisis described in Marcus's scenario is not a niche engineering problem. It is a mission-aligned organizational risk that reaches from the server room to the boardroom. Every organization that operates a hybrid environment, which now includes virtually every enterprise of any scale, is running some version of the same implicit trust model that allowed an attacker to spend 47 days inside Meridian's network without triggering a single containment response. The perimeter-based security model was designed for a world in which users, applications, and data lived inside a defined boundary. The boundary was the network edge. Trust was granted by crossing it. That model made operational sense when offices were the primary workplace, when data centers were physical buildings under organizational control, and when "the cloud" was a marketing term rather than the primary delivery mechanism for critical business applications. That world no longer exists.

The workflow-level impact of implicit trust failures is severe and immediate. When a contractor credential is reused across a flat network, the attacker does not just access one resource; they access every resource the credential can reach, and in a flat network, that reach is determined only by the absence of boundaries rather than by deliberate policy. Finance records, intellectual property, partner credentials, operational system configurations: all of it becomes accessible to an entity that the organization believes it has already verified. The downstream effects cascade through operations, legal obligations, customer relationships, and regulatory standing. Breach notification timelines, measured in days, collide with remediation timelines, measured in months. The cost of a single flat-network lateral-movement event, including incident response, forensic analysis, regulatory penalties, customer notification, and reputational damage, routinely reaches tens of millions of dollars for mid-size enterprises and hundreds of millions for large ones.

The case for zero-trust architecture is not, at its core, a technology-purchasing argument. It is a structural argument about how organizations govern access to their most critical assets. A responsible-by-design security posture is one in which the architecture itself enforces least privilege, requires continuous verification, and limits the blast radius by design, rather than relying on the hope that no attacker finds cracks in the perimeter. This chapter establishes the full scope of the problem: the dissolving perimeter, the explosion of identity types, the economic reality of modern breaches, the regulatory pressure from both federal and commercial frameworks, and the amplifying effects of AI-driven threats and the expansion of IoT. Understanding the problem at this level of specificity is a prerequisite for building solutions that actually address it.

2.2 The Hybrid Enterprise Threat Surface

The term "hybrid enterprise" has become so common in technology discussions that it risks losing its meaning. For this book, a hybrid enterprise is any organization whose computing environment spans at least two of the following five domains: on-premises data centers, public cloud infrastructure, private cloud or colocation facilities, remote or mobile workforce endpoints, and partner or contractor network connections. By that definition, the overwhelming majority of organizations with more than five hundred employees qualify. The practical implication is that the attack surface of a hybrid enterprise is not a single surface but five overlapping ones, each governed by different administrative tools, authentication systems, network controls, and visibility capabilities.

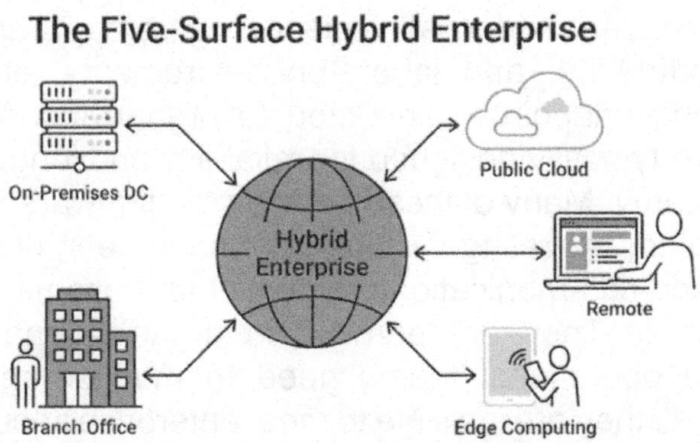

The Five-Surface Hybrid Enterprise

Each of these five surfaces creates attack vectors that the perimeter model was never engineered to address. The on-premises data center is no longer an isolated fortress: it connects to cloud environments via dedicated links, VPNs, and API gateways, and those connections implicitly carry trust in both directions. The public cloud tenant is governed by a shared responsibility model in which the cloud provider secures the infrastructure. Still, the organization must secure its own configurations, identities, data, and workloads. Misconfigured storage buckets, over-permissioned service accounts, and publicly exposed management interfaces are among the most exploited attack vectors in cloud environments today, and they are purely the result of identity and configuration management failures rather than infrastructure weaknesses. Remote workforce endpoints operate entirely outside the organizational network perimeter, connecting through consumer internet services over which the organization has no control, using devices that may be personally owned or that share network connectivity with home automation systems, children's gaming devices, and other organizational endpoints belonging to family members.

The IoT and operational technology surface deserves particular attention because it combines the greatest vulnerability density with the weakest security tooling. Building automation systems, environmental controls, physical access control readers, video surveillance cameras, manufacturing floor controllers, and laboratory instruments all represent computing endpoints connected to organizational networks that were typically designed for reliability and longevity rather than security. Many of these devices run firmware that cannot be updated, operating systems that have been end-of-life for years, and communication protocols that transmit credentials in plaintext. They are enrolled in organizational networks because operational teams need to monitor and manage them. Still, they are invisible to most enterprise security tooling because they do not run standard operating systems or security agents. A compromised building management controller on the same network segment as a financial database exemplifies the structural vulnerability that a flat network architecture enables and that a zero-trust architecture is designed to eliminate.

Partner and contractor connectivity has historically been the weakest link in enterprise network security, and the Meridian scenario illustrates why. Organizations extend network access to hundreds or thousands of external entities whose security posture they cannot directly control, whose personnel change frequently, and whose access is often provisioned for specific projects but deprovisioned inconsistently. The attack surface created by this partner ecosystem is not measured in devices or IP addresses but in trust relationships: how many external parties have been granted some form of access, how granularly that access is defined, and how reliably it is revoked when the relationship ends. In most hybrid enterprises, the honest answer to all three questions is "we do not know with confidence." That uncertainty is the trust crisis in quantifiable form.

The Dissolving Perimeter

The network perimeter was the organizing concept of enterprise security for three decades. Build a high wall around your assets, control who crosses the gate, and treat everyone inside as trusted. The model was coherent when the wall was real: a physical data center with a defined network edge, a workforce that worked inside that edge, and applications that ran within the walls. The perimeter began dissolving before most security teams noticed. The first significant erosion came with web applications that required direct internet connectivity. The second came with mobile devices that carried corporate data outside the walls permanently. The third, and decisive, erosion came with cloud adoption, which moved critical workloads entirely outside any physical perimeter the organization could control. By the time remote work became standard operating practice rather than an exception, the perimeter had not merely dissolved; it had become a conceptual artifact with no physical correlate.

The Dissolving Perimeter

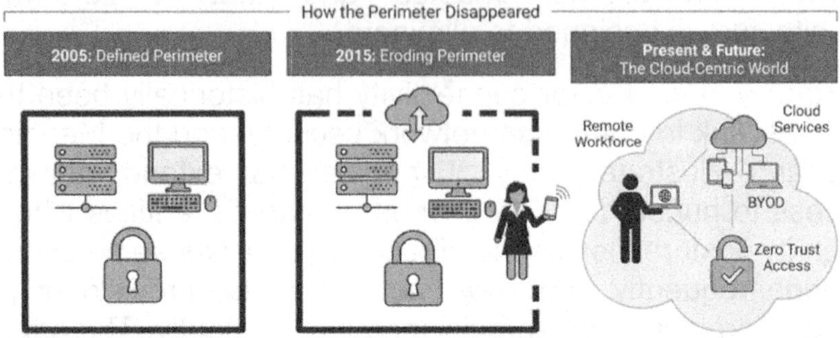

The traditional network perimeter has been replaced by a dispersed, user-centric and cloud-based security model.

The implications of a perimeter-free architecture go beyond technology choices. They require a fundamental reconceptualization of how trust is defined, granted, and maintained. In a perimeter model, trust is binary and location-based: inside the perimeter means trusted, outside means untrusted. In a hybrid enterprise, the concept of "inside" has no consistent meaning. A cloud workload processing sensitive financial data is simultaneously inside the organization's logical environment and outside any physical perimeter. A remote employee connecting from a home office is simultaneously an organizational insider by employment status and an external connection by network topology. An IoT sensor on the factory floor is physically inside the building. Still, it may be running software that is completely unknown to the security team and connected to a manufacturer's cloud service over the public internet. Trust decisions based on network location in this environment are not just inadequate; they are actively misleading.

The Identity Explosion

If the dissolving perimeter is the first dimension of the trust crisis, the identity explosion is the second. The number of distinct identities operating within a modern enterprise has grown by orders of magnitude over the past decade, and most organizations have no comprehensive inventory of how many identities exist, what access rights each one carries, or whether all of them are still legitimate. Human users are the identities that security teams typically consider, but they represent a shrinking fraction of the total identity population in hybrid enterprises. Machine identities, service accounts, workload credentials, API keys, IoT device certificates, and AI agent identities collectively outnumber human identities by a ratio that, in large enterprises, commonly reaches 10:1 or higher.

Consider the identity landscape of a mid-sized enterprise with 5,000 human employees. Those employees generate approximately 5,000 primary user accounts, but each one also creates secondary identities, including application accounts, directory entries, privileged access accounts, shared mailboxes, and service accounts for automated processes. The organization's cloud infrastructure generates hundreds of service accounts, role assignments, and federated identities connecting cloud tenants to on-premises directories. The application portfolio generates API keys, service credentials, and OAuth tokens that enable applications to communicate with one another, cloud services, and external partners. IoT devices generate device certificates and management credentials. Automated workflows, scheduled tasks, and robotic process automation agents generate their own identity footprints. AI workloads operating on behalf of organizational processes require their own access credentials and authorization scopes. By the time a complete inventory is attempted, the total identity count for a five-thousand-person organization frequently exceeds one hundred thousand distinct credentials, each one representing a potential attack vector if compromised, misconfigured, or left unrevoked after its purpose ends.

The Identity Explosion

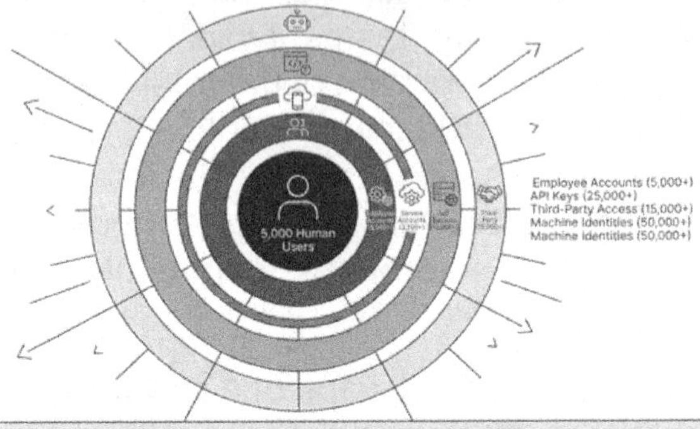

The Identity Explosion refers to the rapid and unmanageable growth of digital identities across human, machine, and service accounts.

The lifecycle management problem compounds the security implications of this identity explosion. Human user identities are typically managed through a joiners-movers-leavers process tied to HR workflows: when someone joins the organization, accounts are provisioned; when they move to a new role, access is adjusted; when they leave, accounts are deprovisioned. This process is imperfect in practice, but at least exists conceptually. Machine identities and service accounts have no equivalent lifecycle management process in most organizations. They are created when a system or application needs them, and they are rarely deprovisioned. API keys are rotated infrequently, if at all. IoT device certificates may have validity periods measured in years. Service accounts created for projects that ended two years ago continue to exist with their original access rights because no process enforces their retirement. Each of these dormant, over-permissioned, unmonitored identities is an opportunity for an attacker who gains access to it: an opportunity that the organization cannot detect or respond to because it does not know the identity exists.

Implicit Trust and Its Failure Modes

Implicit trust is the security assumption that an authenticated entity need not be continuously reverified. Once a credential has been validated, the session or token derived from that validation is treated as trustworthy for its duration, regardless of what the authenticated entity does or where it goes. This assumption was a reasonable engineering compromise in the era of perimeter security. If an employee authenticated at the network boundary and was admitted to the internal network, the overhead of reverifying every subsequent action within that network was not justified by the threat model, because the internal network was assumed to be a trusted environment occupied only by authenticated, authorized organizational members.

Implicit trust fails in hybrid environments for three compounding reasons. First, the authentication event that establishes the session may itself be compromised: stolen credentials, phished passwords, or intercepted session tokens all allow an attacker to begin a session with a legitimate identity. Second, once a session is established, lateral movement through the flat network requires no additional authentication. Hence, a compromise at any point in the network propagates freely to every resource to which the compromised identity has access. Third, in most enterprise environments, session and token durations far exceed what is operationally necessary, meaning that a compromised session has a long window to operate. The Meridian scenario combined all three failure modes: a stolen contractor credential used to initiate a VPN session, a flat network that offered no barriers to lateral movement, and a forty-seven-day session window during which the attacker could operate continuously without reverification.

The operational clarity that zero-trust architecture provides comes precisely from replacing the implicit trust assumption with explicit, continuous, policy-driven access decisions. Every access request, whether from a human user, a machine workload, or an IoT device, is evaluated against a defined policy that considers the identity of the requestor, the health of the device, the sensitivity of the resource being accessed, the behavioral context of the request, and the risk posture of the session at that moment. Trust is not granted once and inherited for the duration of a session; it is re-evaluated continuously, and any change in the session's risk signals can trigger step-up authentication, session limitation, or immediate revocation. This is not a theoretical security ideal; it is the engineering specification for a system that behaves differently at its architectural foundation than the perimeter model does.

2.3 Breach Economics and Business Impact

Security architecture decisions are ultimately business decisions, and the case for investing in zero trust must be made in business terms. The economics of data breaches have shifted dramatically over the past decade, making the cost of architectural inaction far higher than the cost of architectural transformation. Understanding these economics is not a prerequisite for engineering zero trust. Still, it is a prerequisite for funding and sponsoring it at the organizational level, which is why breach economics should be near the beginning of any serious zero-trust discussion.

The Real Cost of a Flat-Network Breach

When security professionals discuss breach costs, the conversation often centers on direct forensic and remediation expenses: the incident response team, forensic investigators, system rebuilds, and additional security tooling purchased in the aftermath. These costs are real and significant, but they represent only a fraction of the total economic impact. The full breach cost model for a mid-size enterprise experiencing a lateral-movement event across a flat network includes at least nine cost categories that compound over an 18- to 24-month period following the incident.

The first category is immediate incident response: the cost of identifying the breach, containing it, and beginning remediation. For a 47-day dwell-time event like the Meridian scenario, this phase alone typically costs between $200,000 and $1,000,000, depending on organizational size and the environment's complexity. The second category is forensic investigation: reconstructing exactly what data was accessed, exfiltrated, or modified. This is often the most time-consuming and expensive phase, because flat networks generate enormous volumes of traffic logs that must be analyzed to establish a precise picture of attacker movement. The third category is notification costs, including legal counsel, notification letter preparation, and credit monitoring services for affected individuals, all governed by mandatory timelines under state breach notification laws, HIPAA, GDPR, and other applicable regulations. Missing these timelines generates additional regulatory penalties on top of the underlying breach penalties.

The remaining categories include regulatory fines and penalties, which in GDPR jurisdictions can reach four percent of global annual revenue; civil litigation from affected parties; operational disruption costs while systems are taken offline for remediation; technology replacement costs for compromised systems; increased cyber insurance premiums in subsequent policy renewal periods; and reputational damage expressed as customer attrition, reduced deal win rates, and difficulty recruiting security talent. When these categories are aggregated into a total cost-of-breach model tailored to an organization's industry, regulatory environment, and data sensitivity, the resulting figure is almost always high enough to justify significant investment in prevention architecture. The pilot-to-production pathway for zero trust architecture, even a multi-year phased implementation, costs a fraction of the breach scenario it is designed to prevent.

Breach Cost Cascade

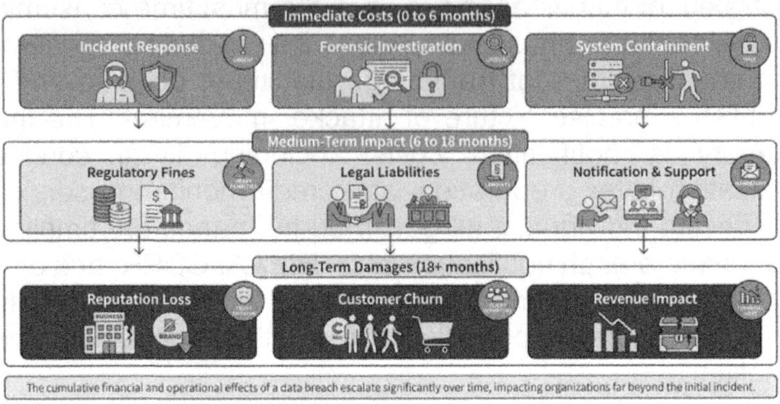

Federal and Commercial Regulatory Pressure

The business case for zero trust architecture is not purely economic; it is also regulatory. In the federal sector, Executive Order 14028, signed in May 2021, directed all federal agencies to develop plans to adopt zero-trust architecture aligned with NIST Special Publication 800-207. OMB Memorandum M-22-09, issued in January 2022, translated that direction into specific technical requirements and a firm deadline: agencies were required to achieve specific zero trust goals by the end of fiscal year 2024. These mandates are not aspirational; agencies that fail to demonstrate progress face budget consequences and heightened oversight from the Office of the Inspector General and congressional appropriators. For federal contractors, CMMC 2.0 requirements create parallel obligations tied to the ability to compete for Department of Defense contracts, making zero trust architecture adoption a business continuity requirement rather than a security preference.

Commercial regulatory frameworks have been moving in the same direction with increasing specificity. PCI-DSS 4.0, effective March 2024, introduced requirements for continuous authentication and behavioral monitoring that align closely with zero trust principles for any organization that processes payment card data. HIPAA enforcement has increasingly focused on network segmentation failures as indicators of inadequate safeguard implementation under the Security Rule, and recent Office for Civil Rights settlements have emphasized the cost of flat-network architectures that allow single compromised credentials to reach protected health information across multiple systems. The General Data Protection Regulation in Europe has led to multi-hundred-million-dollar fines for organizations that failed to implement access controls proportionate to the sensitivity of the personal data processed. This standard is difficult to meet with perimeter-based security. SOX compliance programs increasingly scrutinize logical access controls for financial systems in ways that implicitly favor zero-trust access patterns over legacy network-trust models.

For security architects and managers presenting zero-trust investment to executive leadership, the regulatory landscape provides a second argument that operates independently of the breach-economics case: non-compliance is not just a security risk but also a legal and financial risk, with defined penalty structures and public-disclosure obligations. Framing zero-trust architecture as the intersection of security best practices and regulatory compliance requirements strengthens the business case considerably, particularly for organizations in regulated industries where compliance costs are already visible to the board and the audit committee.

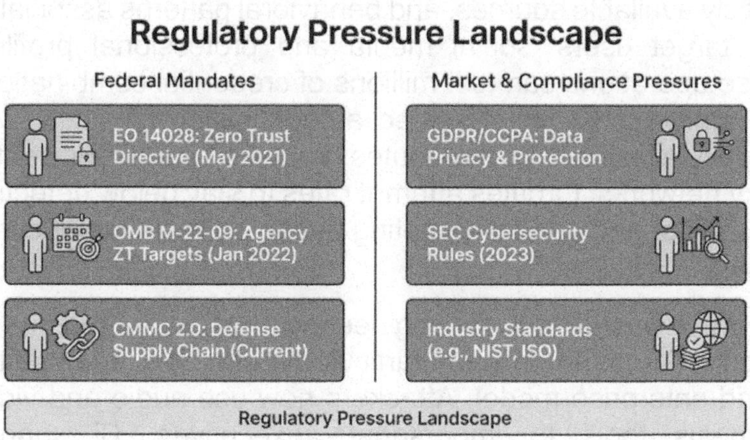

2.4 AI-Driven Threats and the Accelerating Attack Cycle

The threat landscape that zero-trust architecture is designed to defend against has been fundamentally transformed by the availability of AI and machine learning tools to adversaries. This transformation is not a future risk; it is a present operational reality that changes the speed, scale, and sophistication of attacks against hybrid enterprises in ways that static, perimeter-based defenses are structurally incapable of addressing. Understanding how AI amplifies adversarial capabilities is essential context for understanding why continuous verification, one of zero trust's core architectural requirements, is not an optional enhancement but a structural necessity.

Automated Credential Attacks

Traditional credential attacks were constrained by human speed: a human attacker could attempt a limited number of credential guessing or phishing operations per unit time, making detection feasible through simple rate-limiting and anomaly thresholds. AI-powered credential attacks operate at machine speed across machine-scale target populations. Automated credential-stuffing operations use large language models to generate contextually plausible username and password combinations based on correlations with breach databases, organizational naming conventions derived from publicly available sources, and behavioral patterns associated with target users' social media and professional profiles. These operations can test millions of credential combinations per hour against cloud-hosted authentication services, with adaptive evasion logic that routes requests through residential proxy networks, throttles attempt rates to stay below detection thresholds, and adjusts targeting based on real-time success signals.

Deepfake-based social engineering has emerged as a specific AI-driven threat, particularly effective against the hybrid enterprise model. Attackers now use audio and video synthesis tools to impersonate executives, IT support personnel, and trusted colleagues in real-time voice and video calls. A synthesized voice call from what appears to be the Chief Information Security Officer asking a help desk technician to reset a privileged account password, bypassing normal verification procedures because of an "urgent security incident," exploits exactly the kind of human trust that perimeter-based security models relied upon and that zero trust architecture deliberately eliminates from the authentication chain. These attacks are not hypothetical; they have been documented in multiple high-profile wire fraud and credential theft cases, with losses measured in millions of dollars per incident.

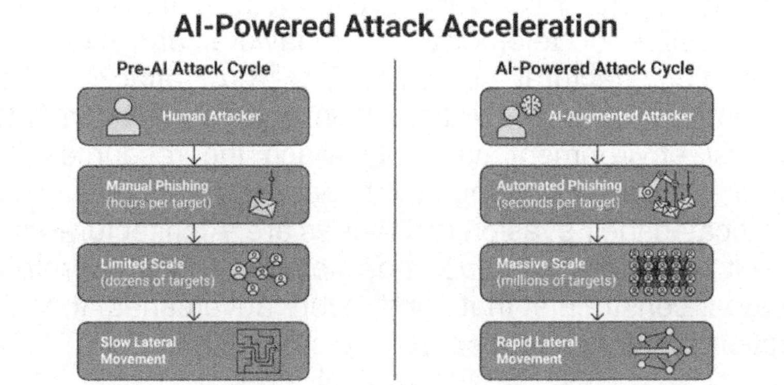

AI tools dramatically increase the speed, scale, and sophistication of modern cyberattacks, outpacing traditional defenses.

Adaptive Evasion and Detection Lag

Beyond the speed and scale of AI-powered attacks, the adaptive evasion capabilities of AI-driven threat actors create a fundamental problem for static detection approaches. Traditional security monitoring relies on signature- and threshold-based detection: known malware signatures are identified, known attack patterns are fingerprinted, and anomalous activity volumes above defined thresholds trigger alerts. AI-powered adversaries can dynamically modify their attack patterns to avoid triggering any of these static rules, observing the detection response to their initial probes and adjusting behavior to operate below detection thresholds or to mimic legitimate traffic patterns. A machine learning model trained on the behavioral patterns of legitimate users in a target environment can generate attacker behavior that is statistically indistinguishable from normal user activity to a rules-based detection system.

The detection lag created by this adaptive evasion is precisely what produced the forty-seven-day dwell time in the Meridian scenario and in many of the high-profile breaches that have defined the security conversation over the past decade. Static detection systems did not fire because the attacker's behavior did not exceed the configured thresholds. Zero-trust architecture addresses this dynamic not primarily through better detection. However, continuous verification does generate superior telemetry for behavioral analytics. Still, through architectural containment, an attacker who compromises a single credential in a properly segmented zero-trust environment can only reach the resources that credential is explicitly authorized to access, regardless of how sophisticated their evasion techniques are. Architectural limits on blast radius do not depend on detection; they operate as structural constraints that bind even adversaries that the detection system cannot see.

2.5 The IoT and OT Expansion

The Internet of Things and operational technology environments represent the frontier of the hybrid enterprise threat surface and pose security challenges qualitatively different from those associated with traditional IT endpoints. Conventional enterprise endpoints, laptops, desktops, and servers, are designed to run security agents, receive software updates, participate in directory authentication, and generate the log data that security monitoring systems consume. IoT devices and OT systems were designed without these capabilities in mind, and retrofitting them is often architecturally infeasible. A programmable logic controller managing a manufacturing process, an HVAC controller managing a building's environmental systems, or a medical device monitoring a patient's vital signs is a computing endpoint. Still, it is not an IT endpoint, and treating it as one creates both operational risk and security blind spots.

The ownership gap between IT security teams and the operational teams responsible for IoT and OT environments compounds the organizational challenge. Facilities departments own building management systems. Industrial control systems are owned by engineering and operations teams. Biomedical engineering departments manage medical devices. These groups have deep domain expertise in their operational systems but typically limited cybersecurity expertise, and they operate under operational constraints that make security controls, which would be standard in IT environments, such as mandatory software updates and security agent installation, potentially disruptive or contractually prohibited by device manufacturer support terms. The result is an expanding population of network-connected endpoints that are invisible to security monitoring, resistant to standard remediation approaches, and potentially connected to physical infrastructure with safety implications that create additional constraints on incident response.

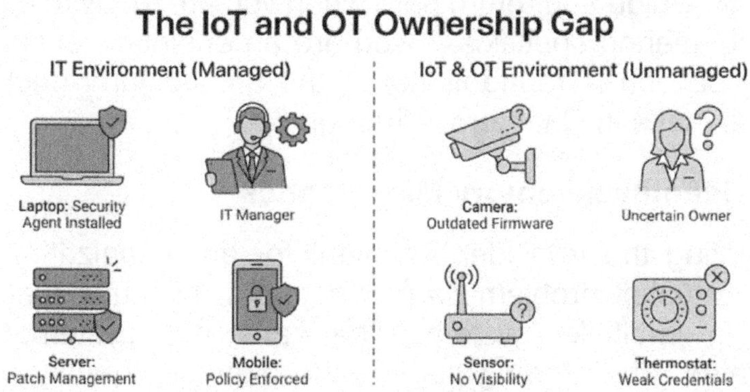

The IoT and OT Ownership Gap creates security vulnerabilities due to a lack of defined **responsibility, visibility,** and management for non-traditional devices.

Zero-trust architecture provides a principled response to the IoT and OT challenges without requiring retrofitting security agents onto devices that cannot support them. By treating every network-connected device as untrusted by default, regardless of its physical location, a zero-trust architecture creates the organizational impetus to inventory IoT and OT devices, classify them by their network behavior and access requirements, and segment them into microsegmented zones that limit their connectivity to only the resources and services their operational function requires. A building management controller that needs to communicate with a central facilities management application does not need to be on the same network segment as financial systems or human resources databases. Microsegmentation creates that separation at the network level without requiring changes to the devices themselves, achieving structural containment even for endpoints that cannot run security software.

2.6 Geeks Only

This section is written for engineers, architects, and technically oriented team leaders who want to understand the quantitative and mechanistic dimensions of the trust crisis before engaging with the zero-trust solutions described in the rest of the book. If you are a manager who has absorbed the preceding sections, you can skip ahead to the Manager's Decision Guide and return here when you are ready to dig into the engineering specifics. If you are an engineer, everything in this section is foundational to the architectural decisions you will make in Chapters 3 through 6.

Identity Inventory Mathematics

Calculating the true identity count for an organization is a more complex problem than it appears, because identities exist in multiple systems that are not systematically correlated. The starting point is the directory identity count: the number of active accounts in the primary identity provider, typically Active Directory or a cloud-based directory service. For a five-thousand-person organization, this number might be 6,000 accounts, including contractor, service desk, shared, and administrative accounts. But the directory count is the smallest of the relevant identity populations.

Machine identities, specifically the X.509 certificates used to authenticate servers, containers, microservices, and automated workloads to each other, are typically managed by a PKI infrastructure that operates independently of the identity provider. A mature cloud-native environment may generate and rotate thousands of short-lived certificates per day via a service mesh with mutual TLS (mTLS) authentication. Each certificate represents an active identity during its validity period, and the aggregate number of concurrent machine identities in a dynamic Kubernetes environment can be 10 times that of human user accounts. API keys and OAuth 2.0 tokens represent a third identity population managed by application-level authorization servers rather than directory services. IAM federation relationships, in which a cloud provider trusts assertions from an on-premises SAML identity provider, represent a fourth population of trust relationships that must be inventoried and governed independently.

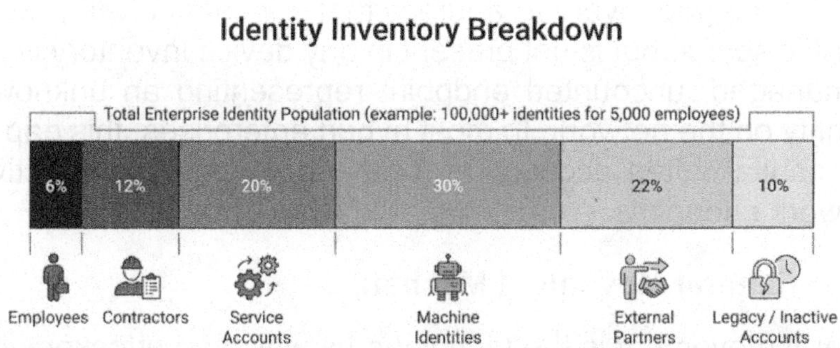

Identity Inventory Breakdown: Comprehensive categorization of all digital identities within an organization provides foundational visibility for security and access management.

The IAM federation architecture is particularly important for security teams to understand because it creates transitive trust relationships that are invisible in any single system's view. When an on-premises Active Directory is federated with a cloud tenant through SAML or OpenID Connect, and that cloud tenant is federated with a partner organization's cloud environment for a joint project, the resulting trust chain means that a compromised credential in the partner organization can, under certain configurations, propagate assertions that are trusted by resources in the on-premises environment. SIEM platforms ingest authentication logs from all three environments, but correlating events across the federated boundary requires specific configuration and is not enabled by default in most deployments. The telemetry blind spot created by this correlation gap is exactly the kind of architectural invisibility that allows lateral movement to proceed undetected for extended periods.

For engineers performing an initial identity inventory, the recommended approach is to enumerate identities from the network layer upward, starting with the directory and enumerating outward. NetFlow data from network infrastructure reveals the actual communication patterns of all devices on the network, including IoT devices and OT systems that do not appear in any directory. Correlating NetFlow source addresses with known device registrations reveals the gap: every IP address that appears in NetFlow as a traffic source but is not present in any device inventory is an unmanaged, uncounted endpoint representing an unknown identity on the network. In most hybrid enterprises, this gap is substantial, often accounting for 20 to 40 percent of active network endpoints.

Lateral Movement Mechanics

Lateral movement is the technique by which an attacker who has gained initial access to one point in a network expands their access to other systems. Understanding the mechanics of lateral movement in a flat network is essential for engineers to build the case for microsegmentation, because it makes concrete and specific what "blast radius" means in engineering terms. In a fully flat network, in which all endpoints can initiate connections to all other endpoints on the same network segment, the blast radius of a single compromised credential is limited only by the access rights associated with that credential. An attacker who compromises a domain user account can use that account to enumerate the Active Directory for high-value targets, attempt authentication to any system that accepts domain credentials, exploit trust relationships between systems to escalate privileges, and exfiltrate data from any system they can authenticate to.

The specific tools and techniques associated with lateral movement in Windows domain environments are well-documented: pass-the-hash attacks that reuse NTLM credential hashes without requiring the cleartext password; pass-the-ticket attacks that forge Kerberos authentication tickets; WMI and PowerShell remoting that use legitimate administrative channels for attacker-controlled command execution; and token impersonation that borrows the authentication context of a higher-privileged process running on the same system. EDR platforms detect many of these techniques through behavioral analysis of process execution chains and memory patterns, but detection alone is not containment. An attacker who is detected but not immediately contained can continue lateral movement faster than a human incident responder can triage alerts. Architectural segmentation, implemented through network policy enforcement rather than detection, stops lateral movement before detection is even required.

Lateral Movement in a Flat Network

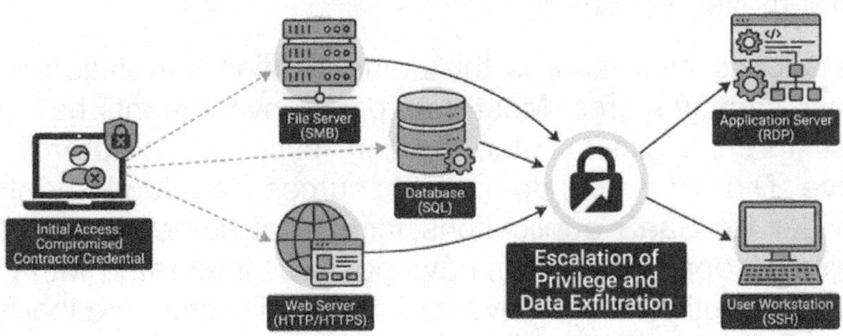

Lateral Movement in a Flat Network

Telemetry Blind Spots in Flat Networks

Effective security monitoring depends on telemetry: the collection of events, logs, and behavioral signals from across the environment that allow a security operations team to detect, investigate, and respond to threats. Flat networks create structural telemetry blind spots distinct from gaps caused by missing security tooling. Even an organization with a comprehensive SIEM deployment, EDR on all managed endpoints, and cloud-native security logging enabled faces specific visibility gaps arising from the flat network's architectural properties.

The first blind spot is the east-west traffic inspection. Security monitoring tools are predominantly designed to analyze traffic entering and leaving the network, typically referred to as north-south traffic. Lateral movement primarily targets east-west traffic: communication between systems within the same network segment. In a flat network, this traffic flows without crossing any inspection point because there are no intermediate controls between systems on the same segment. SIEM platforms receive logs from the endpoints involved. Still, they do not receive network-level flow records for every inter-system connection, as a segmented architecture with embedded inspection points would generate. NetFlow records from core switches provide aggregate traffic volume data. Still, they do not provide the application-layer context needed to distinguish legitimate database replication traffic from a lateral-movement query.

The second blind spot is the authentication correlation gap across identity stores. Most enterprise environments have at least three distinct authentication systems: on-premises Active Directory for domain resources, a cloud identity provider for SaaS applications, and local authentication on servers or applications that have not been integrated with the directory. Authentication events from each system are logged independently. SIEM platforms can ingest all three log sources, but correlating a single attacker session across all three requires custom detection logic that most organizations have not built. An attacker who begins a session with a domain credential, pivots to a cloud application using a service account that shares a common password with the domain account, and then uses the cloud application's API to access data does not generate a single correlated event trail in most SIEM implementations: they generate three separate, unconnected log sequences that individually look legitimate. Only a purpose-built identity analytics layer, informed by the complete identity inventory described in the previous section, can correlate those events into a unified view of the attacker's session.

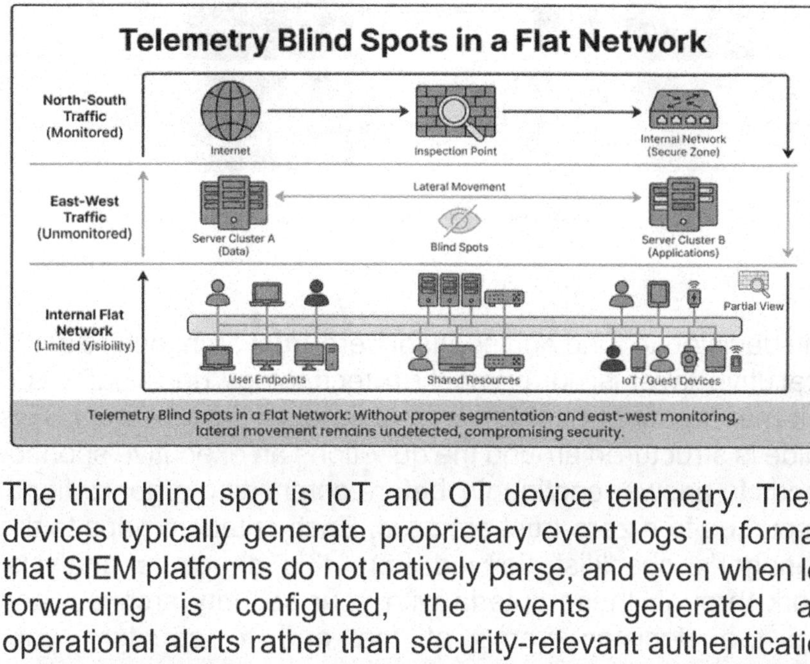

Telemetry Blind Spots in a Flat Network: Without proper segmentation and east-west monitoring, lateral movement remains undetected, compromising security.

The third blind spot is IoT and OT device telemetry. These devices typically generate proprietary event logs in formats that SIEM platforms do not natively parse, and even when log forwarding is configured, the events generated are operational alerts rather than security-relevant authentication and access events. A building management controller may log a door lock failure as an operational event without logging the authentication credential used to initiate the command that failed. Correlating operational events from OT systems with network-level anomalies from NetFlow and authentication events from the identity provider requires custom parsing, custom detection logic, and operational domain knowledge that most security operations centers do not possess for every OT system in their environment. This gap is not a tooling failure; it is a structural consequence of deploying computing devices that were never designed for security visibility into a security-monitored environment without an architectural bridge.

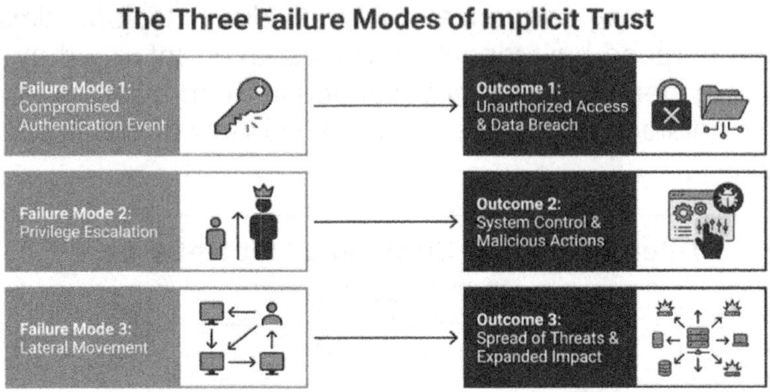

The Three Failure Modes of Implicit Trust

2.7 Manager's Decision Guide

The decision to fund and sponsor zero-trust architecture at the executive level is not primarily a technology decision; it is a risk-management decision with technology implications. This guide is structured around the questions an executive sponsor needs to answer confidently before committing organizational resources to a zero-trust program. Each criterion maps to the manager's checklist derived from this chapter's analysis. Work through these criteria with your security architect and use the decision frame at the end to structure your recommendation to the board or executive committee.

Decision Criteria

- Trust boundary inventory: Have you mapped every network segment, access zone, and connectivity relationship in your hybrid environment and documented which ones rely on implicit trust? Until you can answer yes with evidence, you cannot estimate the scope of the architectural gap you are addressing or the investment required to close it.
- Identity count: Do you know the total number of distinct identities, including human accounts,

service accounts, machine certificates, API keys, IoT device credentials, and AI agent identities, operating in your environment? A zero-trust program cannot be scoped or priced without a credible identity inventory. Commission one before committing to budget figures.

- Blast radius assessment: If a single contractor credential is compromised today, how far can an attacker move before encountering a segmentation boundary? Your security team should be able to answer this with a specific network diagram and a list of resources at risk. If they cannot, your architectural exposure is larger than you realize.
- Breach impact model: Do you have an organization-specific estimate of the full cost of a flat-network lateral-movement breach, including incident response, regulatory penalties, notification costs, litigation exposure, and reputational damage? Generic industry averages understate the cost for organizations with high regulatory exposure. Obtain a model calibrated to your environment.
- Regulatory gap assessment: Have you mapped your current security architecture against the applicable federal mandates (EO 14028, OMB M-22-09, CMMC) or commercial compliance frameworks (HIPAA, PCI-DSS 4.0, SOX, GDPR) to identify where your current controls fall short of the explicit or implicit zero trust requirements in those frameworks?
- AI and IoT threat exposure: Has your security team assessed whether current detection and response capabilities would identify an AI-powered credential attack or a lateral movement campaign originating from a compromised IoT device? If the answer is no, you have specific gap areas that a zero-trust implementation must address as priority workstreams.

- Executive sponsorship alignment: Can you frame the zero trust business case for your board in terms of risk reduction, regulatory compliance, and competitive positioning simultaneously? A zero-trust program framed solely as a security investment is harder to fund than one that is correctly positioned as risk management, compliance enablement, and operational resilience.

Decision Frame

You are ready to sponsor a zero-trust program when you can answer yes to at least five of the seven criteria above, can quantify the financial risk of inaction in terms specific to your organization, and have identified a Security Architect or equivalent technical lead who has a credible phased implementation plan. You do not need to wait until all seven criteria are answered before beginning: the first three criteria, trust boundary inventory, identity count, and blast radius assessment, can be addressed in a six-to-eight-week discovery engagement before a full program budget is committed. The discovery phase generates the evidence base that justifies the full program investment. Funding discovery is not the same as funding transformation, but it is the decision that makes transformation possible. An organization that completes discovery and then decides not to proceed with a full zero-trust implementation will still have generated a materially more accurate picture of its risk exposure than before, which has independent value for insurance, regulatory, and governance purposes.

2.8 Case Study: The Federal Maritime Logistics Administration

The Federal Maritime Logistics Administration, a mid-size federal civilian agency responsible for coordinating port security and maritime shipping data, had completed what its leadership considered a successful modernization in 2014. The agency had migrated from a collection of aging standalone systems to a consolidated hybrid cloud architecture that combined an on-premises data center with two cloud tenants for different operational workloads. The modernization project delivered measurable benefits: faster data access for port operations staff, reduced infrastructure maintenance costs, and improved data sharing with partner agencies. What it did not deliver was a redesigned security architecture to match the new environment. The flat network model that had served the original standalone systems was carried forward into the hybrid cloud environment, and the implicit trust assumptions that had been manageable when the environment was smaller and more homogeneous became structural vulnerabilities at hybrid cloud scale.

The breach began with a routine contractor engagement. A facilities management contractor working on a data center physical access project was issued a VPN credential with access to the agency's internal network. The scope of access was not specifically defined; the credential was provisioned with the same network access as other contractors in that category, which provided broad connectivity to the flat internal network. The contractor completed the engagement and departed. The offboarding process removed the contractor from the contractor management portal. Still, it did not propagate the deprovisioning to the VPN gateway, which maintained its own identity store synchronized weekly from the primary directory. A week after the contractor's portal access was removed, the VPN credential remained active. An adversary with access to that credential, obtained through means that forensic investigation later determined was likely a credential broker operating in specialized underground markets, used it to initiate a VPN session thirty-one days after the contractor's departure.

FMLA Incident Timeline and Blast Radius

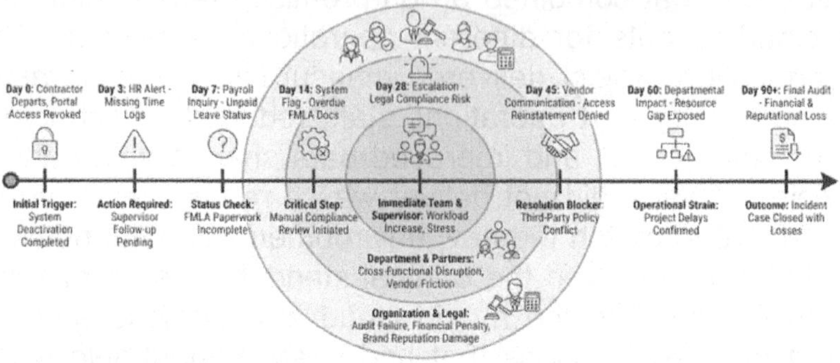

This FMLA incident demonstrates how a single process failure at Day 0 can progressively escalate, expand its impact, and result in significant organizational consequences over time.

Over the forty-seven-day dwell period, the adversary conducted a systematic exploration of the agency's network, moving from the initial contractor access point to file servers, to database systems containing citizen records and vessel tracking data, and ultimately to a shared directory of partner access credentials that the agency maintained for data exchange with affiliated port authorities. The flat network architecture offered no segmentation barriers at any point in this lateral movement chain. The SIEM platform did not flag the adversary's traffic patterns because the VPN session carried a valid credential, and the subsequent access patterns, while broader than typical contractor behavior, did not cross the threshold settings configured for contractor accounts. An automated anomaly-detection module flagged the session on day forty-seven when the access pattern deviated from any known behavioral baseline for the contractor category, but by that point, the attacker had had sufficient time to access 2.3 million citizen records, vessel-tracking data for sixty-three active maritime operations, and forty-seven partner access credentials.

The remediation cost was estimated at 18.6 million dollars, including forensic investigation, mandatory FISMA breach notifications, a congressional inquiry response, partner notifications, credential rotation, system remediation, and a contracted security architecture review that became the foundation of the agency's subsequent zero-trust planning process. A parallel commercial incident at a regional logistics firm during the same period demonstrated that the structural failure is not specific to the federal environment: the logistics firm experienced a nearly identical contractor credential compromise resulting in supply chain data exposure and PCI-DSS non-compliance penalties, following the same flat-network lateral movement pattern and the same deprovisioning gap between separate identity systems. The lesson in both cases is not that the organizations lacked security tools; both had firewalls, VPNs, and SIEM platforms. The lesson is that those tools, deployed in a flat-network implicit-trust architecture, do not limit blast radius. They log it, after the fact, once the damage is done.

2.9 Team Conversation Guide

Run these three discussions in sequence over two to four weeks. The first surfaces the issue, the second works the trade-offs, the third commits the team to action.

Discussion 1. Mapping Our Trust Boundaries

This first discussion asks your team to confront the architecture you actually have, not the architecture you think you have. Before your next sprint planning or architecture review, print the current network topology diagram and bring it to the team. The goal of this conversation is to identify every place where network position alone is being used as a substitute for explicit access policy.

Discussion questions to ask the team:

- If we drew lines on this network diagram for every trust boundary, meaning every point at which an entity must prove its identity before being granted access to a resource, how many lines would we draw? Where are the large trust zones with no internal boundaries?
- Which systems in our environment trust other systems implicitly based on network segment

membership rather than authenticated identity? What would break if we required mTLS or explicit credential verification between those systems?

- Where in our architecture could a compromised contractor credential move laterally without triggering a segmentation boundary or an authentication challenge? How many resources does that reachable zone contain?
- Which of our current security controls are perimeter-dependent? If we assumed today that the perimeter has already been breached, which controls would become ineffective and which would continue to provide value?

What to listen for: Listen for engineers who express uncertainty about where trust boundaries actually are versus where they are supposed to be. That uncertainty is diagnostic: it means the architecture has evolved beyond what the documentation reflects, and the actual trust topology is unknown. Also listen for assumptions that specific systems are "safe" because they are on a particular network segment. Those assumptions are exactly the implicit trust patterns that create blast radius exposure.

Follow-up actions: Document every trust boundary identified and every gap found. Create a working list of implicit trust relationships with a risk rating for each. Schedule a follow-up session to prioritize which implicit trust relationships to address first based on the sensitivity of the resources they expose.

Discussion 2. Understanding Our Identity Landscape

The second discussion shifts from network architecture to identity architecture. Most engineering teams have a good understanding of human user accounts because those are managed through familiar IT processes, but have significantly less visibility into machine identities, service accounts, and IoT device credentials. This conversation surfaces what the team knows and does not know about the total identity population.

Discussion questions to ask the team:

- How many distinct identity types operate in our environment today? Can we enumerate not just human accounts but service accounts, API keys, machine certificates, IoT device credentials, and any AI workload identities we have deployed?
- Do we have a unified inventory of all identities across all identity stores? If a member of this team had to answer the question "how many active credentials exist in our environment right now," how confident would they be in the answer, and how long would it take to produce it?
- Which of our service accounts and API keys have not been rotated in the past ninety days? Which have never been rotated? What is the process for deprovisioning a service account when the workload it served is retired?
- How would an AI-powered credential attack, specifically one that uses machine learning to generate contextually plausible phishing content targeted at our specific users, perform against our current email security and user awareness posture?

What to listen for: Pay particular attention to uncertainty about service account and API key counts. Engineers who are confident about human user counts but uncertain about machine identity counts are reflecting the industry-wide pattern: human identity management processes exist and function, but machine identity management is typically ad hoc. That gap is where most unmanaged credential exposure lives.

Follow-up actions: Assign a team member to run an identity discovery exercise using directory queries, cloud IAM API calls, and network flow analysis to produce the most complete identity count the team can generate in two weeks. The gap between that count and the directory count is the starting point for the identity management workstream in a zero-trust program.

Discussion 3. Quantifying Our Blast Radius

The third discussion is the most operationally concrete and often the most uncomfortable. It asks the team to think like an attacker and trace the specific path that a compromised credential would take through the current architecture. Running this discussion after the first two means the team has already surfaced the trust boundaries and identity landscape; now you are applying that understanding to a specific threat scenario.

Discussion questions to ask the team:

- If a contractor's credentials were compromised today and the attacker had forty-seven days of undetected access, what is the specific list of resources they could reach from the initial access point without crossing any segmentation boundary or triggering any authentication challenge?
- What is the most sensitive data in our environment, and what is the shortest lateral movement path from a typical contractor access point to that data? How many segmentation boundaries does that path currently cross?
- If our SIEM generated a high-confidence alert that a credential was being misused, what is our current mean time to contain the affected session? Is containment a one-click operation, a multi-step manual process, or something that requires escalation through multiple teams?
- What would our architecture look like if we designed it specifically to minimize blast radius? What would we have to break, rebuild, or replace to achieve meaningful segmentation between the highest-risk lateral movement paths?

Jordan O'Neal

What to listen for: The most valuable output of this discussion is a specific, documented blast radius diagram for a realistic compromise scenario. Listen for engineers who can articulate the path in specific terms, naming actual systems and network segments, versus those who describe the risk in general terms. Specific descriptions indicate operational knowledge; general descriptions indicate gaps in documentation and visibility. Both are useful findings, but for different reasons.

Follow-up actions: Produce a formal blast radius assessment document that traces three compromise scenarios through the current architecture: a compromised contractor credential, a compromised service account with cloud access, and a compromised IoT device. This document becomes the primary justification artifact for microsegmentation investment in the zero-trust program business case.

2.10 Takeaway

The trust crisis in hybrid enterprises is not a technology product gap, a talent shortage, or a budget problem, although it has implications for all three. It is an architectural mismatch between the security model most organizations inherited and the environment they actually operate in today. The perimeter that the inherited model was built to defend has dissolved. The identities that the inherited model was built to manage have multiplied by orders of magnitude. The adversaries that the inherited model was built to stop have acquired AI-powered tools that operate faster and at a greater scale than any static defense can match. The regulatory frameworks that organizations must satisfy have begun to reflect these realities in their requirements. The economic cost of architectural inaction, measured in breach remediation, regulatory penalties, and reputational damage, now substantially exceeds the cost of architectural transformation in almost every industry context.

Marcus Reyes's whiteboard session at Meridian Logistics established something that every organization needs to establish before it can build a meaningful zero-trust program:

a clear, specific, evidence-based understanding of where the current architecture fails and what the blast radius of that failure looks like in concrete operational terms. The three team conversations in this chapter, around trust boundaries, identity landscape, and blast radius quantification, are the organizational equivalent of that whiteboard session. They produce the shared understanding between engineering teams and management that makes mission-aligned security investment possible, because both sides of the conversation examine the same evidence and reach the same conclusions about what needs to change.

Chapter 2 takes the problem defined here and translates it into engineering principles. It examines the three foundational axioms of zero trust architecture, never trust and always verify, least privilege access, and assume breach, not as abstract security philosophy but as concrete design constraints that drive specific architectural decisions. It introduces the NIST 800-207 reference architecture as the standards-based framework for building a policy engine, and it reframes "assume breach" as blast radius engineering: the discipline of designing systems so that any single compromise is structurally contained rather than catastrophically propagating. The shift from understanding the problem to designing the solution begins on the next page.

3 Zero Trust Principles Through an Engineer's Lens

Marcus Reyes arrived at Tuesday's architecture review carrying two things: a fresh cup of coffee and a printout of a $4.2 million capital expenditure request. The conference room at Holbrook Financial Services filled steadily with engineers, two junior architects, and Priya Nair, the VP of Infrastructure. On the whiteboard behind him, someone had already written "ZT Phase 2" in green marker, underlined twice, as though enthusiasm alone would make it real.

The CAPex request covered three products: a next-generation firewall suite with a "Zero Trust-ready" badge on the datasheet, an identity governance platform the vendor had been pitching for six months, and a cloud access security broker that promised to wrap every SaaS application in a verification layer. Marcus had read all three proposals the night before. The products were legitimate. The vendors were reputable. The architecture they would sit inside was the problem.

He uncapped a blue marker and drew a rectangle in the center of the whiteboard. "This is our trading platform," he said. Someone compromised a developer's laptop on the corporate network. How many hops to get here?" Silence. Then Priya said, quietly, "Maybe two." Marcus drew a straight line from the edge of the rectangle to a stick figure at the far left of the board—one arrow. No obstacles. He capped the marker. "We are about to spend four million dollars on tools that will sit on top of that line and not change it at all. That is what I want to talk about today."

What followed was not the budget review anyone had prepared for. It was a reckoning with the difference between buying zero-trust products and building zero-trust architecture. Marcus spent the next ninety minutes walking his engineering team through the principles that should drive every design decision before a single purchase order is signed: never trust, always verify; least privilege as a measurable constraint; and assume breach as a primary engineering objective rather than a contingency plan. That conversation is the subject of this chapter.

3.1 Why It Matters

Zero trust is not a marketing category. It is a design philosophy grounded in a specific threat model: that no network boundary reliably separates trusted from untrusted actors, and that perimeter-based access control is therefore structurally insufficient for modern hybrid enterprises. The principles that follow from that threat model are not new. Still, their engineering implications have grown dramatically sharper as workloads moved to cloud platforms, remote access became permanent, and machine identities began to outnumber human ones. Understanding those implications at the level of architectural decisions, not product feature lists, is the foundation of every chapter that follows.

The workflow-level impact of getting the principles wrong is severe and largely invisible until a penetration test or a breach reveals it. Organizations that purchase zero-trust-labeled products without restructuring their underlying architecture end up with expensive telemetry on top of unchanged trust boundaries. Their engineers spend cycles managing new consoles. Their managers receive compliance dashboards. And their core systems remain reachable in two hops from a compromised endpoint. That pattern is so common it has a name, and this chapter addresses it directly.

For teams that get the principles right, the benefits are concrete and mission-aligned: access decisions become auditable and revocable rather than implicit and permanent; blast radius shrinks because lateral movement requires active authorization at every boundary; and security posture becomes measurable because the policy plane produces structured logs of every grant and denial. That operational clarity is what distinguishes genuine zero trust architecture from its imitations.

3.2 The Three Principles as Engineering Constraints

Zero trust is organized around three foundational principles. Security practitioners often state them as axioms, which makes them sound philosophical. Engineers need to read them as design constraints, each of which drives specific decisions about authentication flows, authorization models, and network segmentation topology. The distinction matters because a principle that does not constrain a decision is just a slogan.

Never Trust, Always Verify

The first principle sounds like a paranoid personality disorder applied to network traffic. What it actually means, translated into engineering requirements, is that identity and authorization must be established cryptographically at every access boundary, every time, with no assumption that a prior successful authentication persists as implicit permission across other resources. A user who authenticated to the email system has not authenticated to the financial reporting API. A service that authenticated to the configuration store has not authenticated to the secrets manager. Each boundary is its own verification event.

The engineering implication is that trust cannot be inherited from network position. Being inside a firewall perimeter, on a corporate VLAN, or behind a VPN concentrator grants exactly zero implicit authorization to any resource. Every request carries an identity assertion; every identity assertion is validated against the current policy; and every policy evaluation considers the requesting entity's state at the time of the request, not the state it was in when it first connected. This is not stateless authentication in the REST sense. It is continuous, contextual authorization, and it requires a policy infrastructure capable of evaluating rich attributes in real time.

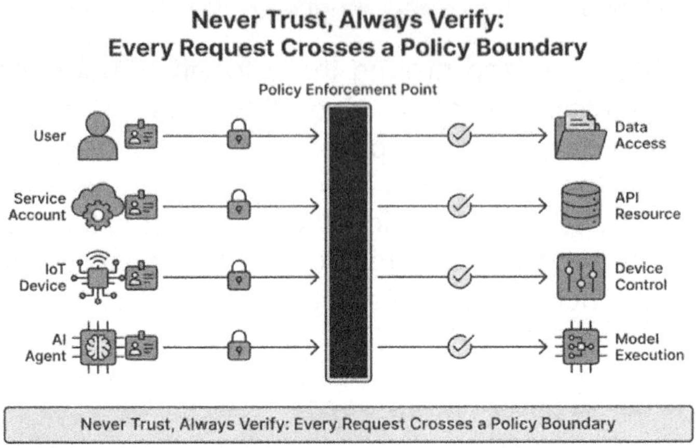

Least Privilege Access

The second principle is frequently misread as "give people the minimum access they ask for." The engineering interpretation is stricter: each identity receives the narrowest set of permissions required to complete the specific task it is performing at the moment it performs it, and those permissions expire when the task completes. Standing access is a liability. Permanent read-write permissions granted at onboarding and never reviewed are not a configuration; they are accumulated debt that compounds with every month the review is skipped.

Operationalizing least privilege requires automated policy enforcement rather than manual access reviews. Human reviewers cannot keep pace with the velocity at which cloud environments change. A service account created for a one-time data migration that still holds production database credentials six months later is not a security failure on the part of the individual who created it; it is an architectural failure of a system that permitted standing access without expiration. Zero-trust least-privilege is enforced through time-bounded credentials, just-in-time access grants, and automated revocation tied to task-completion signals rather than calendar-driven review cycles.

Privilege drift is the accumulation of access rights beyond the current minimum requirement. It is the natural state of any system that adds permissions without removing them, and it is a reliable predictor of the severity of breaches. The blast radius of any compromised identity is bounded by its privilege set. Least privilege access, enforced mechanically and continuously, is therefore not a compliance checkbox but a direct engineering investment in limiting the damage any single compromise can cause.

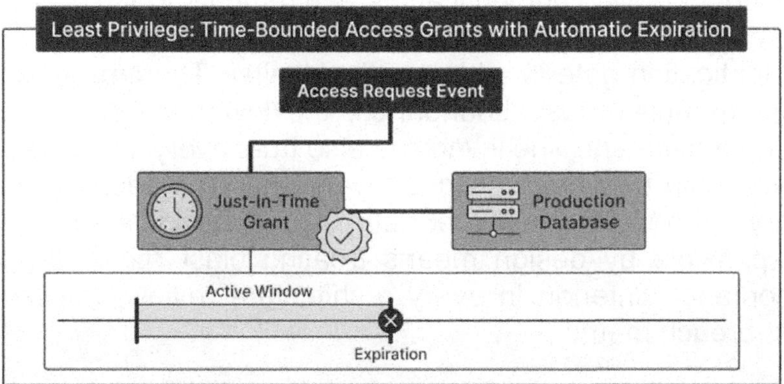

Least Privilege: Time-Bounded Access Grants with Automatic Expiration

Assume Breach: Blast Radius Engineering

The third principle is the most consequential and the most frequently misapplied. "Assume breach" is often treated as a mental posture, an instruction to think pessimistically about attackers. That reading produces a different kind of security theater, where teams run tabletop exercises, update incident response playbooks, and feel appropriately grim, without changing the architecture that determines how much damage an actual breach causes.

The engineering reading of "assume breach" is a design objective: the architecture must be built so that when (not if) a component is compromised, the attacker's reach is bounded by the smallest feasible containment unit. That containment unit is the blast radius, and measuring it is a testable architectural quality metric. A well-designed zero-trust environment can answer, for any given compromised component, exactly which other resources an attacker could reach from that compromise and what additional authentication they would still need to proceed. An architecture that cannot answer that question has not implemented assume breach; it has acknowledged it abstractly.

Blast radius engineering translates directly into microsegmentation topology, identity scope constraints, and service-to-service authentication requirements. Each network segment, each service mesh boundary, and each API authentication gate is a blast radius limiter. The smaller and more granular those boundaries, the lower the ceiling on lateral movement. The inverse is also true: every implicit trust relationship between services, every shared credential, and every open internal network path is a blast radius amplifier. Responsible by design means treating blast radius as an acceptance criterion in every architecture review, not as a post-breach metric.

Blast Radius Engineering: Containing Lateral Movement with Segmentation Boundaries

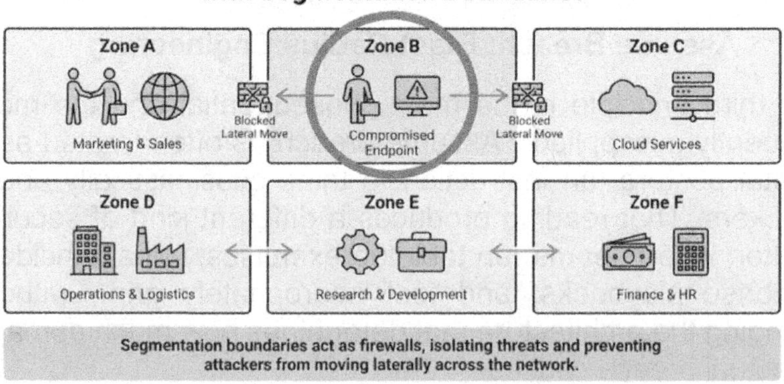

3.3 The NIST 800-207 Reference Architecture

The National Institute of Standards and Technology's Special Publication 800-207 is the closest thing the security industry has to a canonical zero trust reference architecture. Its value is not as a compliance checklist but as a design vocabulary: it names the components of the policy decision plane, describes their relationships, and defines data flows clearly enough that engineering teams can map their existing tools and gaps against it. Every serious zero-trust implementation should be able to answer, for each NIST 800-207 component, what role that component plays in their specific environment.

The architecture defines three primary components and their interactions. Understanding each one as a system design pattern rather than a product category is the starting point for any real implementation.

Policy Engine

The Policy Engine is the decision-making component. It receives a request context, which includes the requesting identity, the requested resource, the device state, the network context, and any other attributes the policy requires. It evaluates that context against the organization's access policy to produce a grant or deny decision. The Policy Engine does not enforce the decision; it makes it. This separation is architecturally important because it allows the decision logic to be centralized, audited, and updated independently of the enforcement points that act on it.

In practice, the Policy Engine is implemented through some combination of identity provider trust signals, device compliance data from endpoint management systems, threat intelligence feeds, and the organization's own policy rules expressed in a policy language. The sophistication of that combination determines how context-aware the decisions are. A minimal implementation makes binary decisions based on identity group membership. A mature implementation incorporates device health scores, behavioral anomaly signals, data classification labels, and time-of-day constraints into a continuous trust evaluation that can escalate authentication requirements mid-session when signals change.

Policy Administrator

The Policy Administrator is the component that translates Policy Engine decisions into access tokens, session credentials, or configuration signals that the Policy Enforcement Points can act on. It sits between the decision layer and the enforcement layer, managing the lifecycle of access grants: issuing short-lived credentials when access is granted, revoking them when the Policy Engine's continuous evaluation produces a deny signal, and logging all grant and revoke events for audit and forensic purposes.

The Policy Administrator is where just-in-time access is operationalized. When the Policy Engine approves a request, the Policy Administrator generates a time-bounded credential scoped to the specific resource and permission level required by the request. That credential is not a session cookie that can be copied and replayed; it is a cryptographically signed assertion with an embedded expiration and scope. When the session ends, or the Policy Engine revokes the underlying grant, the credential becomes invalid, and the enforcement point refuses further access.

Policy Enforcement Point

The Policy Enforcement Point is where access decisions are actually applied. Every resource access passes through a Policy Enforcement Point that checks the requesting credential against the current policy decision for that resource. In a well-segmented architecture, Policy Enforcement Points exist at every boundary: the API gateway between services, the identity-aware proxy in front of internal applications, the network gateway between segments, and the endpoint agent that mediates local resource access.

The key design principle for Policy Enforcement Points is that they should enforce but not decide. An enforcement point that caches decisions locally for performance reasons is making a latency optimization that trades security correctness for throughput; that tradeoff must be made explicitly with a clear TTL, not implicitly through stale cached grants. An enforcement point that contains its own policy logic, rather than consulting the central Policy Engine, creates a shadow policy system that can drift out of synchronization with the organization's actual intent.

NIST 800-207 Reference Architecture: Policy Engine, Administrator, and Enforcement Point

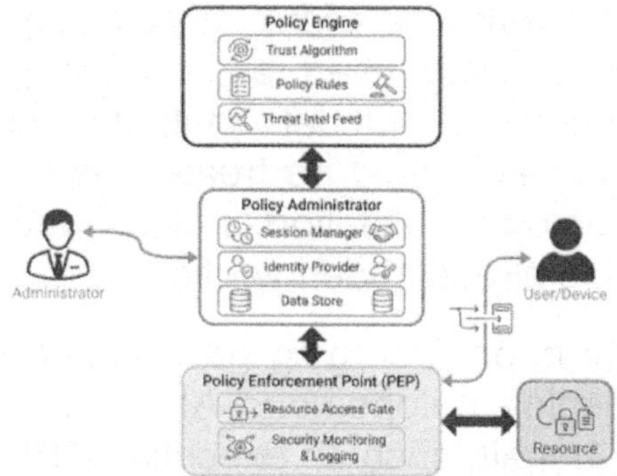

NIST 800-207 Reference Architecture: Policy Engine, Administrator, and Enforcement Point, illustrating the core components and their interactions in a zero trust framework.

3.4 AI and Machine Learning Trust Decisions

Traditional zero-trust frameworks were designed with human-to-resource access as the primary use case. An engineer authenticates, requests a resource, and the policy plane grants or denies access. That model extends naturally to service-to-service authentication and machine identity. Still, AI and machine learning workloads introduce new categories of trust decisions that have no direct analog in the human access model.

An AI model is not simply a service account. It is a computational entity that consumes training data from multiple sources, produces outputs that downstream systems act on, and may run inference workloads that access sensitive data at high velocity. The trust decisions required for an AI workload include: who authorized the model to access the training data it was trained on; whether the training data itself came from sources that meet the organization's data governance requirements; what scope of inference-time data access the model is permitted; and who is authorized to consume the model's outputs and in what contexts.

Model access authorization follows the same least privilege logic as any other identity. An inference endpoint should have read access only to the specific data stores its inference function requires, with no write access and no access to data outside its defined scope. Training pipelines require broader read access across data sources. Still, they should operate under strict network isolation, preventing the training environment from being used as a pivot point to reach production resources. Model outputs consumed by downstream services should be treated as untrusted inputs that require their own validation, not as authoritative data whose source has already been verified.

AI agent identity is an emerging governance challenge. As organizations deploy autonomous AI agents that can take actions on their behalf, such as executing API calls, creating records, or sending communications, each agent requires a verifiable identity that is separate from the human identity of the person who created or configured it. An agent acting on behalf of a user should not inherit that user's full permission set. It should receive a delegated scope limited to the actions the agent is explicitly authorized to perform, with audit trails that associate every agent action with both the agent identity and the delegating human identity.

AI Workload Trust Decisions: Model Identity, Data Scope, and Agent Delegation

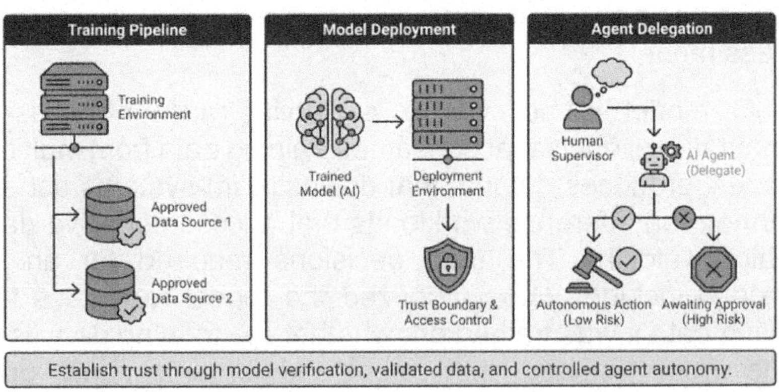

3.5 IoT and Constrained Device Trust

Internet of Things devices present a different category of trust challenge. Where AI workloads require new governance frameworks layered on top of existing authentication infrastructure, IoT devices frequently cannot run that infrastructure at all. A temperature sensor, an industrial control unit, or a medical imaging device may lack the capacity for a full TLS stack, storage for rotating certificates, and an interface for a human to perform MFA enrollment. The trust problem is not how to extend zero-trust authentication to these devices, but how to achieve equivalent security guarantees through different mechanisms.

Device attestation is the primary alternative. Rather than authenticating a device through credentials that the device presents, attestation verifies the device's identity through hardware-rooted evidence: a manufacturer certificate burned into a trusted platform module at the factory, a cryptographic endorsement from the device's hardware root of trust, or a secure boot measurement chain that proves the device is running authorized firmware. The attestation record is not a password the device presents; it is a verifiable proof about the device's physical identity and software state that a relying party can check without requiring the device to run a full authentication client.

Network segmentation is a compensating control when attestation is not feasible. IoT devices that cannot participate in the policy plane should be confined to isolated network segments with explicit, narrowly scoped egress rules that permit only the specific traffic flows required by the device's function. A building automation sensor should not be able to reach the corporate identity provider, the financial systems network, or any segment other than the building management system to which it reports. The segment boundary is the enforcement point, and the segmentation policy is the trust decision.

Firmware integrity monitoring and anomaly detection serve as continuous verification for constrained devices. Because many IoT devices cannot re-authenticate mid-session, the continuous verification function is performed externally: behavioral baseline monitoring detects deviations from expected traffic patterns, firmware scanners compare running images against known-good hashes, and network telemetry identifies devices that are communicating with destinations outside their approved scope. These external verification mechanisms do not replace identity-based authentication, but they provide an equivalent level of ongoing assurance for devices that cannot support it natively.

IoT Trust Architecture: Attestation, Segmentation, and External Continuous Verification

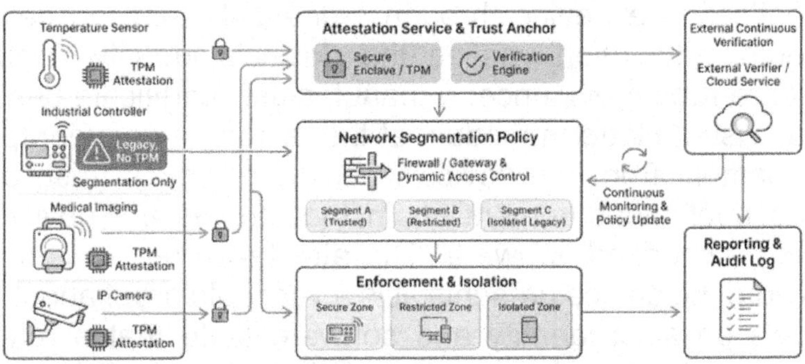

IoT Trust Architecture: Attestation, Segmentation, and External Continuous Verification establishes a resilient security posture by verifying device identities, isolating compromised or legacy assets, and continuously validating trust externally.

3.6 Zero Trust Theater

Zero Trust theater is the organizational failure mode where the language, purchasing, and project management of zero trust proceed without the architectural change that gives them meaning. It is not fraud, and it is not always negligence. It is the natural result of deploying security products into an unchanged architecture, which occurs when procurement precedes design.

The symptoms are consistent. An organization purchases a cloud access security broker and configures it to broker SaaS access; they have not changed the flat network behind the CASB. They deploy an identity governance platform and access certification campaigns; they have not implemented just-in-time access or eliminated standing privilege. They add MFA to the VPN; the VPN still grants full network access upon successful authentication, and the MFA is a stronger lock on a door that opens into an empty field. Each of these is a real improvement. None of them is a zero-trust architecture.

The distinguishing question is whether the underlying trust assumptions changed. Zero trust theater buys new tools and plugs them into old assumptions. Genuine zero-trust architecture first changes assumptions, then identifies tools to operationalize the new model. The sequence is: define the access boundary, define the policy engine and enforcement points, define the identity and authorization model, and then select tools that fill the defined roles. Not: purchase zero-trust-labeled products and hope the architecture assembles itself around them.

Identifying theater in an existing deployment requires asking specific architectural questions rather than reviewing vendor dashboards. Where is the policy engine? Is it a distinct component with its own decision logic, or a feature flag within a firewall console? What is the blast radius from any given compromised endpoint? If the answer requires running a manual simulation rather than relying on continuous measurements, the blast radius is not being managed architecturally. Are credentials time-bounded and automatically revoked, or do they persist until a human revokes them manually? Is microsegmentation enforced at the workload level, or only at the VLAN boundary?

Engineering teams that have inherited Zero Trust theater installations often know it. They see the product stack, they understand the flat network underneath it, and they have been told not to raise the contradiction publicly because the executive team is proud of the investment. Part of the engineer's job in this situation is to translate the architectural problem into business terms, to explain that the $4 million in tools is not protecting what the business thinks it is protecting, and to propose a pilot-to-production pathway that converts the existing investment into genuine architectural change rather than writing it off.

Zero Trust Theater vs. Genuine Architecture: The Underlying Trust Assumption Test

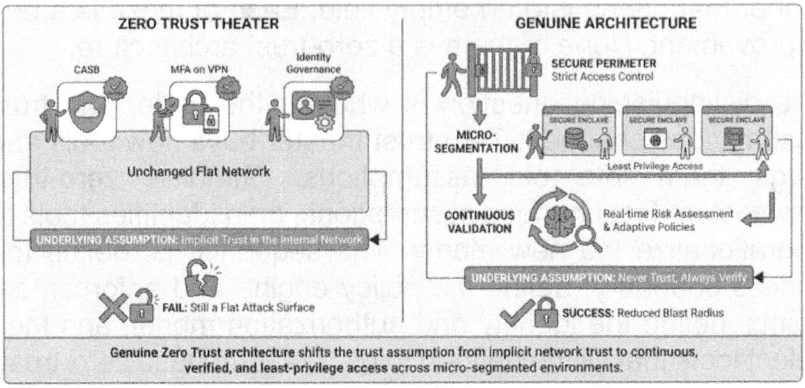

3.7 Geeks Only

The following section goes deeper into the engineering details of the policy decision plane, attribute-based access control, and the design implications of zero trust for microservices and APIs. Readers who are primarily focused on organizational and managerial aspects can move to the Manager's Decision Guide without loss of continuity.

Policy Decision Plane Data Flow

The policy decision plane is the system boundary within which trust decisions are computed. Understanding its data flow at the component level is essential for designing an implementation that is both correct and operationally maintainable. The flow begins at the access request and terminates with an authorized session credential or a deny response with a logged reason.

When a subject initiates an access request, the Policy Enforcement Point intercepts it before it reaches the resource. The PEP does not evaluate the request; it packages the request context into a structured decision request and forwards it to the Policy Administrator. The decision request includes: the subject identity token (typically a signed JWT or SAML assertion from the identity provider), the resource identifier, the requested action scope, the device identity and compliance state from the endpoint management system, the network context (source address, egress path), and any session-level behavioral signals from the SIEM or UEBA system.

The Policy Administrator receives this bundle, enriches it with additional context from its own data sources (threat intelligence, data classification labels, time-based policy rules), and passes the enriched request to the Policy Engine. The Policy Engine evaluates the request against the applicable policy set using a trust algorithm. In a basic implementation, the trust algorithm is a logical rule. If the identity is in group X, the device is compliant, and the time is within office hours, grant access. In a mature implementation, the trust algorithm produces a continuous trust score derived from a weighted combination of signal inputs, where each signal type has a configured weight and a confidence decay function.

The Policy Engine returns a decision (grant or deny), an optional step-up authentication requirement (if the trust score is above the deny threshold but below the grant threshold), and a set of session constraints (maximum session duration, allowed actions, data classification ceiling). The Policy Administrator translates the grant decision into a short-lived, scoped credential: a signed JWT with embedded claims, a SPIFFE SVID for service-to-service authentication, or a time-limited cloud provider role assumption token. That credential is returned to the PEP, which presents it to the resource endpoint. The resource endpoint validates the credential signature and expiration before processing the request.

Continuous evaluation operates in parallel. While the session is active, behavioral signals continue to flow to the Policy Administrator from network telemetry, endpoint detection agents, and the SIEM. If any signal crosses a configured threshold, such as anomalous data transfer volume, access to an out-of-scope resource, or a failed authentication attempt on an adjacent system, the Policy Administrator can revoke the session credential before its scheduled expiration. The PEP on the next request will find the credential revoked and will re-trigger the full decision flow, which may now result in a deny or a step-up authentication challenge.

**Policy Decision Plane Data Flow:
Request, Enrich, Decide, Enforce, Monitor**

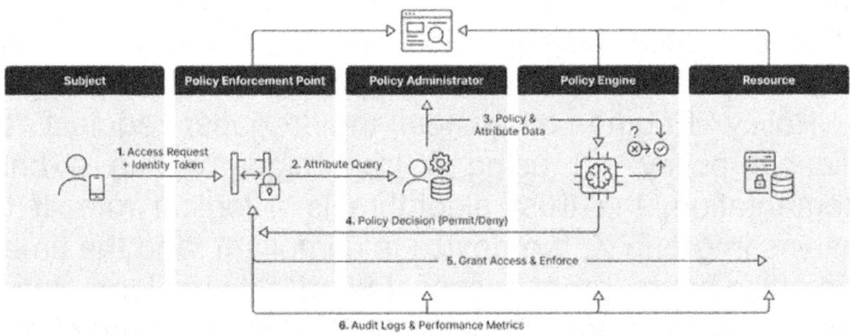

Policy Decision Plane Data Flow: Request, Enrich, Decide, Enforce, Monitor

Attribute-Based Access Control

Role-based access control assigns permissions to roles and users. It is administratively simple and produces predictable, auditable permission sets. It is also fundamentally coarse-grained: a role is a static approximation of a user's access needs averaged across all the tasks the role performs, all the contexts in which those tasks might occur, and all the data classifications the role might touch. That approximation is adequate for many use cases. It is inadequate for zero trust.

Attribute-based access control evaluates access decisions against a policy that can reference any attribute of the subject, the resource, the action, and the environment at decision time. A subject attribute might include identity group, clearance level, department, geographic location, and device compliance state. A resource attribute might include data classification level, owning business unit, retention policy, and regulatory scope. An action attribute describes what the subject wants to do: read, write, delete, or execute. An environment attribute captures context: time of day, network path, session risk score, and concurrent access patterns.

The power of ABAC in a zero-trust context is that it enables policies that are simultaneously more precise and more resilient than RBAC. Precise because the policy can enforce different access levels for the same user based on the context of the specific request: a financial analyst can read aggregated data from any location but can export raw records only from a managed device during business hours. Resilient because adding a new resource or context dimension does not require creating a new role; it requires adding attributes to the resource or policy rule, which is a configuration change rather than an identity management event.

Implementing ABAC requires an attribute store that the Policy Engine can query at decision time, a policy language expressive enough to reference arbitrary attributes in logical conditions, and an attribute governance process that ensures attribute values are accurate and current. XACML is the formal standard for ABAC policy language. However, most modern implementations use more readable alternatives such as Open Policy Agent's Rego language or Cedar, Amazon's purpose-built authorization policy language. The choice of policy language matters less than the commitment to maintaining attribute values: an ABAC policy that evaluates stale clearance levels or outdated device compliance states is making decisions on false premises.

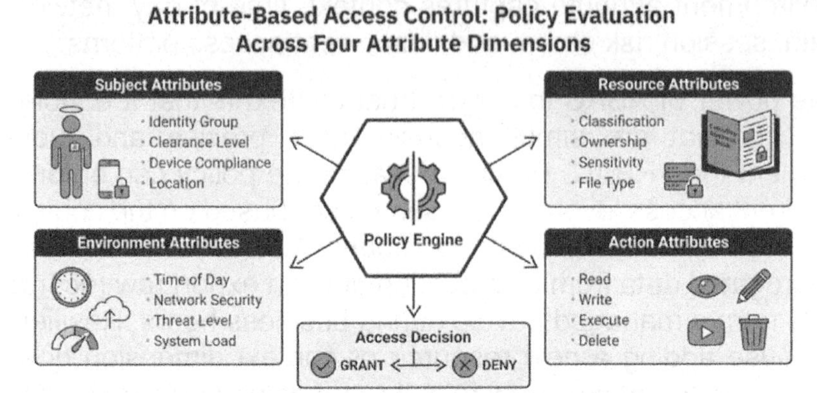

Attribute-Based Access Control: Policy Evaluation Across Four Attribute Dimensions

Microservice and API Design Implications

Zero trust principles change how microservices and APIs are designed, not just how they are secured after the fact. The distinction matters because security controls bolted onto an architecture designed for implicit internal trust will always run counter to the design's grain. Services built to trust any request from within the cluster require extensive middleware to enforce zero-trust policies. Services designed from the start with zero-trust assumptions build verification logic into their request handling.

The first design implication is that every service must have a cryptographic identity. In a microservices environment, service identities are typically implemented through the SPIFFE (Secure Production Identity Framework for Everyone) standard and its runtime implementation SPIRE. Each service receives a SPIFFE Verifiable Identity Document (SVID), a short-lived X.509 certificate or JWT that attests the service's identity to other services it communicates with. The SVID is issued by the SPIRE server, renewed automatically before expiration, and revoked when the service is decommissioned. Service-to-service authentication uses mutual TLS with SVID certificates, so both the client and the server verify each other's identity before any data is exchanged.

The second implication is that API authorization must be enforced at the API layer, not assumed from the network path. An API that accepts requests from any caller with a valid network route is trusting the network. An API that validates a signed authorization token on every request, checks the token's scope against the requested action, and logs the decision is enforcing zero trust at the application layer. This means that API gateway configuration is a security control, not just a routing configuration: the gateway's authorization policy defines which client identities may call which endpoints with which scopes, and that policy must be maintained with the same rigor as firewall rules.

The third implication involves service mesh architecture. A service mesh like Istio or Linkerd implements zero-trust controls at the infrastructure layer, enforcing mTLS between all service pairs, carrying authorization policies as sidecar configuration alongside each service, and producing detailed telemetry for every service-to-service call. The mesh is the Policy Enforcement Point for the microservices layer. Its authorization policy is the policy plane for east-west traffic, and its telemetry is the input to continuous evaluation for service behavior. Teams that have deployed a service mesh have a significant zero-trust foundation; teams that have deployed microservices without a mesh are enforcing zero trust manually at each service, which is brittle and inconsistent.

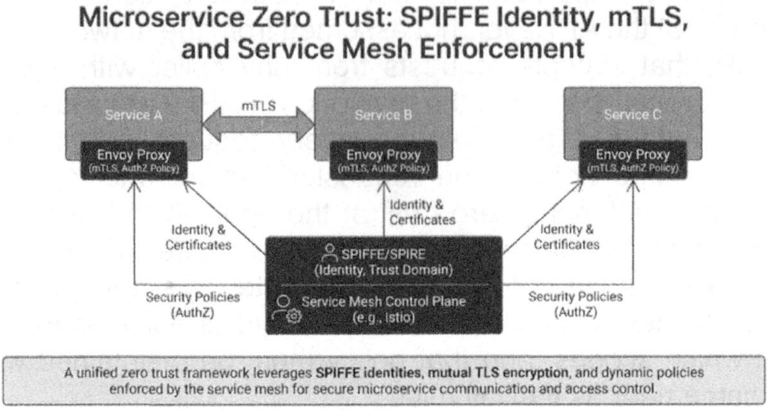

Trust Score Engineering

The continuous trust evaluation model requires a mechanism for producing a numeric trust score from heterogeneous signal inputs. Trust score engineering is the discipline of designing mechanisms that produce accurate, explainable, and appropriately sensitive decisions to signal changes.

A trust score architecture has three layers: signal ingestion, signal normalization, and score computation. Signal ingestion collects raw data from the identity provider (authentication events, MFA outcomes), the endpoint management system (compliance state, patch level, antivirus status), network telemetry (traffic volume, destination patterns, protocol anomalies), and the SIEM (correlated alert context, threat intelligence matches). Each signal source produces data at a different frequency and with different reliability characteristics.

Signal normalization converts raw signals into scores on a common scale, typically 0 to 100, where higher values indicate greater trust. Each dimension has a configured decay function: confidence in a signal degrades over time unless it is refreshed. An endpoint compliance check that ran four hours ago contributes less to the current trust score than one that ran four minutes ago. The decay rate is configurable per signal type because different signals have different staleness profiles. Identity provider signals are nearly real-time; hardware attestation signals may be refreshed only at boot.

Score computation aggregates normalized dimensions into a single trust score using a weighted combination. The weights represent the organization's policy judgment about which signals are most predictive of access risk. A low device compliance score should outweigh a high identity confidence score for privileged access requests, even if the identity is highly trusted, because an unpatched device is a compromised-in-progress risk regardless of who is sitting at it. The output trust score is compared against a threshold policy: above the grant threshold, a grant decision is produced; below the deny threshold, a deny decision is produced; the band in between triggers a step-up authentication challenge.

3.8 Manager's Decision Guide

Use these decision criteria to evaluate zero trust proposals, budget requests, and vendor presentations. Each criterion probes whether the initiative involves genuine architectural change or tool procurement layered on unchanged trust assumptions.

Decision Criteria

- Before approving any zero-trust product purchase, require the engineering team to map the proposed tool to a specific NIST 800-207 component (policy engine, policy administrator, or policy enforcement point). If the team cannot make the mapping, the tool's role in the architecture is undefined.
- Require a blast radius measurement as a precondition for the initiative, not a post-implementation metric. If the team cannot state the current blast radius from the most critical systems, the baseline is unknown, and improvement cannot be measured.

- Ask whether any trust assumptions in the underlying network architecture will change as part of the initiative. New tools on top of unchanged flat networks represent Zero Trust theater regardless of the vendor's marketing claims.
- Confirm that AI and ML workloads and IoT devices are explicitly included in the scope of the initiative. Treating them as exceptions or future phases leaves categories of identity outside the policy plane and creates exploitable gaps.
- Require that least privilege enforcement for the initiative be automated, with time-bounded credentials and automatic expiration. Manual access review cycles are not a zero-trust control; they are an administrative process that does not scale with the velocity of cloud environments.
- Establish a pilot-to-production pathway with measurable security outcomes, specifically blast radius reduction and lateral movement elimination metrics, before committing to full deployment spend. Pilots that cannot demonstrate measurable improvement in these metrics should not proceed to full production.
- For any AI agent deployment, a documented identity governance model is required that separates agent identity from delegating human identity and defines the agent's authorized action scope before deployment.

Decision Frame

The central question in any evaluation of a zero-trust initiative is: when this initiative is complete, which trust assumptions will have changed, and how will we measure that change? If the answer describes only new tools added to the environment, the initiative is theater. If the answer describes specific architectural boundaries that will be moved, specific implicit trust relationships that will replace verified identity assertions, and specific blast radius measurements that will shrink by a defined amount, the initiative has the right architecture. Fund the second kind. Redirect the first kind toward design before procurement.

3.9 Case Study: Meridian Capital Group and the $4 Million Architecture Lesson

Meridian Capital Group manages $14 billion in client assets from offices in three cities. Its technology organization supports a trading platform, a client portal, a risk analytics environment, and the full suite of corporate systems that any financial services firm of its size requires. In 2023, Meridian's board approved a $4 million zero-trust initiative in direct response to two industry peer breaches that had generated significant regulatory attention. The CISO presented a three-product strategy: a next-generation firewall suite with advanced threat detection, an identity governance platform for access certification and lifecycle management, and a cloud access security broker to govern SaaS usage. Eighteen months and $4 million later, a red team engagement by an outside firm revealed that a simulated attacker compromising a developer workstation on the corporate network could reach Meridian's core trading platform in under 9 minutes.

The penetration test report identified the root cause with clinical precision. The firm had deployed three sophisticated products into an architecture that had not changed. The network remained flat between corporate and trading segments, with firewall rules that trusted traffic from the developer VLAN by default. The identity governance platform was running access certification campaigns that confirmed existing access rather than re-evaluating whether the access was still necessary. The CASB governed SaaS usage through a proxy, which users occasionally bypassed by connecting directly via the still-active VPN. The VPN itself granted full network access upon successful authentication, with MFA on the login screen. The MFA was a stronger gate on the same flat field.

Meridian's security team had not been complacent. They had followed vendor deployment guides, completed implementation projects on schedule, and reported green metrics to the board at each quarterly review. The problem was that the metrics measured tool operation rather than architectural security posture. They measured whether the CASB was proxying traffic, not whether the trading platform was reachable from a compromised endpoint. They measured whether identity certification campaigns ran on schedule, not whether standing privilege had been reduced. The distinction between tool operation metrics and security posture metrics is the operational gap that Zero Trust theater exploits.

Meridian Capital Group: Attack Path Before and After Architectural Remediation

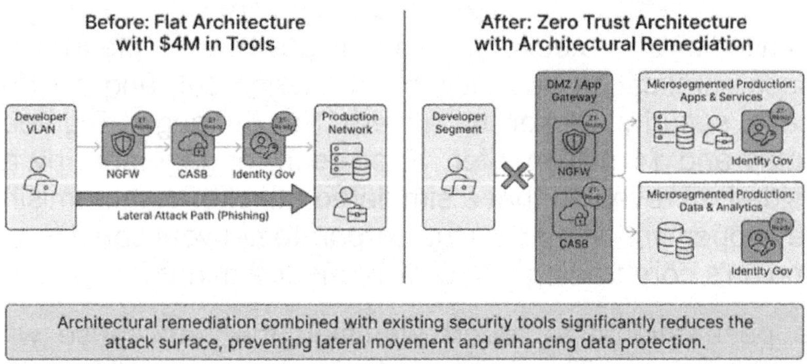

The recovery pathway Meridian adopted after the red team report illustrates the pilot-to-production approach that a genuine zero-trust implementation requires. Rather than writing off the existing investments, the firm's security architects mapped each of the three products to a role in the NIST 800-207 reference architecture and identified gaps. The identity governance platform, reconfigured and integrated with a just-in-time access system, became a partial policy administrator. The NGFW, repositioned as a segment boundary enforcement point rather than a perimeter device, began serving a genuine PEP role once the network was microsegmented. The CASB retained its SaaS governance function. The firm added a dedicated policy engine, an authorization platform that ingested signals from all three products and made centralized access decisions.

The first pilot segment was the trading platform boundary. The team defined a strict microsegmentation policy requiring explicit, identity-verified authorization for any traffic entering the trading zone. They established blast radius measurements at the start: from any developer workstation, an attacker could reach 47 distinct internal resources. After eight weeks of pilot operation with the new policy engine and segmentation controls, the blast radius from the same starting point was four resources, all within the developer zone itself. The trading platform required three verified policy crossings to reach, each requiring a distinct identity assertion. The nine-minute attack path from the red team report had been architected out of existence.

The lesson Meridian's CISO communicated to the board was not that the original products were bad purchases. It was that the products do not produce architecture. The $4 million had been spent on tools before the organization understood what architecture it was trying to build. The recovery required a design-first iteration: define the policy plane, define the enforcement points, define the segmentation boundaries, and then configure existing tools to fill the defined roles while adding purpose-built components where gaps remained. That sequence, design before procurement, is the pilot-to-production pathway that zero-trust implementation requires.

3.10 Team Conversation Guide

Run these three discussions in sequence over two to four weeks. The first surfaces the issue, the second works the trade-offs, the third commits the team to action.

Discussion 1. Mapping Our Architecture to the NIST 800-207 Model

Run this discussion in a whiteboard session with the engineering leads and any architect responsible for infrastructure decisions. The goal is to produce a current-state map, not to plan remediation. Remediation planning comes later. Understanding the gap comes first.

Discussion questions to ask the team:

- Where is our policy engine right now? Is it a distinct, centralized component with its own decision logic, or is it a collection of access rules distributed across firewall consoles, identity provider group memberships, and cloud IAM policies, with no single team having visibility across them?
- Where are our policy enforcement points? Can we name every boundary in our environment where an access request is intercepted and validated before reaching a resource, or are there paths that bypass enforcement entirely?
- If I compromise an endpoint in the corporate network right now, what is the blast radius? How many resources can I reach before hitting an enforcement point that requires a new verified identity assertion?
- Which of our zero-trust-labeled investments map to a specific NIST 800-207 component, and which are operating in a role that is not connected to our policy decision plane?

What to listen for: the team's comfort or discomfort with the policy engine question. If they can answer it quickly and specifically, the organization has architectural clarity. If the answer involves listing products without describing decision logic, the policy plane is likely distributed and ungoverned. Also listen for whether the blast radius question yields a specific number or a hedged estimate; the answer's precision reflects the architecture's precision.

Follow-up actions: Produce a one-page current-state map of the policy plane, including identified gaps, before the next discussion. Assign a specific engineer to own each component mapping. Gaps are not failures; they are the design inputs for the remediation roadmap.

Discussion 2. Identifying Genuine Change Versus Zero Trust Theater

Bring the current-state map from Discussion 1. This conversation is about evaluating which existing investments represent real architectural change and which tools are layered on unchanged trust assumptions. It requires candor from the engineering team and from the manager, because the answer may be uncomfortable.

Discussion questions to ask the team:

- For each security product we have purchased in the past two years, with zero trust in its marketing, what specific trust assumption in our architecture did it change? Not what it monitors or detects, but what implicit trust relationship it eliminated.
- What is still implicitly trusted in our architecture because of network position rather than verified identity? Which services, which segments, which user populations still receive access based on where they are rather than who they are?
- If we turned off our newest zero-trust product tomorrow, would our blast radius change? If the answer is no, what does that tell us about the architectural role that product is actually playing?
- Where are we running manual access review processes as a substitute for automated policy enforcement? Are those reviews producing outcomes different from the access grants they review, or are they primarily confirming existing access?

What to listen for: the team's ability to distinguish between tool operation and architectural security posture. Teams that can articulate the difference between "the CASB is proxying traffic" and "the trust assumption changed" have the conceptual vocabulary for genuine remediation. Teams that conflate the two will need the distinction made explicit before design work can proceed effectively.

Follow-up actions: Categorize each existing security investment as an architectural change or tool layering. For the tool-layering category, identify the architectural change required to convert each investment into a genuine zero-trust control. This becomes the design input for the remediation initiative.

Discussion 3. Defining the Pilot-to-Production Pathway

This is the action-commitment discussion. By this point, the team has a current-state map and a classification of existing investments. The goal is to define a specific, measurable pilot that demonstrates a reduction in blast radius before committing to full-scale remediation spend.

Discussion questions to ask the team:

- Which single boundary in our architecture, if converted to identity-verified enforcement, would produce the largest blast radius reduction? That boundary is our pilot candidate.
- What does success look like for the pilot in measurable terms? Not "zero trust is implemented" but specifically. From a compromised endpoint in zone X, the blast radius shrinks from N resources to M resources, and lateral movement to the trading platform requires three verified policy crossings instead of zero.
- What is the minimum viable policy plane for the pilot? Which combination of policy engine, policy administrator, and enforcement point components can we assemble from existing investments, along with targeted additions, to run the pilot?
- How will we demonstrate to the board that the pilot produced a genuine security posture improvement, not just a successful product deployment? What metrics will we report and how will we collect them?
- What is the rollback plan if the pilot disrupts legitimate workflows? Least-privilege enforcement

that denies legitimate access slows engineering velocity; we need a fast path back to a known-good state while we tune the policy.

What to listen for: Listen for whether the team can define success in blast radius terms rather than product operation terms. A pilot that succeeds when "the policy engine is processing requests" is measuring the wrong thing. A pilot that succeeds when "blast radius from zone X dropped from 47 to 4 verified resources" is measuring the right thing. The ability to define the right metric is the signal that the team understands the architecture, not just the tools.

Follow-up actions: Document the pilot scope, success metrics, and rollback plan as formal acceptance criteria before any pilot spend is approved. Schedule the blast radius measurement as the first activity, not the last. The before measurement is the baseline, without which the after measurement means nothing.

3.11 Takeaway

The three zero-trust principles are not aspirational statements; they are engineering constraints. Never trust, always verify means that identity and authorization must be established cryptographically at every access boundary, with no inheritance from network position. Least privilege means that permissions are time-bounded, task-scoped, and automatically revoked, not granted at onboarding and reviewed annually. Assume breach means that blast radius is a primary design metric, measured continuously and reduced architecturally rather than managed through incident response.

NIST 800-207 gives those principles a system design vocabulary. The policy engine decides. The policy administrator issues credentials and manages the session lifecycle. The policy enforcement point applies decisions at every boundary. Identifiable components must fill those three roles in any genuine zero-trust implementation; an architecture that cannot name what fills each role has not implemented a policy plane; it has implemented a collection of tools.

AI and machine-learning workloads, as well as IoT devices, extend the trust problem beyond human-to-resource access. AI agents require delegated identity governance, scoped inference-time access, and isolation of the training pipeline. IoT devices that cannot authenticate natively require hardware attestation, strict network segmentation, and external behavioral monitoring as compensating controls. Neither category can be treated as a future phase without leaving a significant attack surface unaddressed.

Zero Trust theater is the failure mode that results when procurement precedes design. Products do not produce architecture. The sequence that produces genuine zero trust is: define the policy plane, define the enforcement points, define the segmentation boundaries, measure the blast radius baseline, and then configure tools to fill the defined roles. Every chapter that follows builds on that sequence, going deeper into the three pillars of identity, microsegmentation, and continuous verification with the same design-first discipline that distinguishes real architecture from an expensive collection of dashboards.

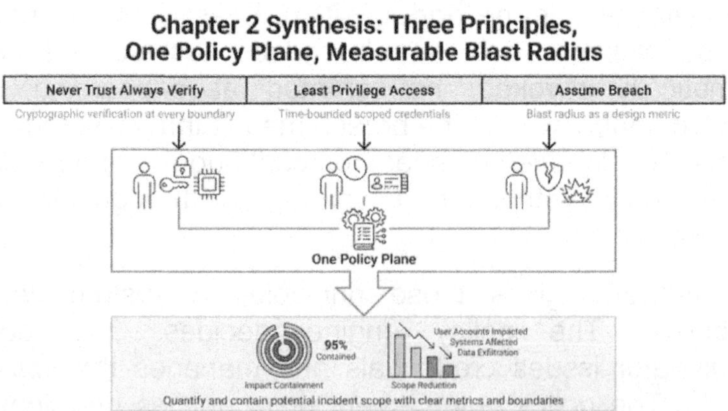

Chapter 2 Synthesis: Three Principles, One Policy Plane, Measurable Blast Radius

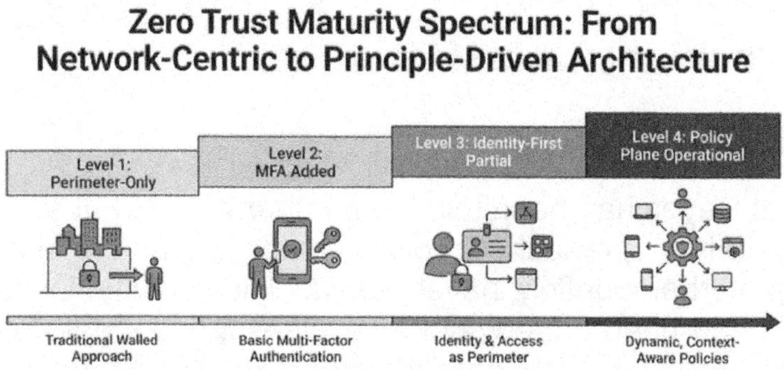

Zero Trust Maturity Spectrum: From Network-Centric to Principle-Driven Architecture

4 Identity Architecture at Scale: Humans, Machines, and Workloads

4.1 Scenario

Marcus Reyes had been told the number before, but seeing it rendered in a spreadsheet column made it concrete in a way that a verbal briefing never could. The identity inventory project had been underway for six weeks, and the team had just delivered the first consolidated count: 4,200 human user accounts and 46,800 non-human identity objects. The ratio was almost exactly eleven to one. For every person who worked at the organization, there were eleven service accounts, API keys, certificates, automation tokens, or device identities running somewhere in the environment, making decisions, opening connections, and reading data on behalf of systems that had been built, extended, patched, and sometimes quietly abandoned over the previous decade.

Marcus walked through the raw numbers with his two senior engineers, Priya and Devonte, on a Tuesday morning. The human side of the inventory was uncomfortable but manageable. About 340 user accounts belonged to contractors whose engagements had ended without formal offboarding, and 88 privileged administrator accounts had not been accessed in more than ninety days. Those were problems, but they were the kind with known solutions. The machine identity side was a different territory entirely. Of the 46,800 non-human identities, roughly 12,400 had no documented owner. Another 8,700 had static credentials, meaning passwords or API keys that had been set at some point and never rotated. Devonte pulled one record at random to illustrate: a service account named svc-legacy-etl-prod, created in 2017, with database administrator rights on four separate data stores, and a password that had not changed in four years. Nobody on the current team knew what it connected to or whether anything would break if it were disabled.

This was the scenario Marcus had been hired to address, but the inventory made clear that the problem was not a simple gap in tooling. It was a governance gap. Nobody had drawn a boundary around machine identity as a category that required the same rigor as human identity. The organization had invested in a mature identity provider for its workforce, had deployed multi-factor authentication for most employees, and had thoughtful joiner-mover-leaver processes for staff. But the parallel universe of machine identities had grown without any of those controls, governed by individual development teams who each made local decisions about how to authenticate services, how long credentials should live, and whether retired credentials needed to be formally retired at all. The answer to that last question, in most cases, had been no.

Marcus opened the chapter of work ahead with a note to himself: identity architecture at scale is not just about the employees. It is about every principal that acts in the environment, whether or not a human being is on the other end of the session. Until every one of those principals has a known owner, a verified lifecycle, and a credential with a defined rotation schedule, the organization cannot make truthful claims about who or what has access to anything.

4.2 Why It Matters

Identity is the foundational control plane of zero-trust architecture. Every other control, whether microsegmentation, continuous monitoring, or device health verification, depends on knowing, with confidence, who or what is requesting access before any access decision is rendered. A network that cannot answer that question reliably is not a zero-trust network; it is a network with additional instrumentation layered on top of implicit trust. The shift from perimeter-based security to identity-based security is not a technology upgrade. It is a rearchitect of the assumption that sits underneath every access control in the environment.

The mission-aligned case for identity investment is straightforward: breaches that exploit identity weaknesses account for the majority of major incidents across every sector. Stolen credentials, orphaned service accounts with elevated privileges, and misconfigured identity federation rules are not exotic attack paths. They are the paths that attackers reliably walk because organizations consistently underinvest in the unglamorous work of identity governance. When an attacker moves laterally through an environment using a legitimate credential, endpoint detection tools treat the session as authorized. Firewall rules see an authorized source. Only an identity-aware control plane that asks continuous questions about whether this credential should do this thing at this time has a chance of detecting the anomaly before data leaves the environment.

The workflow-level impact of poor identity architecture is also significant in both directions. Overly permissive identity controls, such as standing privileges and shared service accounts, create exposure to breaches that grow quietly over time. But overly restrictive controls, implemented without attention to how work actually gets done, create friction that causes teams to route around security processes entirely. The goal of identity architecture at scale is not maximum restriction. It is operational clarity: every principal has exactly the access it needs, for exactly the time it needs it, and it is verified at every step of the session. Achieving that state requires treating identity as a first-class engineering discipline, with the same investment in automation, lifecycle management, and continuous improvement that the organization applies to its application infrastructure.

4.3 The Identity Taxonomy: Who and What Acts in Your Environment

The first instinct when most organizations begin a zero-trust identity project is to think about users: the employees, contractors, and partners who log in to applications and systems. That instinct is understandable but incomplete. In a modern hybrid enterprise, human users typically represent a small fraction of the total identity population. The majority of principals are non-human: service accounts, workload identities, API tokens, machine certificates, AI agent identities, and IoT device identities. Each category has distinct authentication characteristics, distinct lifecycle requirements, and distinct governance challenges. A zero-trust identity architecture must address all of them.

The Complete Identity Taxonomy for Hybrid Enterprises

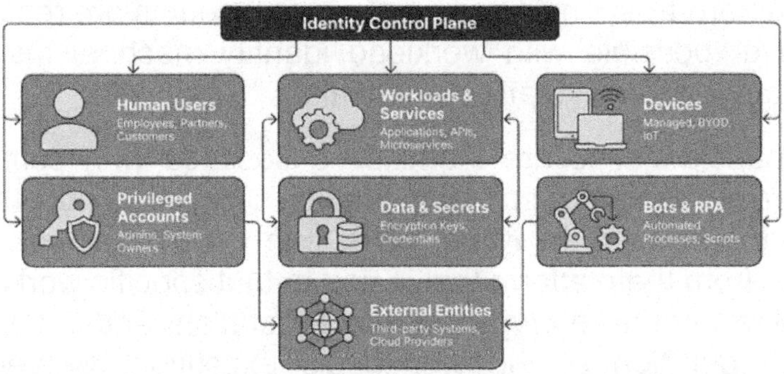

This comprehensive framework organizes **all human and non-human identities** across on-premises and cloud environments, ensuring unified governance, security, and compliance.

Human user identities are the most familiar category and the one most organizations have invested in managing. A human user identity binds a cryptographically verified credential to a real person, with attributes describing their role, organizational unit, clearance or authorization level, and employment status. The authentication strength required for human users scales with the sensitivity of what they are accessing: a standard user reaching a low-sensitivity application might authenticate with a corporate SSO session, while the same user reaching a privileged management console must re-authenticate with a phishing-resistant factor, ideally a hardware security key or a PIV/CAC card in federal environments.

Service accounts are the most numerically dominant identity type in most enterprises and the most poorly governed. A service account is an identity created to represent a software process rather than a person. It authenticates databases, message queues, APIs, and other services on behalf of the application it serves. The governance problems with service accounts are well-documented and consistently exploited: they accumulate over time without systematic deprovisioning, they are often granted excessive privileges because the developer who created them erred on the side of permissiveness, and their credentials are frequently static because rotation requires coordination between the account owner and every system that depends on the credential. In a mature zero-trust environment, service accounts are replaced wherever possible with workload identity mechanisms that eliminate long-lived static credentials.

Workload identities are identities associated with compute workloads rather than with static service accounts. A container running in Kubernetes can receive a workload identity from the platform that is tied to that specific workload, verifiable through a cryptographic attestation, and valid only for the duration of the workload's execution. Serverless functions receive similar short-lived identity tokens from the cloud platform. CI/CD pipelines can authenticate to deployment targets using workload identity federation rather than stored credentials. The key property of workload identity is that it eliminates the static credential: instead of a password or API key that must be managed and rotated, the workload proves its identity through a mechanism rooted in the platform itself.

Machine identities cover infrastructure components that are not executing application workloads but still require a verifiable identity for secure communication. Servers presenting TLS certificates to clients, network devices authenticating to management planes, and endpoint devices enrolling in device management systems all hold machine identities. The certificate is the most common machine identity mechanism: an X.509 certificate issued by a trusted certificate authority that binds a public key to an identity assertion about the device. Certificate lifecycle management at scale requires automation; organizations that manage machine certificates in spreadsheets and with calendar reminders inevitably accumulate expired or misconfigured certificates, disrupting operations or creating security gaps.

AI agent identities constitute an emerging, rapidly growing category. An AI agent is an autonomous software system that makes decisions and takes actions without direct human supervision. Agents interact with APIs, read data stores, write outputs, and sometimes invoke other agents. Each of those interactions requires authenticated access, which means each agent requires a verifiable identity with scoped permissions. The challenge is that agents operate at machine speed, may dynamically compose new action sequences, and may be instantiated in large numbers from a single agent definition. Identity governance frameworks designed for human users and static service accounts are inadequate for agents: an agent session's lifecycle may last seconds, its action scope may vary dynamically, and the number of concurrent agent instances may be orders of magnitude larger than any human workforce.

IoT device identities present a different set of challenges at the opposite end of the computational spectrum. Many IoT devices have limited processing power, no user interface, and firmware that is difficult to update. Enrolling these devices in a zero-trust identity framework requires hardware-rooted identity mechanisms, such as a Trusted Platform Module or a manufacturer-embedded certificate, that can prove device authenticity without relying on user interaction. In healthcare environments like the one described in the Great Lakes case study, connected medical devices may number in the hundreds of thousands, each requiring a verifiable identity to enforce network segmentation and detect anomalous behavior.

Human vs. Non-Human Identity Distribution in a Typical Hybrid Enterprise

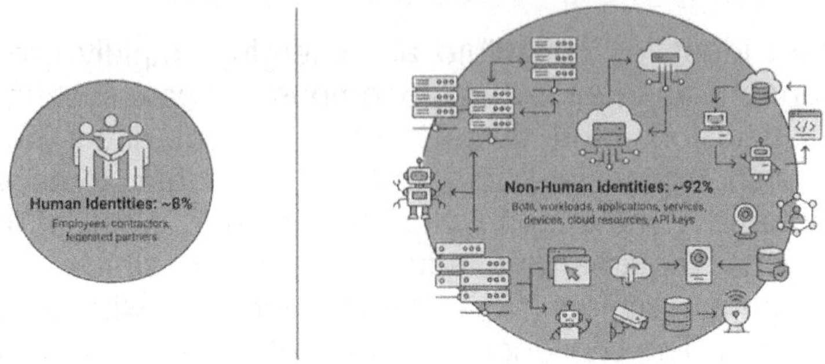

Human vs. Non-Human Identity Distribution in a Typical Hybrid Enterprise highlights the overwhelming majority of machine-based entities managing access and processes in modern business environments.

4.4 Identity Providers, Directories, and Federation

An identity provider is the authoritative system that stores, authenticates, and issues assertions about identities. In a simple environment, a single identity provider serves all principals. In a hybrid enterprise, the reality is almost always a federation of identity providers: a corporate directory for workforce identities, a cloud platform identity service for workload identities, a partner identity provider for B2B users, and potentially separate systems for IoT devices and privileged access. The architecture of how these systems relate to one another and how one system trusts assertions from another is the federation layer.

Identity Provider Federation Architecture

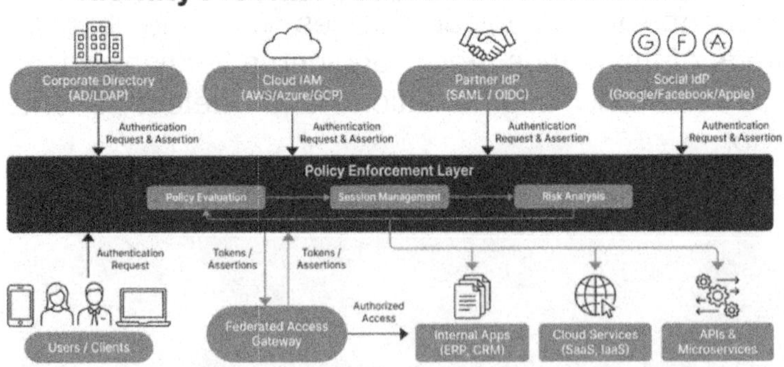

Identity Provider Federation Architecture centralizes authentication and authorization, enabling secure and seamless access across diverse identity sources and protected resources.

The dominant protocols for identity federation in modern enterprises are SAML 2.0 and OpenID Connect (OIDC). SAML is a mature XML-based protocol that remains dominant in enterprise SaaS integration, particularly for browser-based single sign-on. OIDC is a more recent, developer-friendly protocol built on OAuth 2.0 that handles the delivery of authentication assertions via JSON Web Tokens. Most modern enterprises operate with both SAML for the legacy SaaS estate and OIDC for cloud-native applications and API authentication. The practical architecture implication is that the identity team must be fluent in both protocols and must ensure that the policy enforcement layer can consume assertions in either format.

Directory consolidation is a prerequisite for coherent identity governance. Organizations that have grown through acquisition or organic expansion often maintain multiple directories: a Microsoft Active Directory instance from one business unit, an LDAP directory from another, and a standalone cloud identity tenant from a product team that onboarded to a SaaS provider before the enterprise had a standardized approach. Each separate directory represents a separate source of truth, and reconciling them requires either technical consolidation (migrating to a single authoritative directory) or logical consolidation (establishing an authoritative identity graph that aggregates records from multiple directories and resolves conflicts). Neither path is simple, but both are more manageable than accepting ongoing identity fragmentation as a permanent condition.

In federal environments, the identity framework must accommodate PIV and CAC cards as primary authentication mechanisms for human users. PIV (Personal Identity Verification) cards and CAC (Common Access Card) cards are hardware tokens that carry a certificate-based identity credential issued by a federal certification authority. They satisfy phishing-resistant authentication requirements because the private key never leaves the card. The ICAM (Identity, Credential, and Access Management) architecture in federal settings defines how PIV-based identity assertions are federated across agency boundaries and how agency systems establish trust with the federal public key infrastructure. Commercial organizations in regulated sectors often implement equivalent mechanisms using FIDO2 hardware security keys, which provide similar phishing-resistance properties without the federal issuance infrastructure.

FIDO2 Phishing-Resistant Authentication Flow

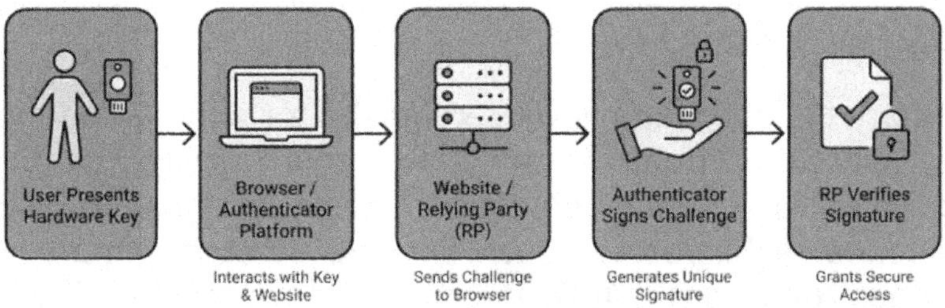

FIDO2 Phishing-Resistant Authentication Flow utilizes **public-key cryptography** to **securely** authenticate users without sharing secrets or passwords, mitigating the risk of credential theft.

Multi-factor authentication at enterprise scale requires more than just enabling an MFA option in an identity provider. The rollout strategy must account for user populations with varying technical capabilities, device access patterns, and risk profiles. Executive populations often have the highest privilege levels and the greatest resistance to authentication friction. Contractor and vendor populations often lack access to corporate-managed devices and may authenticate on personal phones or laptops. Remote workforce users in geographies with unreliable connectivity may need offline-capable authentication methods. A responsible-by-design MFA program defines authentication requirements by resource sensitivity tier and user population segment, then maps available authentication methods to each combination, providing a migration path that moves every population toward phishing-resistant methods over a defined timeline rather than treating MFA as a binary done-or-not-done state.

4.5 Identity Lifecycle Management: Create, Use, Rotate, Retire

The joiner-mover-leaver model is the standard framework for managing the human identity lifecycle. A joiner needs accounts provisioned; a mover needs access adjusted to match a new role; a leaver needs accounts deprovisioned promptly and completely. Most mature organizations have automated the joiner and leaver portions through HR-to-IdP integration. The mover case is typically the weakest link: role changes trigger new grants but do not reliably revoke prior-role access, leading to access accumulation over time.

Identity Lifecycle: Joiner-Mover-Leaver with Machine Identity Extension

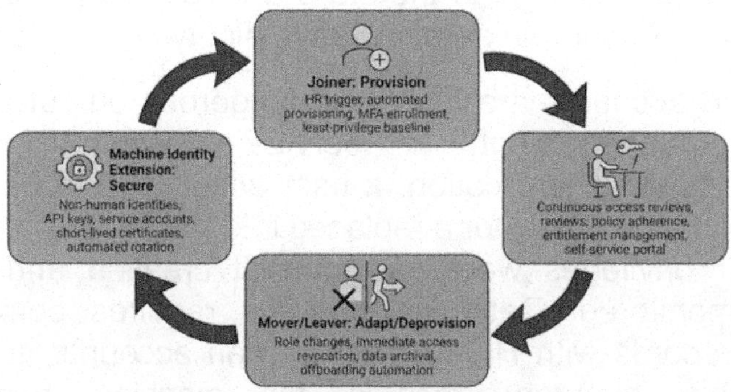

Identity Lifecycle: Joiner-Mover-Leaver with Machine Identity Extension

The machine identity lifecycle requires the same four-phase thinking applied to a fundamentally different operational context. Creation of a machine identity should be triggered by a formal request tied to an application or infrastructure component, with an owner assigned at creation time and a defined expiration. The use phase for machine identities involves authentication events that should be logged and monitored for anomalies, just as human session events are. Rotation is the most operationally challenging phase: rotating a certificate or API key requires coordination between the identity system and every consumer of that credential, which, in practice, means rotation must be automated, or it will not happen reliably. Retirement must be triggered when the associated application or infrastructure component is decommissioned, which requires that the identity system have visibility into the application lifecycle, not just the identity lifecycle.

Access certification is the periodic review by which account owners attest that each identity's access remains appropriate. For human identities, certification campaigns run quarterly or annually and ask managers to review their direct reports' entitlements. For machine identities, the equivalent process asks application owners to review service account permissions. The practical challenge is that ownership of machine identities is often unclear, making it impossible to route a certification request to the right reviewer. Ownership assignment at creation is therefore a functional prerequisite for governance, not an administrative nicety.

Orphaned accounts are the most dangerous output of poor lifecycle management: a service account from a decommissioned application, a user account for a departed contractor, an API key for a replaced integration. Each retains whatever privileges were assigned at creation and is no longer monitored. Detecting orphans requires correlating identity records with HR data for human accounts and with application inventory records for machine accounts. Remediation requires a disabling process with an ownership verification step to confirm that a dormant-looking account is not still serving an infrequent process.

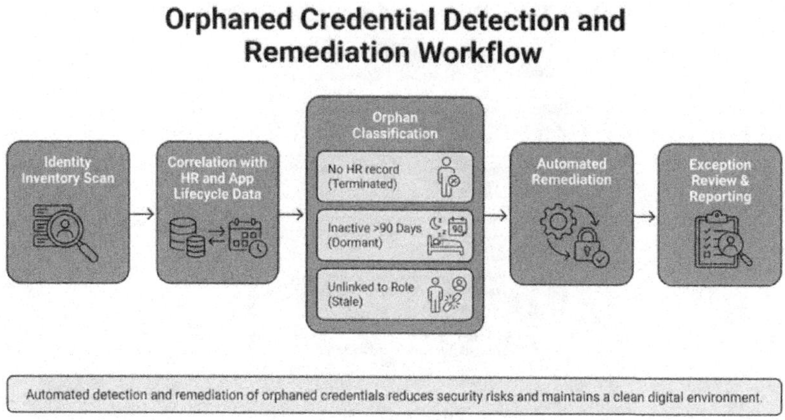

4.6 Privileged Access Management and Just-in-Time Access

Privileged access is qualitatively different from standard user access. A privileged identity can change configurations, read unencrypted data, modify access controls, or shut down systems. The compromise of a privileged identity gives an attacker capabilities that go far beyond what they could achieve with a standard user credential. This asymmetry is why privileged access management deserves dedicated architectural treatment rather than being handled as a variation on standard access controls.

The foundational principle of modern PAM architecture is the elimination of standing privileges. An account with always-on administrative rights is convenient but dangerous: a stolen credential immediately confers full capability. Just-in-time access replaces standing privilege with a workflow: a session is requested, approved (automatically or by a human reviewer, depending on risk level), granted for a defined duration, and automatically revoked when the session ends. The credential used is ideally ephemeral, generated at session initiation and invalid after closure.

Just-in-Time Privileged Access Workflow.

Session recording serves two purposes in privileged access management. During a session, it serves as a deterrent: administrators who know their keystrokes are being recorded are less likely to take shortcuts that violate policy. After a session, it creates an investigation capability: a recording can pinpoint exactly which commands ran under a privileged credential, potentially compressing a week-long investigation into hours. In regulated environments, session recording is often a compliance requirement with defined retention periods.

Break-glass accounts exist for emergency scenarios where the normal JIT workflow is unavailable. They are stored in a secure vault, accessible only through a controlled process that requires dual authorization and triggers an immediate security operations alert. Each use should trigger a post-incident review. Organizations that find break-glass accounts being used regularly have identified a signal that normal privileged access processes are inadequate; the answer is to fix the normal process, not to normalize break-glass use.

Privileged access for automated processes requires a different implementation pattern than for human administrators. A deployment pipeline that must push configuration changes to production infrastructure cannot pause to wait for a human to approve a JIT request. The equivalent architecture for automated privileged access is short-lived credentials issued by the CI/CD platform, scoped to exactly the operations required by the deployment, and tied to the identity of the pipeline run rather than to a standing service account. This pattern, often implemented through a cloud platform workload identity or a secrets manager with dynamic secret generation, achieves the same privilege elimination as JIT for human administrators.

4.7 Machine Identity Governance at Scale

Governing machine identities at scale is one of the most practically challenging problems in zero-trust implementation. The difficulty has three sources. Volume: machine identities typically outnumber human identities by a large margin, and governance workflows designed for humans do not scale to tens of thousands of non-human principals without automation. Ownership: machine identities are created by developers who may not consider governance when they spin up a service account. Lifecycle: machine identity events are tied to application and infrastructure events that may not be visible to the identity team.

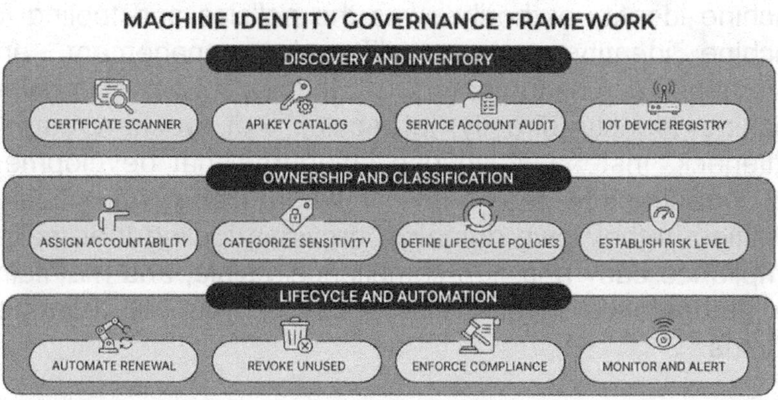

Machine Identity Governance Framework: Ensuring security and compliance throughout the machine identity lifecycle.

Certificate lifecycle management is a particularly acute operational challenge. Expired certificates cause service outages rather than security incidents in the traditional sense, yet they are consistently traced to the absence of automated expiry monitoring. The solution is a certificate lifecycle management platform that inventories all certificates, issues alerts at thirty, fifteen, and seven days before expiration, and automates renewal through integration with the issuing certificate authority. The ACME protocol, which standardizes automated certificate issuance and renewal, should be the default for any certificate not subject to manual review requirements.

API key governance suffers from a different set of problems. Keys created for development or testing are frequently committed to source code repositories or configuration files and persist indefinitely. The governance response has three components: detection, through tools that scan repositories and configuration systems for credential patterns; restriction, through policies requiring keys to be stored in approved secrets management systems; and rotation, through automated expiration windows so that any leaked key has a limited window of utility.

The concept of a machine identity authority is useful for organizations as they scale their governance programs. Analogous to a certificate authority but broader in scope, a machine identity authority owns the policies and tooling for machine identity issuance, lifecycle management, and governance across the enterprise. It does not provision every machine identity directly; at scale, that would create a bottleneck. Instead, it defines standards that development and infrastructure teams follow when managing machine identities in their own domains, provides tooling that makes compliance easy and non-compliance visible, and maintains the authoritative inventory aggregated from all federated sources.

4.8 Geeks Only

This section covers the technical implementation details that underpin the identity architecture concepts in the preceding sections. It is written for engineers who will design, build, or operate these systems and need to understand the protocol-level mechanics, not just the architectural patterns.

OIDC and OAuth 2.0 Flows in Zero Trust

OpenID Connect (OIDC) is the authentication layer built on top of OAuth 2.0. To use it precisely: OAuth 2.0 is an authorization framework that defines how a resource owner can delegate access to a resource server on behalf of a client; OIDC adds an identity layer by defining how the authorization server issues an ID token that asserts facts about the authenticated user. In a zero-trust architecture, both the ID token (for authentication assertions) and the access token (for resource authorization) play distinct roles that must not be conflated.

The authorization code flow with PKCE (Proof Key for Code Exchange) is the recommended OIDC flow for user-facing authentication in modern zero-trust deployments. The flow begins when the client sends an authorization request to the identity provider, including a code challenge derived from a randomly generated code verifier. The user authenticates, and the provider returns an authorization code to the client redirect URI. The client then exchanges that code for tokens by presenting the code verifier, which the provider validates against the earlier challenge. An attacker who intercepts the code cannot complete the exchange without the verifier, which was never transmitted.

OIDC Authorization Code Flow with PKCE in a Zero Trust Context

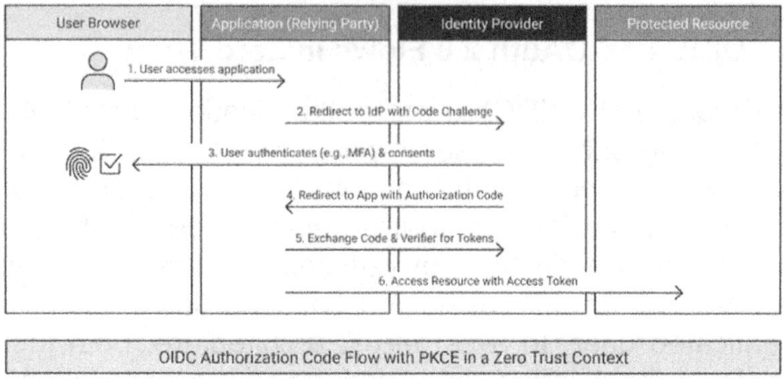

For service-to-service authentication, the client credentials flow is the appropriate OAuth 2.0 pattern. A service authenticates using its client ID and secret (or a signed client assertion from a private key) and receives an access token scoped to the specific APIs it is authorized to call. The zero-trust improvement is achieved by combining short-lived tokens (typically 15 to 60 minutes) with fine-grained scope definitions. A compromise of a short-lived, scoped token gives an attacker a narrow window on a narrow surface, rather than the broad, persistent access that a long-lived shared secret would provide.

mTLS for Workload Identity

Mutual TLS (mTLS) is a transport-layer authentication mechanism in which both the client and the server present X.509 certificates during the TLS handshake and verify the other's certificate against a trusted certificate authority. In standard TLS, only the server presents a certificate; the client authenticates using application-layer credentials such as a password or token. mTLS extends this by requiring the client to prove its identity at the transport layer as well, which means that unauthorized clients are rejected before any application-layer code executes.

In a zero-trust service mesh, mTLS authenticates all service-to-service traffic. Each service receives a short-lived certificate from the mesh control plane at startup. When Service A calls Service B, both sides present and verify each other's certificates against the mesh certificate authority. Neither trusts the other based on network position; they trust based on cryptographically verified identity. The service mesh injects mTLS transparently at the sidecar proxy layer, so application code does not implement it directly.

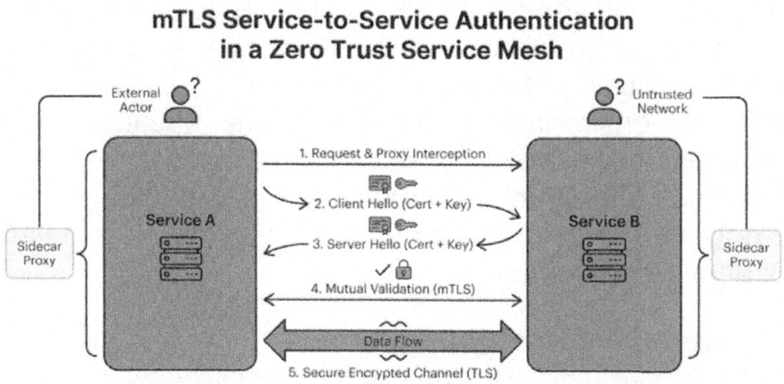

mTLS Service-to-Service Authentication in a Zero Trust Service Mesh ensures mutual verification and encrypted communication, validating identities and securing data in transit between services.

SPIFFE and SPIRE: Workload Identity Standards

SPIFFE, the Secure Production Identity Framework for Everyone, is an open standard defining a uniform identity model for workloads across heterogeneous infrastructure. The core concept is the SPIFFE ID: a URI-formatted identifier of the form spiffe://trust-domain/path that uniquely names a workload within a trust domain. An SPIFFE ID is an identity claim embedded in an authentication credential, most commonly an X.509 certificate (an SVID, or SPIFFE Verifiable Identity Document) or a JWT.

SPIRE, the SPIFFE Runtime Environment, is the reference implementation of the SPIFFE APIs. It provides the machinery for issuing SVIDs to workloads through two components: a SPIRE Server acting as registration and certificate authority for the trust domain, and a SPIRE Agent running on each node to attest the identity of workloads on that node. Node attestation uses platform evidence, such as a cloud instance identity document or TPM attestation, to confirm that the agent is running on the expected infrastructure. Workload attestation uses kernel-level process data to verify the requesting process matches the registered workload definition.

The operational benefit of SPIFFE and SPIRE is a consistent, platform-agnostic workload identity layer. A workload in Kubernetes receives the same SVID style as a workload in a virtual machine, on a bare-metal server, or in a serverless function. Policies referencing SPIFFE IDs work uniformly across all environments. This matters because hybrid enterprises run workloads across multiple cloud providers and on-premises infrastructure, and a workload identity standard that only works in one environment creates segmentation gaps at every boundary.

Credential Rotation Patterns

Credential rotation is the practice of periodically invalidating an existing credential and replacing it with a new one, limiting the window during which a stolen credential remains useful. The rotation interval should be inversely proportional to the credential's exposure risk: a certificate used for external TLS termination that has been logged in thousands of client systems has higher exposure than an internally issued mTLS certificate, and should be rotated more frequently. The practical constraint is that rotation requires coordination between the issuing system and every consumer of the credential, and that coordination cost grows with the number of consumers.

Static credential rotation is the simpler but more operationally demanding pattern. A new credential is generated, distributed to all consumers, and the old credential is revoked after a validation period confirms successful switching. The coordination burden of this pattern means it is frequently deferred, which is precisely the wrong outcome. Automation through a secrets manager that pushes new credentials and manages the switchover window is the only reliable path to consistent rotation at scale.

Dynamic secrets are a more elegant solution. Rather than rotating a long-lived credential on a schedule, the secrets manager generates a new credential for each access request, valid only for the duration of that session. No rotation schedule is needed because the credential was never intended to persist. HashiCorp Vault's dynamic secrets engine and cloud-native equivalents, such as AWS IAM roles for services and Azure Managed Identities, implement this pattern. The prerequisite is that the target system must support dynamic credential issuance, which is not universal in legacy environments.

Secrets Management Integration

A secrets manager is a purpose-built system for storing, distributing, and managing sensitive credentials, certificates, and configuration values. Its core security properties are encryption at rest and in transit, fine-grained access control on which identities can read which secrets, a complete audit log of all secret access events, and versioning that maintains prior secret versions during rotation windows. Secrets managers are not simply secure key-value stores; they are identity-aware policy engines for sensitive material.

Integration between the secrets manager and the deployment pipeline is the architectural seam that determines whether secrets management is operational or theoretical. Applications that retrieve secrets from environment variable files have not adopted secrets management; instead, they add a step before returning to the same insecure pattern. Applications that retrieve secrets at startup from the secrets manager API, authenticated via the workload's identity credential, and receive short-lived or versioned rotating secrets have genuinely integrated secrets management into their operational model.

Containerized workload integration uses one of three patterns. The sidecar injection pattern uses a sidecar container that authenticates with the secrets manager using the pod service account token and writes secrets to a shared in-memory volume, which the application reads. The init container pattern populates secrets before the main container starts. The SDK pattern uses a library within application code to retrieve and refresh secrets at runtime. Each approach carries different trade-offs in complexity, secret freshness, and application coupling.

Attribute-Based vs. Role-Based Access Decisions

Role-based access control (RBAC) is the dominant access control model in most enterprise environments. In RBAC, permissions are assigned to roles, and identities are assigned to roles. An engineer assigned to the database-reader role can read from all databases covered by that role. RBAC is straightforward to reason about and to audit, which is why it became dominant. Its limitation in zero-trust contexts is coarseness: roles aggregate permissions into bundles that are convenient for administration but often grant more access than any individual in the role actually needs for any given task.

Attribute-based access control (ABAC) makes access decisions based on attributes of the requesting identity, the requested resource, and the request context. An access decision in an ABAC system might evaluate: the requester's department attribute, the requester's device health score, the sensitivity label of the requested resource, the time of day, the requester's location, and whether the request matches a known-good behavior pattern for this identity. The decision is a function of all of those inputs, which means the same identity might be granted access under one combination of conditions and denied under another.

The practical architecture for zero trust access decisions is a hybrid: RBAC establishes the coarse-grained permission baseline that determines whether a request is in scope at all, and ABAC provides the fine-grained contextual evaluation that determines whether the request is approved under current conditions. The policy engine receives a structured request containing role assignments, identity attributes, resource attributes, and contextual signals, and evaluates a policy set referencing all of those inputs. Open Policy Agent (OPA), with its Rego policy language, is a widely deployed engine for this hybrid evaluation.

Attribute-Based Access Decision: Input Signals and Decision Logic

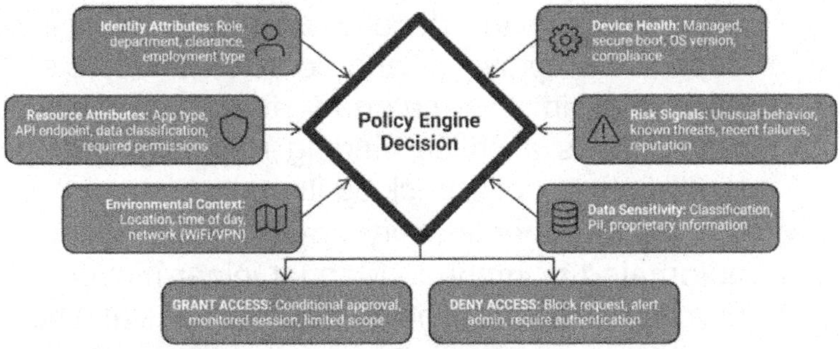

Attribute-Based Access Decision: Input Signals and Decision Logic.

4.9 Manager's Decision Guide

Identity architecture decisions are not made once and revisited annually. They surface constantly, often when a new application is being deployed, a vendor requests access, an audit identifies a gap, or an incident exposes a control failure. The decision framework below helps managers and security architects navigate the most common decision points with a structured rather than reactive approach.

Decision Criteria

- Commission a comprehensive identity inventory before prioritizing any identity tooling investment. You cannot govern what you have not counted. The inventory must cover human users, service accounts, API keys, certificates, workload identities, AI agent identities, and IoT device identities, with ownership and privilege level documented for each.

- Evaluate MFA coverage by population segment, not as a single organizational metric. Identify which populations (contractors, privileged administrators, service accounts, IoT devices) lack phishing-resistant authentication, and build a tiered migration plan that addresses the highest-risk populations first with defined completion dates.
- Measure orphaned and dormant account exposure as a leading risk indicator. Count active credentials associated with decommissioned applications, departed employees, and completed projects. Any credential in this category that holds elevated privileges is a critical finding requiring immediate remediation, not a backlog item.
- Assess whether identity lifecycle management is automated or manual. Manual joiner-mover-leaver processes that depend on ticket queues and human memory will produce systematic gaps. The question is not whether automation is desirable but how far along the automation pathway the organization currently is and what the next automation increment should be.
- Determine whether privileged access uses standing permissions or just-in-time mechanisms. If administrators have always-on elevated access, the standing privilege elimination roadmap should be a near-term workstream, not a future-phase item. The risk exposure of standing privilege compounds over time as credentials age and personnel change.
- Confirm that AI agent identities and IoT device identities are in scope for the identity architecture program. These categories are frequently deferred on the assumption that they will be addressed later. Later typically means after an incident involving one of these unmanaged identity types has forced the conversation.

- Verify that machine identity rotation is automated rather than calendar-driven. If the answer to "how do we rotate this service account credential" is "we schedule a change request and manually update the config files," then the rotation is not reliable at scale. The program needs investment in automation before expanding coverage.

Decision Frame

When evaluating identity architecture investments, apply a pilot-to-production pathway discipline: identify the smallest population or workload that represents the riskiest identity management gap, implement the target-state control for that population, measure the operational outcome, and then scale the pattern. Attempting to solve identity governance for every identity category simultaneously produces a program that is perpetually in design and never in operation. Prioritize by risk exposure first, then by automation feasibility, then by organizational change management complexity. The governance of identity is never finished; the goal is to establish a continuously improving baseline that raises the cost of identity-based attacks across every category of principal in the environment.

4.10 Case Study: Great Lakes Regional Health Network

Great Lakes Regional Health Network is a fictional multi-state healthcare system used throughout this book to illustrate the identity challenges facing large, complex, real-world organizations. At the time its identity program began in earnest, the network operated with 47,000 employees distributed across clinical, administrative, and technical functions; 120,000 connected medical devices ranging from infusion pumps to imaging systems; infrastructure spanning three cloud providers; and four on-premises data centers running a mix of modern and legacy clinical applications. An internal audit, commissioned after a near-miss security event in an unrelated system, revealed 340,000 active identity objects across the environment.

The figure that stopped the CISO in her tracks was not the total count. It was the ratio: 47,000 human identities, 293,000 non-human identities, and no unified lifecycle management system consistently governing any of them. MFA coverage for human users was approximately 68 percent, with notable gaps in the contracted clinical workforce and the overnight technical support teams. For machine identities, the governance picture was far worse. The audit identified 12,000 orphaned service accounts from applications that had been decommissioned over the previous six years. These accounts had not been disabled because no formal decommissioning process had ever required it. Most of them had been created with elevated privileges because the developers who created them had defaulted to broad permissions to avoid troubleshooting access errors during application development, and nobody had ever revisited the permission scope once the application was in production.

The breach that followed was not a sophisticated attack. An adversary who had obtained a credential list from a third-party data aggregator ran a credential-stuffing campaign against the health network's VPN portal over the weekend. One of the credentials in the list matched a service account named svc-clinical-rpt-prod, which had been created to support a clinical reporting application decommissioned two years earlier. The account remained active because the decommissioning project did not include an identity cleanup step. The account retained database administrator rights to the clinical data warehouse because those rights had never been scoped down after the initial development period. The attacker authenticated through the VPN using the compromised credentials and spent three weeks querying the clinical data warehouse, exfiltrating records in batches small enough to avoid triggering the volume-based alerting in place. By the time the activity was detected through an anomaly in query patterns, 890,000 patient records had been accessed.

The post-incident analysis was sobering in a specific way: no technical control in place had been designed to catch this scenario. The VPN authentication succeeded because the credentials were valid. The database queries were executed because the account had valid permissions. The endpoint detection tools detected no malware because none was used. The firewall saw traffic from an authorized network segment. The only control that would have prevented the breach was one that did not exist: a governance process that would have deactivated the service account when the application it served was decommissioned, or a detection process that would have flagged a service account last used two years ago, now suddenly establishing sessions from an external IP address through the VPN.

The remediation program that followed was structured as a phased identity governance initiative. Phase one was the inventory: a six-month effort to enumerate every identity object across all four data centers and three cloud providers, assign ownership, and classify privilege levels. Phase two was orphan remediation: a ninety-day sprint to turn off every identity object in the orphaned category, after a two-week ownership challenge period in which any account that an owner could not claim was turned off automatically. Phase three was lifecycle automation: implementation of automated provisioning and deprovisioning tied to the HR system for human identities and to the application decommissioning workflow for machine identities. Phase four was privileged access modernization: migration of all administrator accounts from standing privilege to just-in-time access, with session recording enabled for all privileged sessions.

The measurable outcomes two years after the program began were notable. The total identity inventory had been reduced from 340,000 objects to 218,000, with the reduction entirely due to orphan remediation and the elimination of duplicate accounts across siloed directories. MFA coverage for human users had reached 96 percent, with the remaining 4 percent concentrated in device categories where interactive authentication was not technically feasible. Orphaned account count had dropped to near zero, maintained by automated detection that flagged new orphans within forty-eight hours of the triggering event. The organization had not eliminated identity risk, but it had replaced an invisible and ungoverned identity landscape with one that was visible, owned, and continuously managed. In a sector where regulatory scrutiny and breach liability are both substantial, the shift from unmanaged to governed identity was a material change in the organization's risk posture.

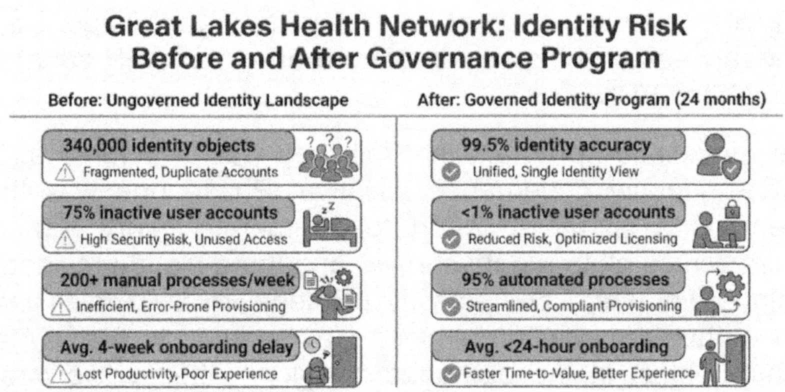

Great Lakes Health Network: Identity Risk Before and After Governance Program

4.11 Team Conversation Guide

Run these three discussions in sequence over two to four weeks. The first surfaces the issue, the second works the trade-offs, the third commits the team to action.

Discussion 1. Surfacing the Identity Inventory Reality

This discussion is designed to surface what the team actually knows about the organization's identity population before making any assumptions about program scope. Many teams discover, in the course of this conversation, that the data they assumed existed does not, or that the counts they had in mind are significantly lower than the actual numbers. That discovery is the point.

Discussion questions to ask the team:

- How many identity objects exist in our environment today, across every category, including human users, service accounts, API keys, certificates, workload identities, IoT device identities, and AI agent identities? Do we have a single authoritative number, or are we estimating?
- Where is the authoritative record for each identity type? Do we have a single source of truth for the whole population, or are we reconciling across multiple disconnected directories, spreadsheets, and platform-specific inventories?
- What is our current orphaned account count, and what is our process for detecting new orphans when an employee departs or an application is decommissioned? Is that process automated or manual?
- Which identity categories do we have the least visibility into? Are there classes of machine identities, for example, developer-created API keys or IoT device credentials, that are essentially untracked?

What to listen for: Listen for uncertainty and estimation. If team members are giving rough numbers with high variance, that is a signal that no authoritative inventory exists and that the actual number is likely higher than anyone expects. Listen also for ownership gaps: if the response to "who owns machine identities" is a reference to whoever created them rather than to a defined governance function, the ownership model is informal and will not scale.

Follow-up actions: Commission a formal identity inventory as the next concrete action, with a defined scope, a completion date, and a named owner. If the organization lacks tooling to enumerate machine identities systematically, that tooling gap should be addressed in the next planning cycle as a prerequisite for any governance work.

Discussion 2. Evaluating Authentication Strength and Coverage Gaps

This discussion moves from existence to the quality of authentication. The goal is to understand not just which identities exist but which of them authenticate with sufficient strength to access the resources they require. The conversation often reveals that MFA coverage metrics reported to leadership measure deployment rather than enforcement, and that significant populations operate with weaker authentication than policy nominally requires.

Discussion questions to ask the team:

- What percentage of our human user identities are enrolled in MFA, and what percentage are actually required to use MFA for every authentication event? Is there a difference between enrollment rate and enforcement rate?
- Which user populations currently authenticate with methods that are susceptible to phishing: SMS OTP, email link, or knowledge-based factors? What is the roadmap for migrating those populations to phishing-resistant methods?
- How would we implement FIDO2 passwordless authentication for our highest-risk user populations? What infrastructure changes, for example, relying party registration, authenticator procurement, and helpdesk process updates, would be required to support it across our federated identity providers?
- What authentication mechanism do our service accounts and machine identities use? Are any using shared passwords, static API keys with no expiration, or other mechanisms that provide no rotation and no individual accountability?

What to listen for: Listen for the gap between policy and enforcement. "Our policy requires MFA" and "MFA is enforced at every authentication event" are different statements, and teams often confuse them. Listen for populations that are technically excluded from MFA requirements through exception lists, legacy application integrations, or geographic workarounds. Those exceptions are likely where the highest-risk authentication gaps live.

Follow-up actions: Map the authentication method to the resource sensitivity tier for every identity population. Any combination in which a high-sensitivity resource is accessible via a phishing-susceptible authentication method should be treated as a remediation item with a defined completion date, not as an accepted risk.

Discussion 3. Designing the Machine Identity Ownership and Rotation Model

This discussion moves the team from diagnosis to design. The goal is to agree on the ownership model for machine identities and the rotation discipline that the organization will adopt. These decisions have direct operational implications for development teams, infrastructure teams, and the secrets management tooling roadmap. Getting the team to a consensus on the model in this discussion makes implementation decisions in subsequent sprints substantially faster.

Discussion questions to ask the team:

- What does our machine identity landscape look like today: how many service accounts, API keys, certificates, and workload identities exist, who owns them formally, and how are they currently rotated? Be specific about what "rotated" means in practice, meaning the last time the credential value actually changed.
- If we needed to assign a named owner to every machine identity in the environment, what process would we use to make that assignment, and what would happen to identities for which no owner could be identified?

- What secrets management tooling do we have available today, and which workload categories are already integrated with it? For the categories that are not integrated, what technical or organizational barriers are there?
- If we needed to assign verifiable identities to every IoT device and AI agent in our environment tomorrow, what identity infrastructure gaps would we need to close first? Which of those gaps are tooling gaps and which are process gaps?

What to listen for: Listen for whether the team distinguishes between "we have a secrets manager" and "our workloads retrieve secrets from it." The tooling exists in many organizations; the integration is what is consistently incomplete. Listen also for resistance framed as application complexity: "we can't rotate that credential because it's wired into six different systems" is a description of a rotation automation problem, not a reason rotation is impossible.

Follow-up actions: Identify three to five machine identity categories with the highest risk exposure (orphaned, elevated privilege, static, and aged credentials) and design the ownership assignment and rotation automation for those categories as a pilot program. Measure the reduction in unowned and unrotated credentials at the sixty-day mark and use that measurement to build the business case for scaling the program.

4.12 Takeaway

Identity architecture at scale is the discipline of extending the same rigor applied to employee accounts to every principal that acts in the environment: service accounts, workload identities, machine certificates, API keys, AI agent sessions, and IoT device credentials. The ratio of machine identities to human identities in a typical hybrid enterprise is 10:1 or higher. Yet most organizations allocate governance effort in proportion to human identities, leaving the population of machine identities largely uncontrolled. That inversion is the gap that attackers exploit most reliably, not through sophisticated techniques, but through the simple observation that orphaned service accounts with elevated privileges and static credentials are consistently available and consistently overlooked.

The path forward is structured and incremental. An authoritative identity inventory is the prerequisite for every other governance action; you cannot deprovision what you have not counted, and you cannot detect behavioral anomalies you have not baselined. Lifecycle automation, applied to both human and machine identity categories, replaces the manual processes that produce systematic gaps with mechanisms that operate at the speed of the infrastructure events that trigger them. Just-in-time access and dynamic secrets replace static credentials and standing privileges with time-bounded, logged, and recoverable access patterns. Phishing-resistant authentication, deployed across every interactive user population on a defined timeline, removes the credential-theft path that remains the most common initial access technique.

The practical sequence for most organizations is: inventory first, orphan remediation second, lifecycle automation third, authentication uplift fourth, privileged access modernization running in parallel with automation. Each phase produces a measurable reduction in the identity attack surface and generates the operational data needed to prioritize the next phase. Responsible-by-design identity architecture does not require solving every problem simultaneously; it requires knowing which problems exist, owning them explicitly, and making continuous progress against each on a defined timeline. The organizations that have done this work report not just improved security outcomes but improved operational clarity: when every identity has an owner, a lifecycle, and a rotation schedule, the environment becomes legible in a way that supports faster incident response, cleaner audit outcomes, and better-informed architecture decisions in the future.

5 Engineering Microsegmentation: Containing the Blast Radius

5.1 Scenario: Wednesday in the Payment Zone

Marcus Reyes had blocked off Wednesday afternoon for what he expected to be a routine review of his team's network segmentation documentation. Three weeks earlier, the CISO had asked every zone owner to produce an up-to-date map of internal traffic flows in advance of the organization's annual risk assessment. Marcus managed the payment processing zone, a cluster of systems that authorizes card transactions, routes settlement batches, and interfaces with the card networks. He knew the zone well, or thought he did.

The documentation his senior network engineer produced was a single page. It showed the DMZ boundary, an ingress load balancer, and a large rectangle in the middle labeled "Payment Processing Segment," with a note reading "240 endpoints, L2 flat segment." Marcus stared at the rectangle for a long moment. He asked his engineer what the note meant in practice. The answer arrived with the quiet certainty of someone who had been waiting for the question: any of those 240 endpoints could initiate a TCP connection to any other endpoint on the same L2 broadcast domain. No firewall rule would stop them. No policy would challenge them. If a point-of-sale integration server were compromised, it could reach the database cluster, the settlement engine, the key management appliance, and the network shares hosting batch reconciliation files, all without crossing a single enforcement boundary.

Marcus had assumed that putting a perimeter firewall around the payment zone meant the zone was segmented. It was not. He had a perimeter around a flat field. Anything that breached the perimeter, whether through a phishing link on an administrator workstation, a vulnerability in a vendor-managed middleware component, or a stolen service account credential, would have unrestricted access to every asset inside that field. The blast radius of a single compromised endpoint was the entire payment zone.

That Wednesday discovery is the starting point for this chapter. Microsegmentation is the discipline of partitioning the network interior so that a breach in one component cannot freely propagate to adjacent components. It is the architectural answer to the question Marcus was forced to ask himself: if something gets inside my zone, how far can it go?

5.2 Why It Matters

The assumption that the network perimeter is the primary security boundary has been eroding for decades and is now functionally obsolete. Attackers who obtain a foothold inside a network, whether through phishing, a compromised vendor, an unpatched vulnerability, or a stolen credential, typically spend weeks or months moving laterally through flat network segments before reaching their target. The dwell time that precedes most major breaches is not a failure of detection alone. It is a failure of containment. A network with no interior segmentation gives an attacker the same freedom of movement that a legitimate administrator enjoys. The only difference is intent.

Microsegmentation creates internal boundaries that limit lateral movement by restricting which workloads can communicate with which others, down to the port and protocol level. When a breach occurs inside a segmented network, the attacker's movement is bounded by policy rather than opportunity. The blast radius, the set of systems reachable from a compromised starting point, shrinks from "the entire zone" to "the workloads in this segment that are permitted to receive connections from this source." This workflow-level impact makes microsegmentation one of the highest-value investments an organization can make: it converts a perimeter breach from a catastrophic event into a contained incident.

Microsegmentation also operationalizes the "assume breach" principle that underpins zero-trust architecture. Chapter 2 established that trust should never be granted based on network location. Chapter 3 built the identity layer that verifies the requester's identity. Microsegmentation enforces the access decisions made by the policy engine at the network and workload level. Without it, a verified identity that has been compromised still has a flat network to traverse. With it, that compromised identity can only reach the specific workloads its policy permits, and every attempted connection to anything outside that permitted set becomes a visible, logged, and alertable event. Microsegmentation is what makes continuous verification, the subject of Chapter 5, operationally meaningful: if the verification engine detects an anomaly, microsegmentation ensures the affected zone is already bounded.

5.3 The Segmentation Maturity Ladder

Organizations rarely arrive at microsegmentation in a single step. Most follow a progression that mirrors the broader maturity of their security architecture. Understanding where an organization sits on that progression is the first step toward a realistic, rather than aspirational, pilot-to-production pathway.

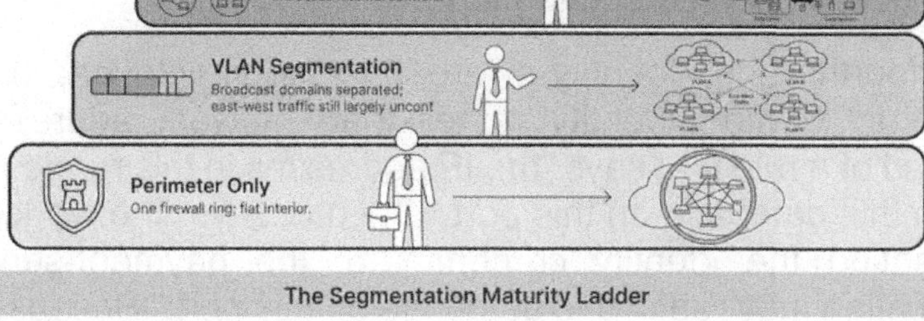

The Segmentation Maturity Ladder

The first rung is perimeter-only segmentation. An organization at this level has a firewall at the boundary between its network and the internet, and possibly a DMZ for public-facing services. Still, the interior is a single flat network. Marcus's payment zone illustrates this perfectly. A perimeter exists; the interior does not. An attacker who breaches the perimeter faces no further obstacles.

The second rung is VLAN-based segmentation. VLANs separate broadcast domains and are widely deployed in enterprise networks. They provide a meaningful improvement over a completely flat network: a compromised workstation in the finance VLAN cannot trivially send broadcast traffic to workstations in the HR VLAN. However, VLANs are a layer-two construct. East-west traffic between VLANs typically flows through a router or layer-three switch, and the routing rules at that device are often permissive by design, configured to support application connectivity rather than restrict it. An attacker who compromises a device in the finance VLAN can frequently reach the HR VLAN simply by routing through the inter-VLAN gateway, because no firewall policy inspects or restricts that routing path.

The third rung is microsegmentation. At this level, policies are enforced between individual workloads or small groups of workloads, and the default posture is deny-all with explicit permits. A web server in the application tier can reach the database on port 5432 because there is an explicit permit rule for that specific flow. It cannot reach the database on any other port, cannot reach the key management appliance, and cannot initiate connections to other application servers in the same tier. Every flow is explicitly authorized. Anything not authorized is denied and logged.

The fourth rung, identity-aware microsegmentation, binds policy to verified identity rather than network addresses. Instead of a rule that says "the IP addresses in this range may reach this database on this port," the rule says "the workload presenting the identity credential of the payment-service application may reach this database on this port." Because IP addresses can be spoofed and are frequently reused by dynamic workloads in cloud environments, identity-aware policy is both more precise and more durable. It is the natural convergence of the microsegmentation layer with the identity layer built in Chapter 3.

Marcus's organization is on rung one. His immediate goal is to reach rung three for the payment processing zone while designing the architecture to support rung four as identity infrastructure matures.

5.4 East-West Traffic Visibility: Seeing Before Enforcing

The most common mistake organizations make when beginning a microsegmentation project is attempting to enforce policy before they understand traffic. A policy that is written without an accurate model of existing communication flows will break applications. In complex environments with legacy workloads, undocumented middleware dependencies, and organic network growth over years or decades, the actual traffic flows frequently differ substantially from what any documentation describes.

East-west traffic is traffic between workloads inside the network, as opposed to north-south traffic between the network and external entities. Traditional security architectures instrumented the perimeter extensively, with firewall logs, IDS alerts, and proxy records capturing everything crossing the boundary. The interior was often invisible. Security teams that have spent years analyzing north-south traffic frequently discover, when they first instrument east-west flows, that their internal network resembles a dense graph with hundreds of undocumented edges rather than the clean tiered architecture their diagrams depict.

East-West vs. North-South Traffic Visibility

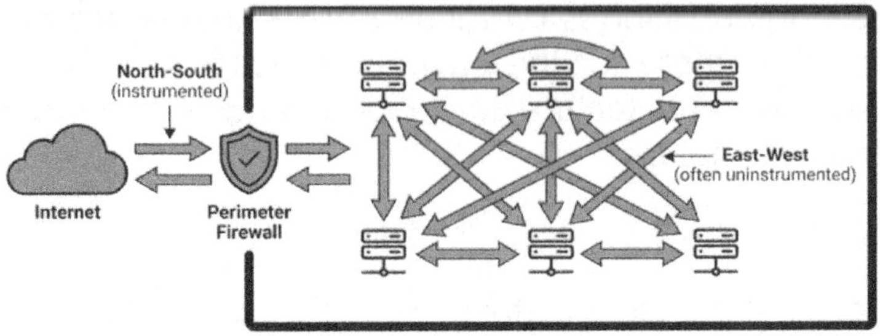

Traditional security architectures often lack visibility into lateral (East-West) traffic, focusing primarily on perimeter (North-South) flows.

Achieving east-west visibility requires instrumentation at the workload level, not at the perimeter. The primary approaches are network flow collection, agent-based telemetry, and network taps at the virtual switch layer.

Network flow protocols such as NetFlow, IPFIX, and sFlow can be enabled on core switches and routers to capture metadata for every internal flow: source and destination IP addresses, ports, protocols, byte counts, and durations. Flow records do not capture packet content, which makes them suitable for policy discovery even in regulated environments where full packet capture is prohibited. The limitation of flow-based telemetry is that it reflects traffic as seen by the switch, which may aggregate flows across multiple workloads behind a load balancer or miss flows that never traverse a monitored segment.

Agent-based telemetry, deployed on individual hosts, captures process-level context that flow records cannot. An agent knows not only that a connection was made from this IP to that IP on port 443, but also which process initiated the connection, which user account the process ran under, and what certificate the connection presented. This context is invaluable for writing precise policy: it allows a policy author to permit the payment-service process to connect to the database, rather than permitting the entire host to do so.

The observation period before enforcing policy is not a formality. It should span a complete business cycle, including month-end batch processing runs, quarterly reporting periods, and any seasonal traffic patterns. Many application breakages during microsegmentation rollouts trace back to flows that were present in the traffic baseline but missed because the observation window was too short to capture infrequent but legitimate communication.

5.5 Application Dependency Mapping

East-west visibility produces raw flow data. Application dependency mapping transforms that data into a structured model of which workloads serve which functions, which workloads depend on which other workloads, and which flows are necessary for application availability versus which flows are opportunistic or unauthorized. The dependency map is the design input for microsegmentation policy.

Application Dependency Map: Payment Processing Zone

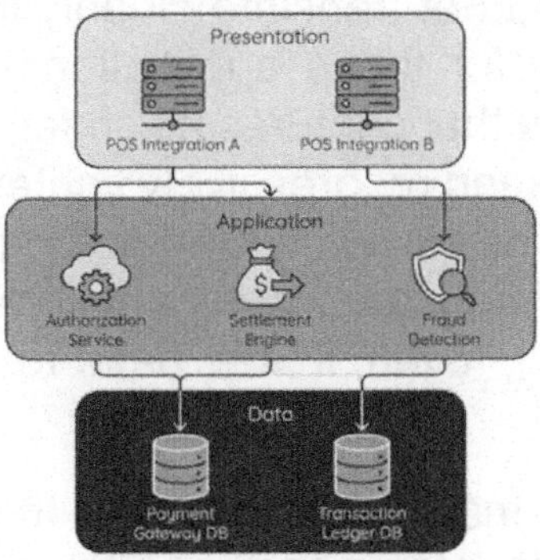

This map visualizes the critical dependencies and data flow between components within the secure payment processing environment.

Building a dependency map begins with automated discovery from the flow data collected during the visibility phase. Tools that perform dependency mapping ingest flow records and agent telemetry, group flows by application tier or workload identity, and produce a visual graph of communication relationships. The initial automated output is a starting point, not a final product. It contains noise: backup agent connections to storage targets, monitoring agent connections to collection endpoints, and vendor license validation calls that are not documented in any architecture diagram but are essential for application operation.

The refinement process requires application owners to walk through the automated map and categorize each flow as required, conditional, or unexpected. Required flows are those without which the application fails. Conditional flows are those triggered only by specific operational events, such as disaster recovery failover or batch processing. Unexpected flows are those that no application owner can explain and often represent either unauthorized lateral movement or undocumented legacy behavior that must be investigated before policy is written.

Legacy applications present a particular challenge for dependency mapping. Applications written before network segmentation was a design consideration often make broad connection attempts, probe the network for services they need, and rely on broadcast-based service discovery mechanisms that do not translate cleanly into explicit permit rules. The strategy for these applications is to start with a permissive segment that captures all observed flows from the legacy application, then progressively tighten the policy as the application's actual behavior is understood. Responsible by design means building the constraint into the architecture from the beginning, even if the path to full constraint is incremental.

The dependency map also reveals the communication structure that should inform segment boundaries. Workloads that communicate frequently and necessarily with each other belong in the same segment. Workloads that have no documented communication requirement belong in separate segments with a default-deny policy between them. The goal is not to create one segment per workload, which would maximize security but also maximize operational complexity. The goal is to find the segmentation granularity that meaningfully reduces blast radius while remaining operationally manageable.

5.6 Policy Authoring Without Breaking Flows

Policy authoring is the translation of the dependency map into enforceable rules. The central challenge is writing policy that is restrictive enough to meaningfully contain lateral movement but permissive enough to allow every legitimate application flow to continue without interruption. These goals are in tension, and the resolution of that tension is the operational art of microsegmentation.

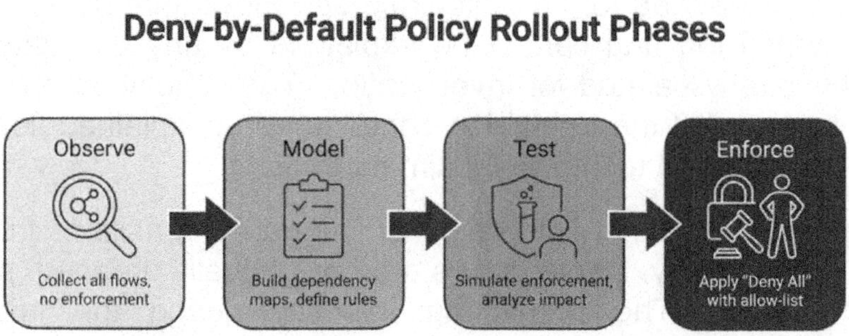

Deny-by-Default Policy Rollout Phases are designed to gradually transition network security from permissive to restrictive by first observing, then modeling, testing, and finally enforcing a strict access control policy.

The rollout pattern that reliably avoids application breakage follows four phases. The first phase is observation, during which the segmentation tooling or agent is deployed in monitoring-only mode. No traffic is blocked. The goal is to capture a comprehensive baseline of every flow in the environment. The observation period should be long enough to capture all regular operational cycles.

The second phase is modeling. The collected flow data is used to build the dependency map and draft the initial policy rule set. Policy rules should be written at the level of specificity supported by the tooling: if the tooling supports process-level or identity-level rules, write at that level. If only IP and port rules are supported, use the most specific IP ranges available. Every rule in the initial draft should correspond to at least one observed flow in the baseline data. Rules that permit flows that were never observed are not policy; they are guesswork, and guesswork in a deny-by-default architecture quickly becomes a source of operational problems.

The third phase is simulation, often called log-only or audit mode. The policy rules are activated, but instead of dropping traffic that does not match a permit rule, the enforcement engine logs a denial record. Operations teams review the denial log daily. Flows that appear in the denial log but are known legitimate, because they were missed during the observation period or triggered by an infrequent operational event, are added to the policy rule set. Flows that appear in the denial log and cannot be explained by any application owner are escalated for investigation. The simulation phase continues until the denial log contains no unexplained flows that correspond to known legitimate traffic.

The fourth phase is enforcement. The policy transitions from log-only to deny. Denied flows are now actually dropped, not just logged. The monitoring intensity during the initial enforcement period should be high: daily review of denial logs, application performance metrics, and helpdesk tickets for application failures. Most enforcement-phase breakages occur within the first 48 hours and are resolved by adding rules for flows that were present in the environment but not captured during observation.

One principle that must be maintained throughout the rollout is that policy changes should be tracked in version control. Each rule addition, modification, or deletion should correspond to a documented change with an owner, a justification, and a reviewer. Policy drift, the gradual accumulation of undocumented rule changes that erode the original design intent, is the primary cause of segmentation architectures that look secure on paper but are permissive in practice. Operational clarity requires that every rule in the policy set can be traced to a documented justification.

5.7 Software-Defined Segmentation vs. Host-Based Agents

The two primary implementation approaches for microsegmentation are software-defined networking overlays and host-based agent enforcement. Each has architectural trade-offs that determine which approach is appropriate for a given environment. Many organizations deploy both, using network overlays for environments where host agents cannot be installed and host-based agents for environments where process-level policy granularity is required.

Microsegmentation Implementation Approaches

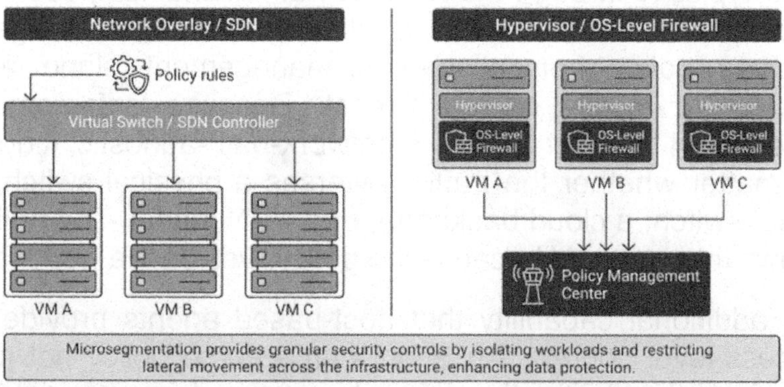

Software-defined networking overlays implement segmentation policy at the virtual switch or SDN controller layer. Traffic between workloads passes through a software-controlled switch that applies permit-or-deny decisions based on policy rules distributed by a central controller. The primary advantage of the overlay approach is that it requires no changes to the workloads themselves. A virtual machine or bare-metal server does not require an agent; the policy is enforced in the network fabric beneath the workload. This makes the overlay approach well-suited to environments with a large number of workloads that cannot accept agent installation, including legacy operating systems, appliances, and systems running under vendor support agreements that prohibit host modifications.

The limitation of the overlay approach is that policy is bound to network addresses rather than workload identity. When IP addresses change, as they frequently do in dynamic cloud environments, policies must be updated. When a workload moves between hosts or availability zones, the policy must follow it by mapping workload identity to the current network address. This mapping layer adds operational complexity and can introduce gaps if workloads outpace the policy distribution mechanism.

Host-based agent enforcement installs a software component on each workload that intercepts network calls at the kernel level, typically through the host firewall mechanism, such as iptables, nftables, or Windows Filtering Platform. The agent receives policy from a central management plane and enforces it locally on each host. Because enforcement happens on the host itself, it is network-path-agnostic: it does not matter whether the traffic traverses a physical switch, a virtual switch, a cloud backbone, or a VPN tunnel. The policy follows the workload because the policy lives on the workload.

The additional capability that host-based agents provide is process-level context. An agent can enforce policy not just based on the destination IP and port, but based on which process on the host is initiating the connection. A rule can permit the payment-service process to connect to the database while denying the same connection from any other process on the same host, even if both processes share the same IP address. This level of granularity is not achievable with network-layer enforcement.

5.8 Container and Kubernetes Namespace Isolation

Container orchestration platforms introduced a new segmentation surface that traditional network controls were not designed to address. In a Kubernetes cluster, hundreds of pods sharing the same physical host infrastructure can communicate with each other through the cluster's overlay network by default. Without an explicit network policy, any pod in the cluster can initiate a TCP connection to any other pod, regardless of namespace boundaries. This default-open posture makes a freshly provisioned Kubernetes cluster architecturally similar to Marcus's payment zone: a perimeter exists around the cluster, but the interior is flat.

Kubernetes Namespace Isolation with Network Policy

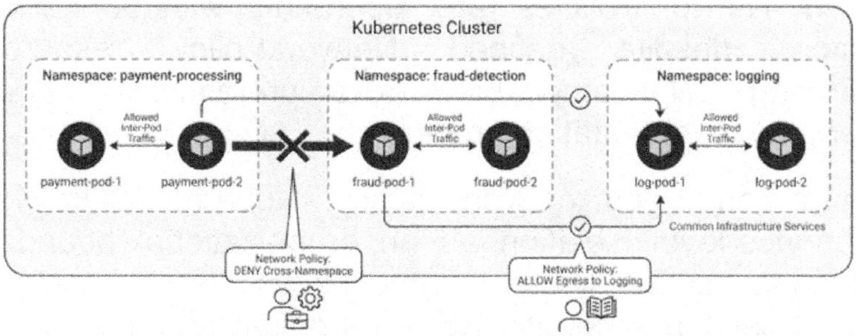

Kubernetes Namespace Isolation with Network Policy ensures secure and controlled traffic flow between application components by defining clear communication rules.

Kubernetes provides a native mechanism for network-level isolation: the NetworkPolicy resource. A NetworkPolicy object specifies, for a set of pods selected by labels, which ingress and egress connections are permitted and which are denied. NetworkPolicy resources are namespaced, meaning they apply within a specific namespace and can reference pods in other namespaces by namespace selector. The default behavior in the absence of a NetworkPolicy resource is to permit all traffic.

The operational pattern for implementing Kubernetes microsegmentation mirrors the broader rollout pattern described earlier: observe traffic first, build a dependency map for pod-to-pod communication, then apply NetworkPolicy resources in a sequence that progressively restricts traffic without breaking services. A common starting point is to apply a default-deny NetworkPolicy to each namespace that blocks all ingress and egress, then add specific permit rules for each documented flow. This approach surfaces undocumented dependencies quickly because any flow that was not captured during the observation period will be blocked immediately, generating a log event and often a service failure that triggers rapid remediation.

A critical architectural note: Kubernetes NetworkPolicy is only enforced if the cluster's Container Network Interface plugin supports it. Not all CNI plugins support NetworkPolicy. Organizations that deploy clusters with CNI plugins that do not enforce NetworkPolicies are operating without network isolation, despite defined NetworkPolicy resources. Confirming CNI compatibility is a prerequisite for any Kubernetes segmentation project.

Beyond pod-to-pod network policy, Kubernetes provides namespace-level isolation as an organizational boundary. Workloads across different namespaces can be assigned different service accounts, RBAC permissions, and network policies. For multi-tenant clusters or clusters hosting workloads with different data-sensitivity classifications, namespaces should be used as the primary segmentation boundary, with NetworkPolicy resources enforcing permitted flows between namespaces.

5.9 Network Security Groups and Cloud-Native Segmentation

Public cloud platforms provide native segmentation primitives that, when properly configured, deliver microsegmentation capabilities without requiring additional tooling. The most widely deployed of these primitives is the network security group, which exists in equivalent form across the major cloud providers under different names and with slightly different behavioral models.

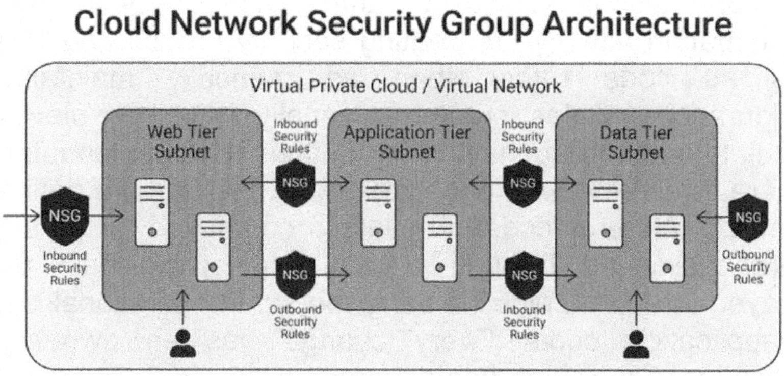

Cloud Network Security Group Architecture

In AWS, the equivalent construct is a combination of security groups and network access control lists. Security groups are stateful, instance-level firewalls that control which traffic can reach an EC2 instance or other resource. Network ACLs are stateless, subnet-level rules that apply to all traffic entering or leaving a subnet. The distinction matters architecturally: security groups are the appropriate tool for workload-to-workload microsegmentation because they operate at the level of individual resources and their rules reference other security groups by ID rather than by IP address. This means that when a new instance is added to the application-tier security group, it automatically inherits the rules that permit it to reach the database-tier security group and is automatically subject to the rules that prevent it from reaching anything else.

In Azure, network security groups serve a similar function and can be associated with either subnets or individual network interface cards. Azure also provides application security groups, which allow policies to reference groups of VMs by logical name rather than by IP prefix, enabling policy rules that read as "the payment-service application group may reach the database application group on port 1433" rather than "these five IP addresses may reach these three IP addresses on port 1433." This abstraction is more durable in dynamic environments where IP addresses change frequently.

The operational discipline that makes cloud-native segmentation effective is treating security groups and NSG rules as code rather than as manually maintained configurations. Rules created manually in a console are difficult to audit, frequently drift from their documented intent, and are rarely reviewed once created. Rules that are defined in infrastructure-as-code templates, committed to version control, reviewed through a pull request process, and deployed through a pipeline carry the same operational rigor as application code. Every change has an owner, a justification, a reviewer, and an audit trail.

5.10 Geeks Only

This section provides the engineering depth that security architects, network engineers, and platform engineers need to implement microsegmentation in practice. It assumes familiarity with networking fundamentals, Linux host internals, and cloud infrastructure concepts.

Application Dependency Discovery Techniques

Automated dependency discovery at scale requires a combination of passive observation and active interrogation. The passive layer collects flow data without disturbing the environment. The active layer queries running processes on hosts to correlate flow data with application context.

For flow-based passive discovery, enable IPFIX or NetFlow v9 on all core switches and virtual switches. Use a collector such as ntopng, Elastiflow, or a SIEM with flow ingestion capabilities to aggregate and index flow records. Configure the collection interval at no more than 60 seconds for internal flows; shorter intervals improve accuracy for short-lived connections but increase collection overhead. Store at least 90 days of flow history before beginning policy authoring to capture monthly and quarterly patterns.

For host-based agent discovery, deploy the agent in discovery-only mode before activating enforcement. Most commercial microsegmentation platforms provide this capability. The agent captures process name, process ID, user account, listening ports, active connections, and the destination of each outbound connection. This data produces a richer model than flow records alone because it includes the application context behind each network flow.

For Kubernetes environments, use a network observability tool such as Cilium Hubble or Calico's flow logging capability to capture pod-to-pod traffic metadata. These tools integrate with the CNI layer and can produce flow records tagged with the pod name, namespace, and service account, allowing dependency maps to be built at the pod identity level rather than the IP address level.

Dependency discovery should produce a structured output, typically a JSON or YAML graph, which serves as input to policy generation. Tools that automate the translation from a dependency graph to a policy rule set reduce manual effort and the risk of transcription errors. Review the generated policy carefully against the dependency graph before advancing to simulation mode; automated policy generation can produce overly broad rules if the input data contains noisy flows.

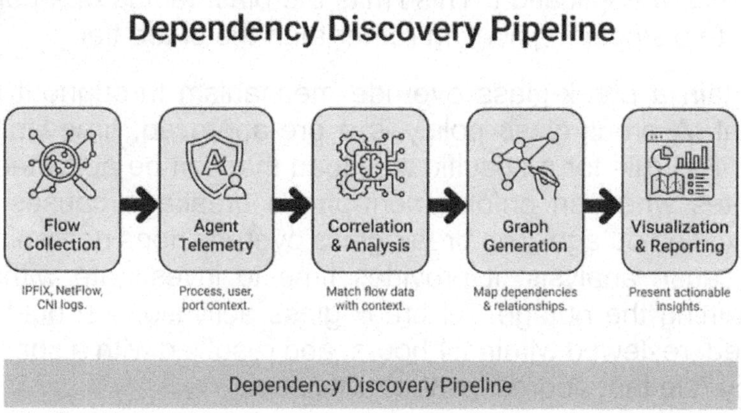

Deny-by-Default Rollout Without Breaking Production

The rollout sequence described in the policy authoring section above can be made more precise by applying a few engineering-level disciplines, which reduce the risk of enforcement-phase breakages.

Segment the rollout by application tier. Begin enforcement in the highest-isolation tier first, which is typically the data tier. Database servers have the most constrained legitimate connection patterns: application servers connect to them on specific ports, backup agents connect to them on specific ports at specific times, and monitoring agents connect to them on specific ports. The set of permitted flows for a database server is small enough that the deny-all-plus-explicit-permits model is easy to validate. Starting enforcement at the data tier also produces the highest security value per rule, because databases are the primary target of lateral movement.

Use a canary enforcement approach when rolling out to application tiers. Select one representative application in the tier, complete the observe-model-simulate-enforce cycle for that application, validate stability for 2 weeks, then move on to the next application. This limits the blast radius of a policy error to a single application rather than the entire tier.

Maintain a break-glass override mechanism throughout the rollout. A break-glass policy is a pre-approved, time-limited permit-all rule for a specific workload that can be activated in minutes when an enforcement-phase breakage causes an application outage. The break-glass override does not replace root-cause analysis; it provides time to investigate without sustaining the outage. All break-glass activations should be logged, reviewed within 24 hours, and resolved with a specific policy rule that addresses the root cause.

Track the ratio of deny events to permit events over time. In a newly enforced segment, the deny-to-permit ratio will be high initially as legitimate flows surface that were missed during observation. This ratio should decrease rapidly in the first week of enforcement. If it remains elevated after two weeks, the dependency model is incomplete, and enforcement should be paused while additional flows are investigated. A declining deny rate, accompanied by stable application performance, is the primary signal that enforcement is healthy.

Network Security Policy as Code

Infrastructure-as-code pipelines for network security policy should follow the same engineering practices as application code pipelines: version control, peer review, automated testing, and deployment automation.

For AWS environments, security group rules should be defined in Terraform or CloudFormation templates and stored in a version-controlled repository. Each security group resource should include metadata tags identifying the owning team, the application, and the review date. Security group changes should require a pull request with at least one approver who is not the author. Automated policy drift detection, using tools such as AWS Config or Terraform Cloud's drift detection capability, should alert when the live configuration diverges from the declared state.

For Kubernetes environments, NetworkPolicy resources should be defined as YAML manifests stored in the same repository as the application they protect and deployed through the same GitOps pipeline. Using a policy-as-code framework such as OPA Gatekeeper or Kyverno allows enforcement of organizational standards: every namespace must have a default-deny NetworkPolicy; every NetworkPolicy must include a label identifying its author and review date; no NetworkPolicy may use a pod selector that matches all pods in the namespace without explicit approval.

For on-premises firewalls, tools such as Batfish provide network-level configuration analysis that can validate firewall rule sets against a policy model before the rules are deployed to production devices. A Batfish analysis can answer questions like "Does any rule in this configuration permit a host in the POS integration subnet to reach the key management appliance?" before the configuration is committed to the device. This type of pre-deployment validation is the network equivalent of unit testing for application code.

Security Policy as Code Pipeline

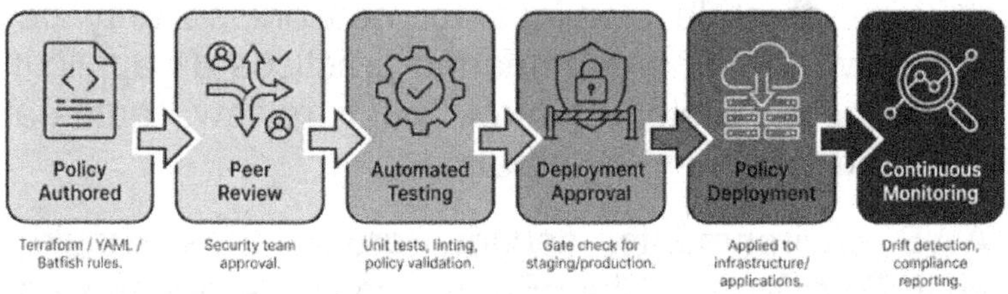

Security Policy as Code Pipeline integrates security best practices directly into the software development lifecycle for automated and compliant infrastructure management.

eBPF and Host-Based Enforcement

Extended Berkeley Packet Filter (eBPF) has become the dominant technology for host-based network policy enforcement in Linux environments. eBPF allows programs to run in the Linux kernel without modifying kernel source code, with safety guarantees enforced by a kernel-level verifier. A network policy enforcement engine built on eBPF can intercept every network call, apply policy rules, and either permit or drop the connection, all within the kernel without copying packet data to user space. This architecture delivers significantly lower latency and lower CPU overhead than enforcement mechanisms that require packets to be processed in user space.

Cilium, the most widely deployed eBPF-based networking solution for Kubernetes, uses eBPF programs attached to each pod's network interface to enforce NetworkPolicy resources. Because Cilium's enforcement is eBPF-based, it can inspect not only L3 and L4 headers (source IP, destination IP, port, protocol) but also L7 protocol content: it can enforce a policy that permits HTTP GET requests to a specific URL path while denying POST requests to the same destination, without requiring a sidecar proxy. This L7 awareness without a sidecar is one of the most significant architectural advantages of eBPF-based enforcement.

For traditional Linux hosts outside Kubernetes, eBPF-based host agents from commercial microsegmentation vendors provide the same kernel-level enforcement with a management plane that integrates into the broader policy framework. These agents replace or augment iptables-based enforcement, which becomes a performance bottleneck at scale due to iptables' linear rule traversal model and lock contention under high connection rates. nftables, the successor to iptables, provides better performance through a map-based lookup architecture. Still, eBPF-based enforcement typically outperforms nftables at high packet rates because eBPF programs can be JIT-compiled and execute with minimal kernel-context overhead.

Service Mesh for L7 Segmentation

Microsegmentation at the network layer controls whether a connection can be established between two workloads. Service mesh segmentation controls what can happen within that connection after it is established. These are complementary layers, not alternatives. An organization that implements both achieves defense in depth: the network layer prevents unauthorized connections, and the service mesh layer prevents unauthorized operations within permitted connections.

Service Mesh L7 Policy Enforcement

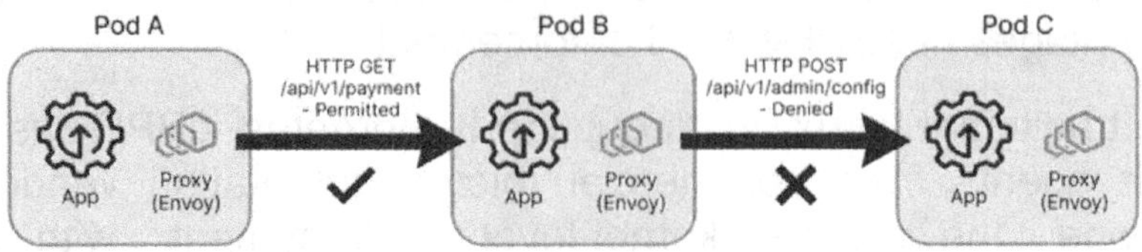

Service mesh proxies enforce fine-grained Layer 7 access policies at the application level.

A service mesh deploys a proxy sidecar alongside each application container. All network traffic to and from the application is intercepted by the sidecar, which operates at L7 and can inspect the full content of each request for HTTP, gRPC, or other application-layer protocols. The control plane distributes authorization policy to each sidecar: rules that specify which services may call which APIs, which HTTP methods are permitted, and which response codes should trigger circuit breaking or traffic rerouting.

Istio is the most widely deployed service mesh for Kubernetes. Its authorization policy model uses the same identity system as Kubernetes RBAC, binding L7 policy to Kubernetes service accounts. A policy that says "the payment-service service account may call the fraud-scoring service on the /api/v1/score endpoint using GET" is enforced by the Envoy sidecar on the fraud-scoring pod. Any call from the payment service that uses a different HTTP method, path, or service account credential is denied by the sidecar without the request ever reaching the application container. The application is insulated from unauthorized requests; it never processes them.

Linkerd takes a simpler approach to service mesh, focusing on mTLS between all services by default with a lighter-weight proxy implementation. Linkerd's authorization policy model is less expressive than Istio's at the L7 level but provides lower operational overhead and simpler debugging. Organizations that need mTLS across all services, without complex L7 authorization logic, may find Linkerd's simpler model easier to operate and maintain.

The mutual TLS capability that service meshes provide as a baseline is valuable independent of L7 authorization policy. When all service-to-service communication uses mTLS, each connection is authenticated using a short-lived cryptographic certificate issued by the mesh control plane. This means that the fraud-scoring sidecar will reject a connection from a compromised pod claiming to be the payment service if the compromised pod lacks a valid payment service certificate. The network layer may permit the connection; the TLS handshake at the sidecar will refuse it.

Troubleshooting Policy Denials

The most operationally intensive phase of a microsegmentation deployment is the period immediately following enforcement activation, when unexpected denials surface application dependencies that were missed during observation and modeling. A systematic approach to denial investigation reduces resolution time and prevents ad-hoc policy additions that erode the intended design.

For each unexpected denial, the investigation should establish four facts: the source workload identity (not just IP address), the destination workload identity, the specific port and protocol of the denied connection, and the process on the source workload that initiated the connection. IP addresses alone are insufficient because multiple workloads may share an IP in NAT environments and because DHCP lease expiry can transfer an IP from one workload to another between the time of denial and the time of investigation.

Most commercial microsegmentation platforms provide a denial visualization that shows denied flows on the dependency map, highlighting flows that were permitted during observation but are now denied under the active policy.

This view immediately clarifies whether a denial represents a missed legitimate flow or a new, unexpected connection attempt. The former requires a policy addition; the latter requires investigation.

For Kubernetes environments, kubectl commands provide the primary diagnostic interface. The command "kubectl get networkpolicy -n <namespace>" lists all active NetworkPolicy resources. A useful diagnostic pattern is to temporarily apply a logging-only NetworkPolicy that mirrors the deny rules and generates events for each denied connection, allowing denial events to be correlated with pod logs without dropping the traffic. Once the root cause is identified, the logging policy is replaced with either a permit rule (if the flow is legitimate) or is left in place as a detection signal (if the flow represents anomalous behavior).

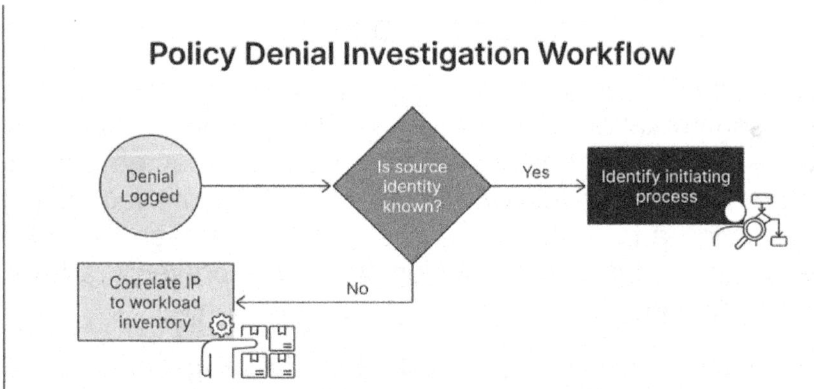

Policy Denial Investigation Workflow provides a structured approach for analyzing and resolving security policy violations.

5.11 Manager's Decision Guide

Microsegmentation decisions require balancing security outcomes against operational risk and engineering capacity. The following criteria and decision frame give managers a structured basis for moving from assessment to commitment.

Decision Criteria

- East-west visibility: Before committing to enforcement, confirm that the team has collected at least one full business cycle of internal flow data and has reviewed it with application owners. Enforcement without visibility leads to outages; visibility without enforcement leads to dashboards. The goal is both.
- Blast radius assessment: Use the dependency map to calculate the current blast radius from each critical asset. Express this as the number of endpoints reachable from a single compromised workstation within the segment. This number should decrease measurably after each enforcement milestone and should be reported to leadership as a security outcome metric.
- Application owner readiness: Microsegmentation cannot succeed without active participation from application owners. If application teams are unavailable to participate in dependency map review or are unwilling to accept brief validation windows during the simulation phase, the rollout will stall. Secure this commitment before the project begins, not during it.
- Tooling selection: The choice between network-overlay and host-based agent approaches should be driven by the mix of workload types in the environment and the policy granularity required. Environments with a large legacy workload footprint that cannot accept agents favor the overlay approach. Environments with modern operating systems and a requirement for process-level policy granularity favor host-based agents. Most large environments require both.
- Policy-as-code readiness: If the organization does not have a version-controlled infrastructure-as-code pipeline for network policy, establishing one should be part of the microsegmentation project

scope. Manually managed firewall rules will drift from the intended design within months of enforcement activation. Code-managed policy will not.

- Cloud and container coverage: Confirm that the segmentation plan covers cloud workloads and Kubernetes clusters, not just on-premises data center infrastructure. Organizations that segment the data center but leave cloud workloads in default-open virtual networks have reduced their overall blast radius by a fraction of what a comprehensive plan would achieve.

Decision Frame

The central decision in a microsegmentation project is not which tool to buy; it is how to sequence the rollout across a complex, mixed environment without causing application outages that erode organizational confidence in the project. The recommended frame is: start with the highest-value, lowest-complexity segment (typically the data tier of the most sensitive application), achieve demonstrable success within 90 days, measure the reduction in blast radius, and use that result to justify the next phase. A pilot-to-production pathway that produces a visible security outcome within 90 days is more likely to sustain executive support than a comprehensive plan that promises results in 18 months.

5.12 Case Study: Vanguard Defense Systems

Vanguard Defense Systems is a fictional Tier 1 defense contractor supporting classified weapons programs. The organization's red team, an internal group tasked with simulating adversary behavior, conducted a two-week exercise to test the corporate network's resilience against a workstation-compromise scenario. The exercise began by simulating the compromise of a single employee workstation on the corporate network through a phishing link. The workstation was in the general administrative subnet, physically and logically separate from the weapons systems design enclave, which was connected to the corporate network through a dedicated network link for file transfers and collaboration workflows.

The results were immediate and alarming. The red team traversed 14 network hops from the compromised workstation to the weapons systems design repository in 11 minutes. They crossed four subnets without triggering a single firewall rule denial, because every inter-subnet routing path was configured to permit traffic for operational convenience. They reached the design repository's file share and exfiltrated a sample of documents. The only detection during the exercise came from a file access audit log that flagged an unusual access pattern. Still, the alert was generated 47 minutes after the initial exfiltration and was routed to an unmonitored queue.

The post-exercise debrief revealed that the organization's leadership found it difficult to present to its defense agency customers: the network had a perimeter, but the perimeter protected a flat interior. Any attacker who breached the perimeter by any means had unrestricted access to all network assets, including the most sensitive classified repositories. The blast radius from a single compromised workstation was the entire corporate network plus the connected enclaves.

Vanguard's security leadership commissioned a microsegmentation program with a 12-month implementation timeline. The first phase, lasting three months, focused exclusively on observation: deploying host-based agents in discovery mode across both classified and unclassified network segments, collecting flow data, and building dependency maps for every application tier. The discovery phase revealed over 400 east-west flows that were not documented in any network architecture diagram, including automated backup processes, monitoring agent calls, and several undocumented peer-to-peer synchronization processes between workstations that had accumulated over years of organic growth.

The second phase, months four through eight, implemented enforcement for the classified enclave first. The weapons systems design repository and its associated compute infrastructure were placed in a microsegment with a deny-by-default policy and explicit permit rules for the 23 flows identified by the dependency map as necessary for the application function. The enforcement rollout used the canary approach: enforcement was activated for one system at a time, stability was validated for 48 hours, then the next system was added. The rollout produced two enforcement-phase breakages. One was a license validation call from the CAD application to a vendor server that was missed during observation because it ran only once every 30 days. The second was an undocumented synchronization process between the design repository and a backup vault that ran on a non-standard port. Both were resolved with specific permit rules within hours of detection.

By month 12, the classified enclave had a blast radius of 12 endpoints, down from the entire corporate network. A subsequent red team exercise began from the same compromised workstation as the original exercise. The red team reached the classified enclave network boundary in six minutes, the same as before. But the boundary now reflected actual policy enforcement. The first lateral movement attempt within the enclave was blocked at the host-based agent on the target workstation, 90 seconds after the initial connection attempt. A deny event triggered an alert in the monitored security operations queue within 15 seconds. The red team terminated the exercise. The weapons systems design repository was never reached.

The measurable outcome of Vanguard's program illustrates the mission-aligned impact of microsegmentation: the reduction in blast radius from "entire corporate network" to "12 endpoints in a bounded segment" is not an abstract security improvement. It is the difference between a breach that compromises classified program data and a breach that terminates at a containment boundary without reaching the asset it was designed to protect.

5.13 Team Conversation Guide

Run these three discussions in sequence over two to four weeks. The first surfaces the issue, the second works the trade-offs, the third commits the team to action.

Discussion 1. Do We Know What We Cannot See?

Run this discussion before any tooling decisions or vendor evaluations. Its purpose is to establish a shared, honest picture of the team's current east-west visibility. Many teams discover during this conversation that they have assumed visibility they do not actually have.

Discussion questions to ask the team:

- Can we produce a complete map of all internal traffic flows between our workloads in the payment processing zone, including flows between workloads in the same subnet?
- If a workstation in our zone were compromised right now, how many other workloads could it reach without crossing a firewall policy boundary? Does anyone know this number with confidence?
- What tooling do we currently use to observe internal network traffic? Does it give us flow-level metadata, or only perimeter logs?
- Have we ever discovered an unexpected internal communication flow, one that no application owner documented, during a security review or incident investigation?

What to listen for: Listen for uncertainty about the perimeter-versus-interior distinction: team members who conflate "we have a firewall at the boundary" with "we have microsegmentation inside." Also listen for confidence that turns out to be based on documentation rather than observed data. The goal of this discussion is to surface the gap between what the team believes is true about internal visibility and what is actually instrumented.

Follow-up actions: After the discussion, assign a specific owner to assess the east-west visibility gap and produce a written inventory of what internal flows are currently observable and what observation period would be needed to build a complete dependency map. This assignment should have a two-week deadline and result in a brief written output that serves as input to the next discussion.

Discussion 2. How Do We Segment Without Breaking Things?

Run this discussion two to three weeks after the first, once the team has assessed the visibility gap and understands what a dependency mapping exercise would require. This discussion focuses on the operational risk of the rollout and how the team would manage it.

Discussion questions to ask the team:

- Given what we now know about our east-west flows, what is the minimum observation period we need before we can write policy with confidence? What events or operational cycles would we need to capture?
- Which application owner relationships do we need to secure before we can begin dependency map review? Are there application owners who are likely to resist participating, and how would we address that?
- What is our break-glass plan if enforcement causes an application outage? How quickly could we identify the affected flow, suspend enforcement for that workload, and restore service?
- Should we start enforcement at the data tier, where the permitted flow set is smallest and most constrained, or is there a different starting segment

that would produce a demonstrable security outcome while managing outage risk?

What to listen for: Application owners described as "hard to reach" or "unlikely to prioritize security reviews." These are the relationships that need management attention before the project begins, not after the first enforcement-phase outage. Also, listen for whether the team has a realistic assessment of the observation period required: teams often underestimate how long it takes to capture infrequent flows and how much time application owner reviews consume.

Follow-up actions: Agree on the pilot's starting segment, the required observation period, and the list of application owners to engage. Assign the relationship management tasks to named individuals. Schedule a follow-up session for three weeks out to review the dependency map produced during the observation period.

Discussion 3. How Do We Measure Whether It Is Working?

Run this discussion after the first enforcement milestone is reached, once at least one segment is operating under the deny-by-default policy. The purpose is to establish the metrics that the team will use to demonstrate segmentation effectiveness to leadership and to identify gaps in the current measurement approach.

Discussion questions to ask the team:

- What is the blast radius, expressed as the number of reachable endpoints from a single compromised workstation, before and after enforcement in the pilot segment? Can we calculate this number with confidence?
- How would we detect lateral movement within a segment that is now enforced? What events would be generated, where would they go, and who would review them?

- How often should we run a red team or automated lateral movement simulation to validate that enforcement is working as intended? What would a positive result look like, and what would a failure look like?
- What reporting would give leadership confidence that the microsegmentation program is producing the security outcomes we committed to? How do we translate technical enforcement metrics into mission-aligned language?

What to listen for: Listen for whether the team can articulate blast radius as a quantitative metric, not just as a conceptual improvement. Organizations that cannot express segmentation effectiveness in measurable terms will find it difficult to sustain executive support through a multi-year implementation. Also, listen for whether the denial log review process is actually staffed: a denial log that no one reviews is not a security control.

Follow-up actions: Establish a regular cadence for blast-radius reporting, denial-log review, and red-team validation. Define what each metric looks like when the program is healthy and when it needs attention. Assign metric ownership to specific individuals rather than to the team, because generic ownership yields no accountability.

5.14 Takeaway

Microsegmentation is not a product that gets installed and forgotten. It is an ongoing engineering discipline that begins with visibility, proceeds through modeling, advances through careful phased enforcement, and matures through continuous validation. The organizations that implement it successfully share three characteristics: they observe before they enforce, they treat policy as code from the first day, and they measure blast-radius reduction as a concrete outcome rather than assuming that deploying the tooling achieves the goal.

Marcus's payment zone with 240 endpoints on a flat L2 segment is not an unusual starting point. It is the modal starting point for most enterprise segmentation projects. The gap between that starting point and a fully enforced microsegmented architecture is real and requires real engineering effort. What changes with microsegmentation is not the probability of a breach, which remains a function of many other factors. What changes is what the attacker can do after the breach. A flat network gives an attacker the entire zone. A microsegmented network gives the attacker a bounded segment, a set of deny logs, and a race against detection time measured in seconds rather than months.

The segmentation maturity ladder has four rungs. Most organizations are standing on the first one. Each rung above it reduces the blast radius of any given compromise and increases the cost, complexity, and detectability of lateral movement. That progression is not academic. It is the engineering expression of the zero-trust principle: no single breach should become an enterprise-wide catastrophe. Operational clarity about where the organization stands on that ladder and what it would take to climb one rung is the honest foundation of every credible zero-trust implementation plan.

6 The Continuous Verification Engine: Trust Nothing, Verify Everything, All the Time

Marcus Reyes was three hours into a Thursday afternoon review of Meridian Logistics Group's SIEM and UEBA dashboards when the realization arrived not as an alarm but as a question he could not answer. He had been auditing authentication events across the hybrid environment, cross-referencing MFA completions against active session durations, when he noticed the gap that had always been there but had never been framed quite this way: every security control he was looking at operated at the moment of login. The MFA challenge fired once, at the front door. The SAML assertion was issued once, at the front door. The session token was granted once, at the front door. Everything that happened afterward, the file access, the database queries, the lateral API calls between services, was governed by a trust decision made at a single point in time and then extended indefinitely until the session expired or the user logged out manually. Eight hours of unmonitored activity running on a single credential event.

He pulled up the session records for a subset of accounts and ran a simple calculation. The average session for a Meridian knowledge worker lasted six hours and forty minutes. During that window, the account sent an average of 340 API calls to internal services, executed an average of 14 database queries, accessed an average of 23 distinct file shares, and changed network context at least once as the user moved between office, VPN, and cloud applications. Not a single one of those 340 API calls was evaluated against the user's current context. Not a single database query was assessed against the user's behavioral baseline for that hour. Not a single context change triggered a re-evaluation of whether the trust extended at login still reflected reality. The session token was a six-hour blank check drawn on a single verification event.

The risk calculus was stark. If an attacker acquired a valid session token after the MFA gate, through session hijacking, a malicious browser extension, or a compromised endpoint that passed the initial device health check before its posture degraded, they inherited that blank check entirely. They could operate as the user for the remainder of the session without additional verification. Marcus had spent months building the identity and microsegmentation pillars of Meridian's zero trust program. He had reduced implicit network trust. He had implemented least-privilege access policies. But all of that architecture was protecting a perimeter around a session model that treated trust as a binary state: authenticated at login, trusted until logout. The architecture needed a third pillar, one that made trust a continuous question rather than a one-time answer.

That Thursday afternoon, Marcus opened a new document and began mapping what a continuous verification engine would need to do. It would have to ingest signals from device posture systems, behavioral analytics platforms, network context monitors, and threat intelligence feeds. It would have to aggregate those signals into a trust score that updated on a cadence short enough to detect meaningful changes mid-session. It would have to connect that score to enforcement actions that could challenge suspicious sessions, reduce access scope, or terminate sessions that crossed a risk threshold, all without disrupting the six thousand legitimate users whose sessions would look completely normal. By the time he left the office that evening, he had the outline of an architecture. The rest of this chapter is the engineering and decision-making work that fills in the outline.

6.1 Why It Matters

Authentication and trust are not the same thing, and confusing them is one of the most expensive architectural mistakes an enterprise can make. Authentication confirms that an entity presented valid credentials at a specific point in time. Trust is a forward-looking assertion that the entity will continue to behave in ways consistent with the access it has been granted. In a static environment where users log in from the same device, the same network location, at the same hours, accessing the same resources, the gap between authentication and trust is narrow enough to be manageable. In a hybrid enterprise where users move between networks, share devices, operate from dozens of countries, and access resources spanning on-premises and cloud environments, the gap is enormous. A credential that was authentic at eight in the morning may be in an attacker's hands by ten. The authentication event at eight does not help you detect that transition.

The workflow-level impact of this gap shows up in breach timelines. Credential compromise incidents consistently show dwell times measured in hours to days, not seconds, because the attacker acquires a valid credential and then operates within the session model as though they were the legitimate user. The authentication gate has already been passed. The SIEM generates no alert because the credential is valid. The behavioral anomalies, the unusual access volume, the off-hours activity, and the atypical resource sequence accumulate without triggering containment because no system is evaluating those signals against the trust decision in real time. Continuous verification closes this gap by connecting behavioral signals to trust decisions throughout the session, not just at its beginning. When anomalies accumulate past a risk threshold, the session faces a new challenge, regardless of whether the original authentication was valid.

The regulatory environment reinforces the mission-aligned case for continuous verification. Federal frameworks derived from NIST 800-207 and OMB M-22-09 explicitly require that trust be re-evaluated continuously, not assumed based on a prior authentication event. Commercial frameworks, including PCI-DSS 4.0 and emerging HIPAA guidance, have moved in the same direction. The pilot-to-production pathway for continuous verification is also among the most concrete in the zero-trust program: organizations can implement session monitoring and risk scoring against existing authentication infrastructure without requiring a full rearchitecture, generating immediate security value. In contrast, the broader zero-trust deployment matures. This is not a future-state capability. It is a currently deployable control that addresses the highest-frequency attack vector in enterprise security today.

6.2 Trust as a Continuous Variable, Not a Binary State

The foundational shift in continuous verification is conceptual before it is technical. Traditional authentication models operate on a binary state machine: a user is either authenticated or not, and authentication at login persists until the session ends. Zero trust replaces this binary with a continuous variable, the trust score, that reflects the current probability that the entity requesting access is who they claim to be and is behaving consistently with that identity. That probability changes with every new signal the environment produces. A device health check that degrades because a signature update fails. A user who begins accessing resources that fall outside their behavioral baseline. A network location that shifts from the corporate office to an unusual geographic region. Each of these events adjusts the trust variable, and the access controls the user experiences adjust in response.

This reframing has significant implications for system architecture. If trust is a binary state, a single authentication system can manage it. If trust is a continuous variable, an entire signal processing pipeline is required to compute and update it throughout the session. The pipeline must ingest data from multiple sources at different cadences, normalize and weight the signals, compute a composite score, compare that score against policy thresholds, and route enforcement actions to the appropriate control plane, all within a latency budget short enough to be operationally effective. Building that pipeline is the engineering challenge at the center of this chapter. Understanding what it is doing and why is the management challenge that determines whether the organization's resources and governance are correct.

Responsible-by-design principles apply directly to how the trust score is used. An opaque, unauditable, or disconnected trust score in high-consequence situations creates its own operational and legal risk. Organizations that implement continuous verification must build governance structures around the score: documentation of which signals feed it and with what weights, audit logs of score changes and the signals that drove them, human review processes for automated enforcement actions that affect business-critical operations, and exception procedures for legitimate users whose behavioral patterns diverge from baselines for legitimate reasons. Continuous verification, when done well, is not a black box. It is an auditable, tunable, and explainable system that earns operational trust through transparency.

Trust as a Continuous Variable

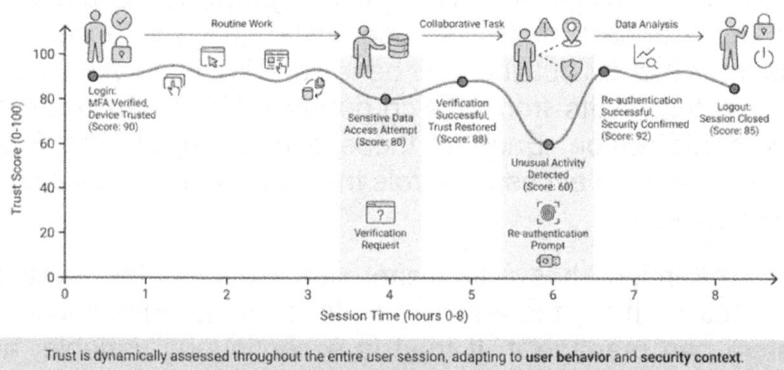

6.3 Signals That Feed the Trust Score

A trust score is only as good as the signals that compose it. The engineering challenge is not computing a score; it is identifying which signals are meaningful, collecting them reliably at the cadence required for real-time decisions, normalizing them so they can be combined coherently, and weighting them in proportion to their actual predictive value for the threat scenarios the organization faces. Four signal categories form the foundation of most continuous verification architectures, and each brings its own collection-engineering requirements, latency profile, and false-positive characteristics that must be understood before the score can be trusted.

Device Posture

Device posture signals represent the current health state of the endpoint the user is operating on. At login, a device posture check evaluates whether the device meets a defined compliance baseline: operating system version, patch level, presence and health of endpoint detection and response agents, disk encryption status, and screen lock configuration are common examples. In a continuous verification model, that check does not end at login. The posture agent on the device continues to report status throughout the session, and changes in posture feed the trust score in real time. A device that passes the initial posture check and then receives a detection event from its EDR agent, or whose patch level drops out of compliance because an update fails to apply, or whose endpoint agent is terminated by a process running on the device, generates a posture degradation signal that the trust engine must process and reflect in the score.

The collection engineering for device posture signals requires a posture assessment agent that operates continuously and communicates status to the verification engine on a defined cadence. The cadence is a design decision with security and performance trade-offs: a posture check every 30 seconds provides near-real-time detection of posture changes but generates significant agent and network overhead at enterprise scale. A posture check every 5 minutes reduces overhead while extending the window during which a compromised endpoint operates before the trust score reflects the change. Most enterprise implementations land between 60 and 180 seconds for the routine posture polling interval, with event-driven immediate reporting for high-severity posture changes such as EDR detection events or agent termination.

Behavioral Baselines

Behavioral signals represent deviation from the patterns that characterize a specific user's normal activity. Where device posture is largely objective (the patch level either meets the threshold or it does not), behavioral signals require a baseline model to be meaningful. That model is built over time from historical observations of the user's access patterns: the hours at which they typically access resources, the categories of resources they access most frequently, the volume of data they transfer in a normal session, the sequence of applications they typically use, and the geographic and network contexts from which they normally operate. Deviation from that baseline is not itself an indicator of compromise; it is an indicator that something has changed, and the trust engine must weigh the significance of the deviation against other available signals to determine whether it warrants action.

Behavioral baseline models must account for the legitimate variability in user behavior. A financial analyst who normally accesses accounting systems suddenly accessing a large volume of HR records is a meaningful anomaly. The same analyst accessing accounting systems at 6:00 PM instead of 9:00 AM during quarter-close is not, because quarter-close off-hours work is a documented pattern for that role. UEBA platforms that handle this well maintain role-adjusted baselines that normalize for known behavioral patterns associated with job function, department, and organizational calendar events. They also maintain peer group baselines that allow an individual's behavior to be assessed against similar users, which is particularly valuable for detecting behaviors that are unusual for the peer group, even if the individual has no prior history of that behavior.

Behavioral Baseline and Anomaly Detection

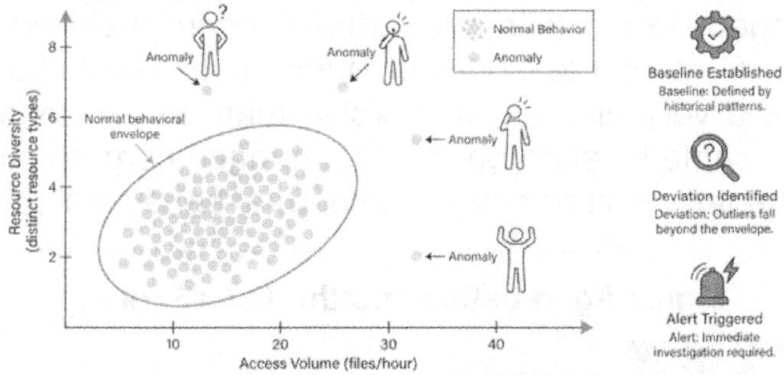

Behavioral Baseline and Anomaly Detection identifies deviations from normal user activity patterns to flag potential security threats.

Geographic and Network Context

Geographic and network context signals capture where the user is connecting from and whether that location aligns with their established patterns and physical plausibility. Geographic context is one of the most reliable continuous verification signals because human beings cannot be in two places simultaneously. If a user authenticates from New York at 9:00 AM and a session request arrives from Singapore at 9:45 AM, the geographic signal is a near-certain indicator of credential compromise: no legitimate user can have traveled from New York to Singapore in 45 minutes. This class of anomaly, often called an impossible travel event, is among the highest-confidence signals available to a continuous verification engine and typically warrants immediate step-up authentication or session termination rather than a score adjustment.

Network context signals go beyond geographic location to include the character of the network from which the connection originates. A user connecting from a known corporate office network presents a different risk profile than the same user connecting from a residential ISP, which in turn presents a different profile than a connection from a data center IP range, a known VPN exit node, or a Tor exit node. Network context signals are most powerful when they are combined with behavioral signals and device posture data. A user connecting from a new geographic location on an unmanaged device with no endpoint agent installed, while accessing resources outside their behavioral baseline, presents a very different risk profile than a user connecting from the same location on a fully managed corporate device with a healthy posture check and normal behavioral patterns.

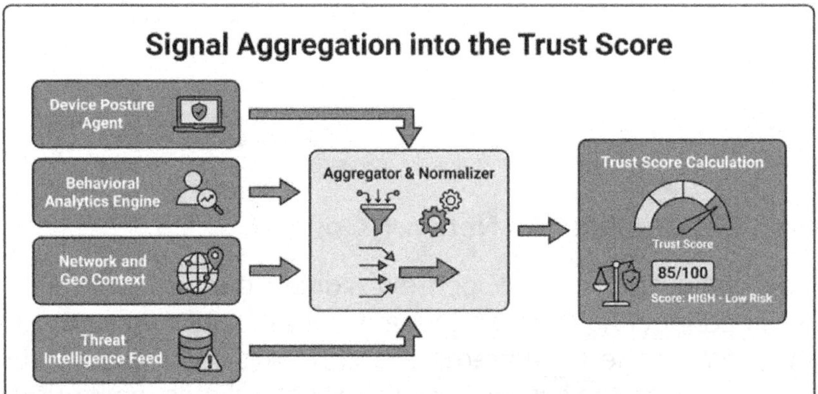

The central system collects, processes, and analyzes signals from various sources to compute a unified, dynamic trust score for access decisions.

Risk Indicators from Threat Intelligence

Threat intelligence feeds add an external dimension to the trust score by incorporating knowledge of known malicious indicators into the real-time evaluation. When a user's session network characteristics match a known command-and-control infrastructure IP range, or when the device posture system identifies a file hash associated with a known malware family, or when an email address associated with the account appears in a credential breach database published after the user's last password rotation, those external signals should feed the trust engine as high-weight negative indicators. Threat intelligence integration requires a normalization layer that maps diverse feed formats to the internal signal schema the trust engine expects, and a freshness management system that ensures the engine operates on current intelligence rather than stale data.

The weighting of threat intelligence signals relative to behavioral and posture signals is a policy decision that reflects the organization's assessment of feed reliability. High-confidence feeds from well-maintained sources, such as government threat-sharing platforms and reputable commercial threat-intelligence providers, warrant higher weights than lower-confidence feeds. False-positive rates in threat intelligence feeds can be significant, particularly for IP reputation signals, where legitimate infrastructure is sometimes miscategorized as malicious. Organizations should evaluate feed quality empirically by tracking how often threat intelligence hits correlate with confirmed malicious activity in their own environment before assigning maximum weight to any single feed.

6.4 Step-Up Authentication: Dynamic Escalation Without Disruption

Step-up authentication is the mechanism that translates a drop in trust score into a user-facing verification event. When the continuous verification engine detects that the trust score has fallen below a defined threshold, rather than terminating the session immediately, it can challenge the user to perform an additional authentication step, typically one that is stronger than the method used at initial login. A user who authenticated with a password and an SMS one-time code at login might receive a push notification to their authenticator app for a step-up, be required to insert a hardware security key, or be prompted to complete biometric verification on their device. The step-up challenge is designed to be resolved in seconds by a legitimate user but to block an attacker who lacks access to the user's out-of-band authentication factor.

The user experience design of step-up authentication is as important as its security design. A step-up mechanism that generates too many challenges per session trains users to approve challenges reflexively without scrutiny, a pattern known as MFA fatigue that has been exploited in major breach events. A step-up mechanism that generates challenges at infrequent but appropriate risk moments trains users to treat each challenge as meaningful and to pay attention to its content. The design goal is that a legitimate user should encounter a step-up challenge only when something genuinely unusual has happened in their session, and the challenge should feel commensurate with the anomaly. An impossible travel detection, for example, might warrant a challenge with explicit messaging: "We have detected a login attempt from a new geographic location. Please verify your identity." That context helps the user understand why the challenge is appearing and makes it more likely they will recognize a malicious challenge for what it is.

Step-up authentication must be integrated with the session management layer so that a successful completion updates the trust score appropriately without resetting it to the initial maximum. A user who completes a step-up challenge after genuinely moving to a new geographic location should see their score recover from the geographic anomaly signal. Still, other active signals, such as a device posture degradation that occurred independently of the travel event, should continue to contribute to the score. The step-up event is an attestation about identity, not a blanket certificate of normality. Treating it as a full trust reset would allow an attacker who steals a session and passes a step-up challenge to clear all pending anomaly signals from the score, which is the opposite of the intended behavior.

Step-Up Authentication Workflow

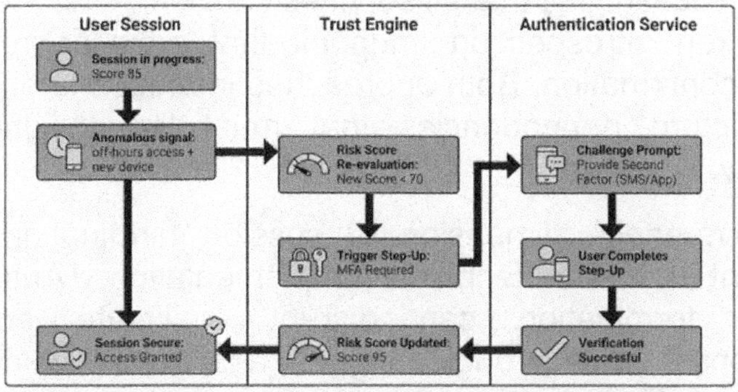

Step-Up Authentication Workflow ensures enhanced security by requiring additional verification when anomalous activity or high-risk signals are detected during an active user session.

6.5 Session Monitoring and Termination

Continuous verification is not only about detecting risk signals mid-session. It is also about having the infrastructure and authority to act on those signals up to and including terminating an active session when the risk level warrants it. Session monitoring means that every access event, every API call, every data transfer within an active session is logged and evaluated against the current trust context. Session termination occurs when the trust score falls below the termination threshold, and step-up authentication either fails or is not applicable. The system can then revoke the session token, terminate the connection, and log the event with sufficient detail for forensic analysis.

The technical implementation of session termination requires careful integration with the token management infrastructure. In a system that relies on short-lived JWT tokens with a one-hour expiration, terminating a session is relatively straightforward: the token can not be renewed at expiration, and the user is re-challenged within 60 minutes. In a system that uses long-lived session tokens with an eight-hour or longer lifespan, termination requires a token revocation mechanism that actively invalidates the token before its natural expiration. Building a real-time revocation capability that scales to enterprise session volumes is a non-trivial engineering problem, because it requires either a centralized token revocation list that all services check on every request, or a token introspection endpoint that provides real-time validity confirmation. Both approaches introduce latency and infrastructure dependencies that must be designed for reliability.

The governance dimension of session termination is as important as the technical implementation. Automated session termination can disrupt legitimate business operations if the trust engine triggers a false positive during a critical workflow. An executive who is mid-presentation and whose session is terminated because a network context change temporarily drops their trust score below the threshold will have a visceral, negative response to the continuous verification system. Organizations that deploy session termination capability must define the specific score thresholds and signal combinations that justify automated termination without human review, versus those that require a human decision before the session is terminated. For most commercial environments, the appropriate threshold for fully automated termination is set conservatively to capture only the clearest evidence of compromise: impossible travel events, active threat intelligence matches, or concurrent sessions from different geographic locations.

Session Termination Decision Flow

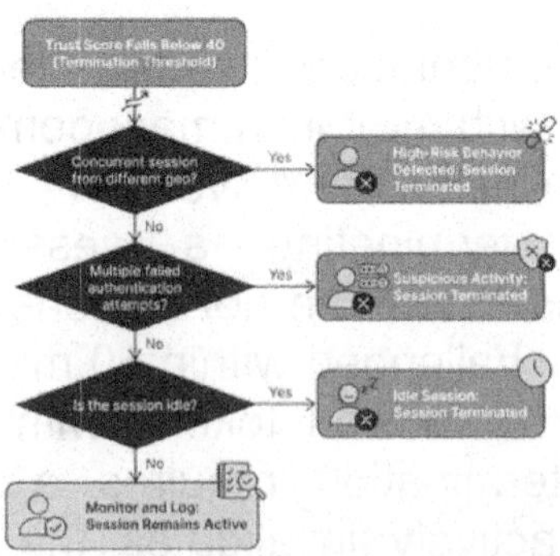

Session Termination Decision Flow

6.6 Behavioral Analytics and UEBA

User and Entity Behavior Analytics (UEBA), commonly abbreviated UEBA, is the technology discipline dedicated to building behavioral baseline models and anomaly-detection capabilities that feed the continuous verification engine. A UEBA platform ingests event telemetry from multiple sources, applies statistical and machine learning methods to characterize normal behavior patterns for each user and entity, and generates alerts or scores when observed behavior deviates from those patterns in ways that exceed a defined sensitivity threshold. In the context of continuous verification, the UEBA platform is not a standalone detection system; it is a signal source that feeds the trust engine alongside device posture, network context, and threat intelligence signals.

The behavioral dimensions that enterprise UEBA platforms typically model include access timing patterns, which capture the hours and days during which a user or entity typically accesses resources and flag access outside those windows; resource access patterns, which model the set of resources a user typically accesses and flag unusual resource categories or sequences; data transfer volumes, which establish baselines for how much data a user moves in a normal session and flag significant volume deviations; authentication event patterns, which model the cadence and success rate of authentication attempts and flag unusual rates or patterns; and privilege usage patterns, which model how often a user exercises elevated privileges and flag unusual privilege escalation events. Together, these dimensions create a behavioral fingerprint for each user that is both sensitive enough to detect genuine anomalies and robust enough to tolerate the normal variation in legitimate user behavior.

Entity behavior analytics extends the same approach to non-human entities: service accounts, application workloads, IoT devices, and API clients. Service accounts are particularly valuable targets for entity behavior analytics because their behavioral patterns are typically narrow and consistent, making genuine deviations highly conspicuous. A service account that normally performs a specific set of database queries against a specific schema at regular intervals from a specific source IP will generate a very clear anomaly signal if it begins performing queries against different schemas, from different source IPs, or at irregular intervals. That signal is the early indicator of a compromised service account that might be invisible to a system monitoring only human user behavior.

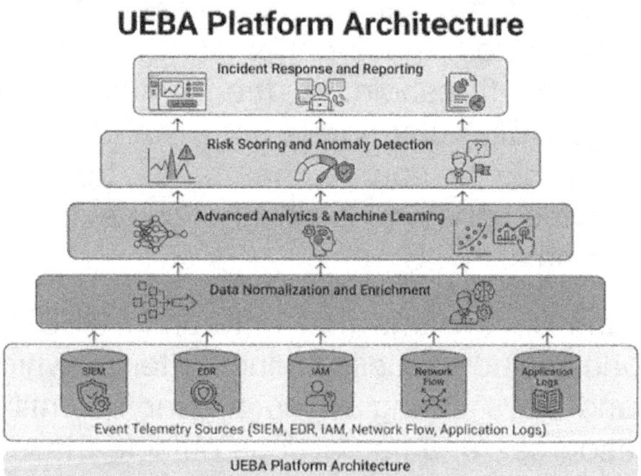

6.7 AI and Machine Learning in the Continuous Verification Engine

Machine learning is not an optional enhancement to continuous verification; it is a foundational requirement. The behavioral baseline models that underpin UEBA cannot be built and maintained at enterprise scale using manually configured rules. An enterprise with ten thousand users and several hundred thousand distinct entities generates billions of behavioral events per day. Building meaningful baseline models for each user and entity from that telemetry volume requires statistical methods that automatically learn patterns from data, update those models as behavior changes over time, and flag deviations without generating an operationally unmanageable volume of alerts. The two primary ML paradigm families deployed in verification engines are unsupervised learning methods for anomaly detection and supervised learning methods for classification and risk prediction.

Unsupervised anomaly detection is particularly well-suited to behavioral baseline modeling because it does not require labeled training data. Methods such as Isolation Forest, Local Outlier Factor, and autoencoder neural networks learn the statistical structure of normal behavior from historical data and identify events that deviate significantly from that structure, without requiring a pre-defined rule for what constitutes deviation. This is critical because insider threats and credential compromise attacks often take forms that security teams have not specifically anticipated. A rule-based system that only flags the patterns an analyst has thought to codify will miss novel attack behaviors; an unsupervised model that characterizes normality will flag anything that deviates from that normality, regardless of whether the specific deviation pattern has been seen before.

Supervised learning methods play a complementary role by incorporating labeled examples of confirmed malicious behavior into the model training process. When an organization has confirmed breach events, insider threat incidents, or red team exercise results that can be labeled as positive examples of malicious behavior, those labeled examples can be used to train classifiers that assign higher risk scores to behavioral patterns similar to the labeled examples. Supervised methods tend to achieve higher precision on the attack patterns they were trained on, at the cost of reduced coverage for novel patterns outside the training distribution. A mature continuous verification deployment typically uses both approaches in an ensemble: unsupervised methods for broad anomaly coverage and supervised classifiers for high-confidence detection of known-pattern attacks.

ML Model Types in the Verification Engine

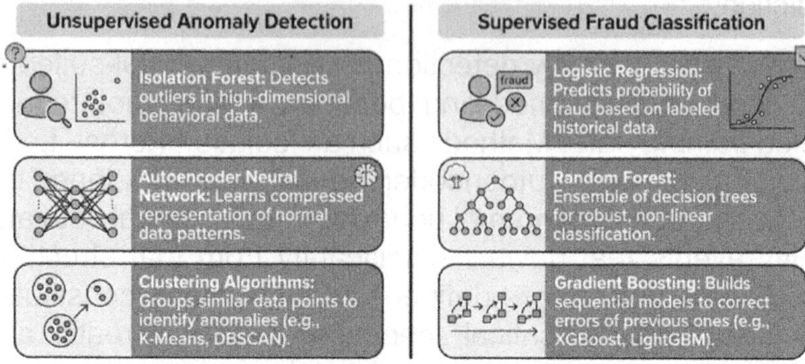

ML Model Types in the Verification Engine employ both unsupervised and supervised machine learning techniques to detect anomalies and classify fraud risks.

6.8 The Trust Score as a Dynamic Value

The trust score is the operational interface between the continuous verification engine and the access control infrastructure. It translates a complex, multi-signal reality into a number that policy engines, microsegmentation platforms, and identity systems can act on, without requiring each system to interpret raw behavioral and posture telemetry independently. But for the trust score to function as a reliable operational interface, the organization must make and document a set of architectural decisions about how the score is computed, what it represents, and how it connects to enforcement.

Score composition decisions define which signals contribute to the score, in what proportions, and with what temporal decay. A device posture event from thirty seconds ago should carry more weight than a behavioral pattern observation from six hours ago. A confirmed threat intelligence match should carry more weight than a mild behavioral deviation. The relative weighting of these signals is not a mathematical question with a single correct answer; it is a security policy question that reflects the organization's threat model and risk tolerance. Organizations in high-security environments may treat threat intelligence hits as near-maximum impact, so that any threat intelligence match immediately triggers step-up authentication or session termination. Commercial environments with higher user-experience sensitivity may use more graduated weighting, so that a single threat-intelligence hit from a lower-confidence feed results in a moderate score reduction rather than an immediate challenge.

Score thresholds define the bands of the trust score that map to specific enforcement actions. A common three-band model defines a green zone (score 70-100) in which the session continues normally, a yellow zone (score 40-69) in which step-up authentication is triggered, and a red zone (score below 40) in which automated enforcement actions up to and including session termination are initiated. The threshold values that define these bands must be calibrated against the organization's false-positive tolerance. Setting the step-up threshold too high generates frequent challenges for legitimate users and creates MFA fatigue. Setting it too low allows attackers operating just above the threshold to persist in sessions without challenge. Threshold calibration is an ongoing operational task, not a one-time configuration decision, and it requires continuous monitoring of the false-positive rate alongside the detection rate to ensure the score remains operationally calibrated.

Trust Score Bands and Enforcement Actions

Trust Score Bands and Enforcement Actions

6.9 Geeks Only

This section is written for security engineers, platform architects, and technically oriented team leads who want to understand the mechanistic and implementation-level dimensions of continuous verification before making decisions on system design and tool selection. If you are a manager who has absorbed the preceding sections, the Manager's Decision Guide on the next page covers the questions you need to answer at the organizational level. If you are an engineer designing or evaluating a continuous verification pipeline, the subsections below address the technical decisions that determine whether your implementation actually performs at production scale.

Anatomy of a Verification Engine Pipeline

A production-grade continuous verification engine is a real-time data processing pipeline with a well-defined set of functional stages. Understanding those stages, their dependencies, and their failure modes is a prerequisite to designing a system that meets the latency, reliability, and scale requirements of an enterprise deployment.

The ingestion stage accepts telemetry events from source systems, typically via message-queue infrastructure such as Apache Kafka or a cloud-native equivalent, and writes them to a processing buffer. The ingestion stage must handle event bursts, out-of-order events from sources with variable latency, and schema variations between different source systems. Designing for schema evolution from the beginning is important because source systems will change their event schemas over time, and a verification engine that cannot tolerate schema changes without downtime is operationally fragile.

The normalization stage transforms ingested events into a canonical internal representation that the downstream processing stages can operate on consistently. Normalization includes schema mapping, timestamp normalization to a common time zone and precision, entity resolution that maps diverse identifiers, such as email addresses, UUIDs, and certificate hashes, to a canonical identity, and field enrichment that adds context not present in the raw event, such as geographic location derived from IP addresses. Normalization errors are among the most common sources of false positives in verification engines because incorrectly normalized events may appear anomalous simply because their representations diverge from the baseline model.

The scoring stage applies the behavioral models and weighting logic to the normalized events and computes or updates the trust score for the affected session. The scoring stage must be completed within the latency budget defined for the overall pipeline. For access control integration, where the trust score must be available before an access decision can be made, the end-to-end pipeline latency budget is typically 200-500 milliseconds. Scoring stages that exceed this budget cause access decisions to revert to a default policy, which, in a well-designed system, should be deny-by-default, but in practice is often allow-by-default to avoid operational disruption. That fallback behavior is a significant security design consideration.

The enforcement stage routes score changes to the appropriate control plane for action. When the score drops below the step-up threshold, the enforcement stage must notify the session management system to issue a step-up challenge. When the score drops below the termination threshold, it must invoke the token revocation API and log the event with full signal context. The enforcement stage must be designed as an idempotent consumer: if the enforcement action is delivered more than once due to message-queue semantics, processing the action a second time must not produce harmful results, such as issuing two simultaneous step-up challenges to the same user or terminating an already-terminated session.

Verification Engine Pipeline Stages

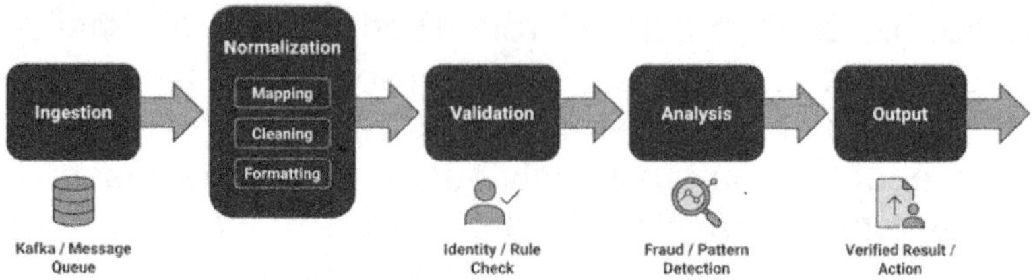

The verification engine pipeline transforms raw data through sequential stages into trusted, actionable intelligence for business decision-making.

Signal Ingestion Patterns

Signal ingestion is the first engineering challenge in continuous verification, and it is harder than it appears. Source systems emit events in different formats, at different cadences, with different reliability characteristics. Device posture agents may emit events every 60 seconds with a reliable heartbeat. Behavioral analytics events may arrive in near-real-time from some sources and in batches from others. Threat intelligence feeds may update hourly, daily, or irregularly. The ingestion architecture must handle this heterogeneity without creating integration debt that makes the pipeline brittle as source systems change.

A common pattern is to define a canonical event schema for the verification engine and implement source-specific adapters that transform each source system's native event format into the canonical schema before writing to the message queue. The adapter pattern isolates the pipeline from schema changes in source systems: when a source system changes its event format, only the corresponding adapter needs to be updated, not the downstream pipeline stages. Adapters should also handle backpressure, the condition in which the message queue is processing events more slowly than they are being produced, by implementing configurable shedding policies that drop lower-priority events when the queue depth exceeds a threshold. Behavioral telemetry events from low-risk sessions are reasonable candidates for shedding; device posture events signaling EDR detections are not.

Building Behavioral Baselines

Behavioral baseline construction is the most time-intensive phase of a continuous verification deployment. A meaningful baseline requires a minimum observation period during which the system accumulates sufficient behavioral history to reliably characterize normal behavior. For most enterprise user populations, a minimum of 30 days of behavioral telemetry is required before anomaly detection becomes operationally useful, with 90 days providing materially better baseline quality. The cold-start problem, the inability to detect anomalies for new users or entities until a baseline has been built, must be handled explicitly in the system design.

Common approaches to the cold-start problem include peer group baseline initialization, in which a new user is initially assessed against the baseline for their peer group rather than an individual baseline. Peer group baselines are derived from the aggregated behavioral history of users with similar roles, departments, and access patterns and provide reasonable anomaly-detection coverage until the individual baseline matures. The transition from peer group baseline to individual baseline should be gradual: a weighted blend of peer group and individual observations during the baseline maturation period, shifting progressively toward individual observations as the history grows. Most UEBA platforms implement a variant of this approach natively, but the specific parameterization of peer group definition and transition timing requires tuning for each deployment.

Baseline models must also handle legitimate behavioral change over time. A user who changes roles, takes on a new project, or moves to a different office will exhibit behavioral patterns that diverge from their prior baseline. If the baseline model does not adapt to these changes, the user will generate persistent false-positive anomaly signals, degrading both operational efficiency and user experience. Most production UEBA implementations use adaptive baselines that update continuously as new observations accumulate, with a configurable decay rate that determines how quickly old observations lose influence relative to new ones. Decay rates must be balanced: too fast and the baseline adapts to attacker behavior during a slow-burn compromise; too slow and the baseline fails to adapt to legitimate behavioral change.

False-Positive vs. False-Negative Trade-offs

No continuous verification system can simultaneously minimize false positives and false negatives. These are competing objectives, and the system design must make explicit choices about where to position the operating point on the precision-recall trade-off curve. A system tuned for maximum recall, catching every possible anomaly, will generate a high volume of false-positive step-up challenges and session terminations, disrupting legitimate user workflows and eroding trust in the system. A system tuned for maximum precision, challenging only the most certain anomalies, will miss genuine threats that fall below the confidence threshold and provide weaker security guarantees.

The appropriate operating point depends on the security context. High-security federal environments may accept a higher false-positive rate in exchange for maximum detection sensitivity, particularly for privileged account sessions and access to classified systems. Commercial environments serving large consumer-facing workforces may prioritize low false-positive rates to protect the user experience, accepting a somewhat higher false-negative rate for low-risk session types. A tiered approach that applies different sensitivity settings to different session categories, based on the sensitivity of the resources being accessed, is often the best practical solution: aggressive detection for sessions accessing high-value data, and more permissive thresholds for sessions accessing lower-sensitivity resources. This approach is consistent with the least-privilege principle applied to verification overhead: apply the most intensive verification where the cost of a false negative is highest.

False-Positive vs. False-Negative Operating Point

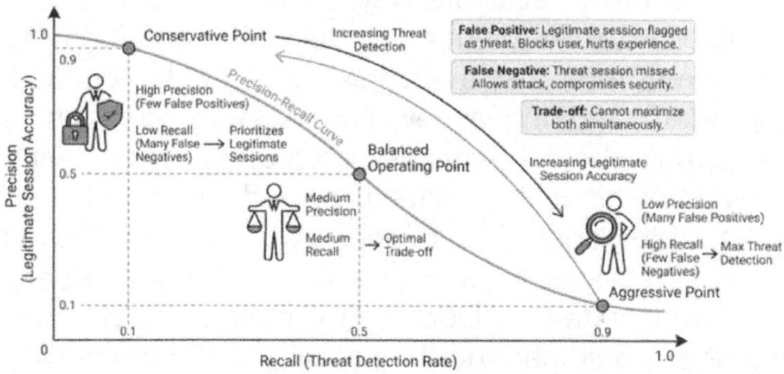

Choosing the operating point determines the trade-off between missing threats (false negatives) and blocking legitimate users (false positives), impacting security effectiveness versus user experience.

Real-Time vs. Near-Real-Time Evaluation

Continuous verification pipelines operate across a spectrum from true real-time evaluation, where access decisions wait for the current trust score before completing, to near-real-time evaluation, where the trust score is computed asynchronously and used for subsequent requests. The choice between these modes has significant implications for pipeline latency requirements, infrastructure complexity, and the enforcement actions that are practically achievable.

True real-time evaluation requires the trust engine to complete its scoring computation within the latency budget of the access control check. For systems using a policy enforcement point that blocks access requests pending a trust score evaluation, this means the entire ingestion-normalization-scoring pipeline must complete in under 200 milliseconds at the 99th percentile. Achieving this latency target at enterprise scale requires in-memory scoring, pre-computed baseline models loaded into memory at session initialization, and a careful engineering approach to the normalization stage that avoids external service calls in the critical path. True real-time evaluation provides the strongest security guarantees because every access request is evaluated against the current trust score.

Near-real-time evaluation relaxes the latency constraint by computing the trust score asynchronously and caching the most recent score for use by the policy enforcement point. Access requests are evaluated against the cached score rather than requiring a new computation for each request. This approach can handle enterprise scale with more conventional pipeline architectures and more forgiving latency targets, at the cost of a staleness window: the cached score may be several seconds to several minutes old, depending on the refresh cadence. During that staleness window, a risk signal that would change the score below a threshold will not affect access decisions until the cached score is refreshed. For most commercial deployments, a cached score refreshed every 30 to 60 seconds provides a reasonable balance between security and infrastructure cost.

Model Drift Detection

Machine learning models embedded in continuous verification systems are not static artifacts. They are trained on historical data that reflects the organization's environment at a point in time. As that environment evolves, the models' assumptions about what constitutes normal behavior may become increasingly inaccurate. This phenomenon, known as model drift, is a critical operational concern for verification systems deployed over months and years. A model that was accurately calibrated during initial deployment may begin to generate higher false-positive rates as legitimate user behavior evolves, or may miss new attack patterns that did not exist in the training data. Both failure modes degrade system performance without generating obvious error signals in standard monitoring.

Detecting model drift requires explicit monitoring of model performance metrics over time. For unsupervised anomaly detection models, the most accessible drift indicator is the anomaly score distribution: if the distribution of scores assigned to the user population shifts significantly over time without a corresponding change in confirmed incident rates, the model baseline may have drifted. Tracking the percentage of sessions that receive anomaly scores above the alert threshold, the mean and variance of the score distribution, and the correlation between high-anomaly scores and confirmed malicious events provides an empirical basis for detecting drift. When drift is detected, the response options include retraining the affected model on more recent data, adjusting the model's sensitivity thresholds to compensate for the drift, or deploying a new model architecture better suited to the changed behavioral landscape. Most production deployments implement scheduled retraining on a quarterly or semi-annual cycle, with continuous monitoring to detect drift that warrants an unscheduled retraining event.

Model Drift Detection and Response Cycle

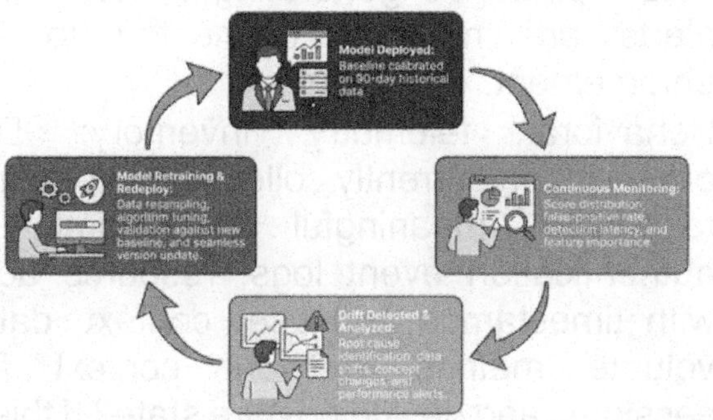

Continuous vigilance ensures AI model performance and reliability over time through this structured response mechanism.

6.10 Manager's Decision Guide

Continuous verification is the most operationally complex component of a zero-trust program to deploy and govern. It simultaneously touches authentication infrastructure, behavioral analytics platforms, session management systems, and enforcement integrations. The decision criteria below help you assess organizational readiness, identify the highest-priority gaps, and structure a deployment approach that delivers security value without creating unacceptable operational disruption. Work through these criteria with your security architect and SIEM or UEBA platform owner before committing to a deployment timeline or budget.

Decision Criteria

- Binary vs. continuous authentication audit: Has your security team assessed whether the organization's current authentication model is binary, granting trust at login and persisting it for the session duration, or whether any continuous evaluation is already in place? This assessment identifies the baseline from which you are starting and the scale of the required architectural change. Organizations with eight-hour session tokens and no mid-session behavioral monitoring have a larger gap to close than organizations that already have UEBA platforms generating alerts, even if those alerts are not yet connected to automated enforcement.
- Behavioral telemetry inventory: Does the organization currently collect the telemetry required to build meaningful behavioral baselines: authentication event logs, resource access logs with timestamps and user context, data transfer volume metrics, network context for active sessions, and device posture state? If this telemetry is being collected in a SIEM today, the data foundation for continuous verification may already exist. If it is not being collected consistently, the telemetry pipeline must be built before behavioral baseline models can be developed, which extends the deployment timeline by six to twelve weeks.
- Step-up authentication capability: Is the organization currently capable of issuing a mid-session re-authentication challenge to a user, and does the user population have the out-of-band authentication factor required to satisfy such a challenge? Organizations that have deployed

hardware security keys or push-based authenticator apps to all users have the strongest step-up authentication capabilities. Organizations relying on SMS-based one-time codes have a meaningful capability gap for step-up use cases because SMS one-time codes are susceptible to SIM-swapping attacks, allowing an attacker who has compromised the session also to satisfy the step-up challenge.

- Session termination authority and infrastructure: Does the security operations team currently have the technical capability to terminate an active user session, and is that capability fast enough to be operationally useful? A manual session-termination process that requires 30 minutes and two escalation steps provides minimal value against an attacker operating in real time. Automated session termination tied to trust score thresholds requires that the session management infrastructure support a revocation API that the trust engine can call with sub-minute latency. Assess whether that API exists and whether it covers all session types in your environment, including cloud application sessions managed by SSO and on-premises application sessions managed by legacy authentication systems.
- User experience impact assessment: Has the team modeled the expected frequency of step-up authentication challenges under the proposed trust score configuration, and is that frequency acceptable to the user population and business leadership? An implementation that is technically correct but generates step-up challenges five times per day for the average user will face organizational resistance that undermines the program. The pilot phase of the deployment should include explicit measurement of step-up frequency, false-positive rate, and user satisfaction to validate that the

threshold configuration is operationally viable before broad deployment.

- Enforcement integration completeness: Are the continuous verification outputs, specifically the trust score and the enforcement actions it triggers, integrated with both the microsegmentation enforcement plane and the identity-based access control plane? A verification engine that detects anomalies and issues alerts but does not trigger automated enforcement provides a fraction of the security value of a fully integrated system. The integration path from the trust engine to enforcement must be mapped and tested before the system is declared production-ready, because gaps in enforcement integration are invisible until an incident occurs in which detection worked but containment did not.

Decision Frame

The organization is ready to deploy a production continuous verification capability when the behavioral telemetry foundation is in place, the step-up authentication infrastructure covers the full user population, and at least one enforcement integration, either session termination or microsegmentation isolation, is operational and tested. You do not need to wait for perfect baseline model maturity or complete enforcement coverage before deploying: a phased approach that begins with monitoring-only mode, advances to step-up authentication challenges, and then adds automated termination allows the system to generate security value at each phase while operational confidence builds. The critical gate before enabling automated session termination without human review is a demonstrated false-positive rate below the threshold your organization defines as acceptable, measured over a minimum 30-day monitoring period. Operating automated enforcement with an unvalidated false-positive rate is the most common cause of organizational rejection of continuous verification programs. Validate before automating, but do not let the pursuit of perfection delay the deployment of a monitoring-only capability that begins collecting the baseline data you will need.

6.11 Case Study: Atlas Commerce International

Atlas Commerce International, a fictional global retailer operating across 14 countries with a workforce of 200,000 employees, built its authentication infrastructure around the prevailing commercial standard of the time: MFA at login, followed by an 8-hour session. The decision was defensible on its own terms. MFA at login represented a meaningful security improvement over password-only authentication, and an eight-hour session duration minimized user friction for a workforce operating across multiple time zones and needing sustained access to retail operations systems throughout their working day. The security team monitored authentication events, tracked MFA completion rates, and investigated failed authentication attempts. What they were not doing was evaluating whether the trust established at login still reflected reality as sessions progressed.

The credential compromise that exposed this gap began in a way that the authentication perimeter was not designed to detect. An employee's credentials were obtained through a spear-phishing campaign targeting the employee's personal email account, in which a password matching their corporate credentials was reused. The attacker waited. Rather than using the credential immediately after acquiring it, the attacker waited eleven days before initiating the first corporate access session, a deliberate delay designed to minimize correlation between the phishing event and the subsequent access. When the attacker did connect, they connected during the employee's normal working hours from a geographic location the employee had not previously used, a residential ISP in a neighboring city, close enough to avoid a dramatic geographic anomaly but different enough that a behavioral context model would have flagged it.

The session proceeded for six hours. During that time, the attacker navigated through the retail operations systems, progressively expanding their access to customer data tables. The access volume grew steadily: the first hour involved reconnaissance queries against schema metadata, and the second and third hours involved selective data sampling across multiple customer record tables. Hours four through six involved systematic extraction of payment card records from the transactions database. By the time the employee's legitimate session token expired at the end of the business day, the attacker had exfiltrated 340,000 payment card records. The SIEM had logged every access event. None of the logged events individually crossed the threshold for an alert under the organization's rule-based detection configuration, because the access patterns, while broader than the employee's normal behavior, did not match any specific rule the team had configured.

The post-incident analysis identified four signal categories that a continuous verification engine would have processed into an anomalous trust score within the first 90 seconds of the session: a network context anomaly because the connection originated from a new ISP not previously associated with the employee account, a geographic proximity anomaly because the IP geolocation was 180 miles from the employee's registered work location, a behavioral deviation because the first action taken in the session was a schema metadata query rather than the employee's typical start-of-session application access pattern, and a data access volume anomaly because the session began accumulating query volume at a rate 340% above the employee's baseline within the first 15 minutes. Any one of these signals would have been insufficient to trigger action under a conservative threshold. The composite trust score computed from all four signals would have fallen to 31 within 90 seconds, below the step-up authentication threshold. It would have triggered a push notification challenge that the attacker, lacking access to the employee's registered authenticator application, could not satisfy.

Following the implementation of a continuous verification engine over the twelve months after the incident, Atlas Commerce ran a simulated replay of the attack scenario using the same credentials, network context, and behavioral sequence. The trust engine detected the composite anomaly within 87 seconds and issued a step-up challenge. The simulated attacker, unable to satisfy the push notification challenge, was locked out within 92 seconds of session initiation. Given the reconnaissance-first behavioral pattern of the original attack, the maximum number of payment card records accessible in that window was fewer than 200. The operational impact of the step-up challenge on the legitimate user population during the same 30-day simulation period was measured at 2.3 additional authentication interactions per user per week, a number the organization's user experience team assessed as within the acceptable friction budget. The case for continuous verification at Atlas Commerce was not abstract. It was 340,000 customer records versus fewer than 200, and it was an 87-second detection window versus a six-hour exfiltration session.

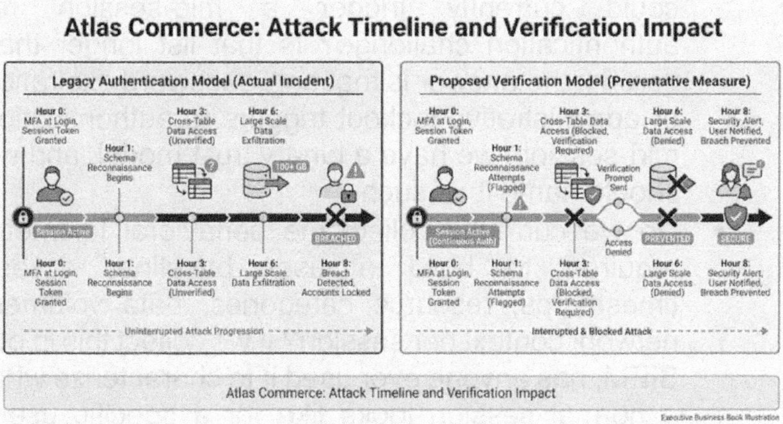

6.12 Team Conversation Guide

Run these three discussions in sequence over two to four weeks. The first surfaces the issue, the second works the trade-offs, the third commits the team to action.

Discussion 1. Auditing Our Current Session Trust Model

Before your team can design a continuous verification capability, they need to characterize the current session trust model honestly. This discussion surfaces what happens today when a user authenticates, how long that trust lasts, and what, if anything, currently evaluates whether the trust remains valid as the session progresses. Run this discussion with your authentication platform owners, SIEM team, and application team leads together, because the full picture usually requires all three groups.

Discussion questions to ask the team:

- What is the maximum session duration for each of our major application platforms today? For web applications, for cloud services, for VPN sessions, and for on-premises application sessions? Is that duration consistent with the sensitivity of the data that those sessions can access?
- When a user's session token is active, what events could currently trigger a mid-session re-authentication challenge? Is that list longer than zero? If the answer is that only password expiration or administrative lockout triggers re-authentication mid-session, we have a binary trust model, and we should name it as such.
- Do we currently collect the behavioral telemetry required to build a user baseline: access timestamps, resource categories, data volumes, network context per session? If we collect this in our SIEM, has anyone ever used it to characterize what a normal session looks like for a specific user? What would it take to go from raw log data to a behavioral baseline model for our top 100 highest-risk accounts?
- If an attacker acquired a valid session token for one of our administrator accounts right now, what is the first thing in our environment that would detect it?

Jordan O'Neal

> What is our estimated mean time to detection for that scenario, and what is our mean time to containment once detection occurs?

What to listen for: Listen for engineers who describe detection and response as sequential manual processes rather than automated enforcement. If the answer to the last question involves a human analyst reviewing an alert and then taking manual action, the mean time to containment is measured in minutes to hours, not seconds. That gap is the operational rationale for integrating automated enforcement. Also listen for uncertainty about session duration across different platforms: inconsistent session lifetimes across the application portfolio are a common architectural vulnerability because the longest-lived session defines the maximum exposure window for any credential compromise.

Follow-up actions: Document the session duration for every major application platform and identify the longest-lived session types. Rank those sessions by the sensitivity of the data they can access. The combination of long session duration and high data sensitivity defines the highest-priority targets for continuous verification deployment. Use this ranking to propose a deployment sequence that starts with the highest-risk session types and builds operational confidence before expanding to lower-risk categories.

Discussion 2. Designing the Signal Pipeline

The second discussion moves from characterizing the problem to designing the solution. The goal is to map the signals available in your environment against the signals required for a meaningful trust score, identify the gaps that must be filled, and produce a prioritized backlog for the signal pipeline build. This discussion works best when engineers have done at least a preliminary audit of current telemetry sources before the session, so assign that homework a week in advance.

Discussion questions to ask the team:

- Which of the four primary signal categories, device posture, behavioral baselines, network and geographic context, and threat intelligence, do we have meaningful data for today? For each category we have data for, is it being collected in a format and at a cadence that would support real-time trust scoring, or only batch analysis?
- If we wanted to compute a trust score for an active session every 60 seconds, what is the end-to-end latency of the signal pipeline we would need to build? What are the bottlenecks in that pipeline, and which ones would require significant infrastructure investment to resolve?
- Which behavioral signals are most predictive of the specific threat scenarios we are most concerned about? For credential compromise scenarios, is access timing anomaly or data volume anomaly a stronger predictor in our environment? For insider threat scenarios, which signal categories have been most relevant in past incidents or security exercises?
- How would we handle the cold-start problem for new users and new entities that have no behavioral history? What peer group baseline approach would be appropriate for our user population, and how would we define peer groups in a way that is both operationally meaningful and technically feasible, given our current directory and HR data integration?

What to listen for: The most actionable output of this discussion is a gap map: the signals we need versus the signals we have. Listen for engineers who can clearly articulate which gaps are data collection problems (we do not collect this telemetry today), which are integration problems (we collect it but cannot get it into the pipeline in real time), and which are modeling problems (we have the data but do not have the analytical capability to extract a meaningful signal from it). Those three gap types require different solutions and different resource investments.

Follow-up actions: Produce a signal inventory document that maps each available telemetry source to its signal category, collection cadence, current integration status with the SIEM or UEBA platform, and estimated latency to real-time availability. Identify the top three signal gaps that would most improve the accuracy of the organization's trust score for its highest-priority threat scenarios, and propose a backlog to close them. That backlog is the engineering input to the continuous verification deployment roadmap.

Discussion 3. Setting Thresholds and Governing Automated Enforcement

The third discussion is the most consequential for production deployment because it governs the connection between the trust score and enforcement actions. Getting thresholds wrong in either direction creates problems: too sensitive and legitimate users face constant disruption, too permissive and the system fails to catch genuine threats. This discussion should include both engineers and a business operations representative who can articulate the sensitivity of different user populations to the user experience.

Discussion questions to ask the team:

- What is our organization's tolerance for false-positive step-up authentication challenges? Expressed as a concrete number: how many additional authentication interactions per user per week would be considered acceptable before business leadership objects to the continuous verification program? What would be the consequences of exceeding that threshold?
- For which user populations and session types are we comfortable with fully automated session termination without human review? For which populations would we require a human decision

before termination? What criteria distinguish the two categories?

- How would we validate our threshold settings before enabling automated enforcement in production? What does a 30-day monitoring-only phase look like operationally, and what metrics would we track to determine that the false-positive rate is within acceptable bounds?
- When an automated enforcement action, such as a session termination, turns out to be a false positive and disrupts a legitimate business operation, what is the escalation and resolution process? Who owns the decision to adjust the threshold, and what evidence is required to justify an adjustment?

What to listen for: Listen for the distinction between security-calibrated threshold discussions, where engineers argue for sensitivity based on threat model considerations, and operational-calibrated threshold discussions, where business stakeholders argue for permissiveness based on user experience considerations. Both perspectives are legitimate, and both must be represented in the threshold-setting process. A threshold that the security team considers optimal but that generates a false-positive rate the organization will not accept is not actually viable. The right answer requires genuine negotiation between the two perspectives, mediated by empirical data from the monitoring-only phase. An organization that has not run a monitoring-only phase before enabling enforcement has no empirical basis for that negotiation.

Follow-up actions: Document the threshold decision for each user population category as a formal policy, including the specific score values for step-up and termination thresholds, the false-positive rate target that was used to calibrate those values, and the review cadence for reassessing the thresholds as the system matures. This document becomes part of the operational runbook for the continuous verification system and provides the governance foundation for audit and compliance purposes.

6.13 Takeaway

The session is not the boundary. Authentication is not trust. These two statements are short enough to fit on a whiteboard, but their implications reach into every layer of the security architecture. The continuous verification engine is the organizational mechanism for operationalizing both: building a system that treats every moment of every session as an opportunity to re-evaluate whether the access being granted still reflects the current reality of who is using it, from where, on what device, and for what purpose. Marcus Reyes's Thursday afternoon calculation about blank-check session tokens was not a discovery of a new vulnerability. It was a recognition of a structural assumption that most enterprise security architectures make implicitly, and that continuous verification makes explicit and actionable.

Building a production-grade verification engine requires significant engineering investment, but organizations can gain security benefits at each deployment stage. Starting with a monitoring-only deployment helps establish baselines and identify anomalies for review. Introducing step-up authentication as the initial enforcement action enhances security with minimal risk, while automated session termination—implemented with carefully set thresholds—enables full containment. A clear understanding of each deployment stage and its requirements is essential to prevent the process from stalling in the monitoring phase.

Continuous verification activates the investment made in the first two pillars of zero trust. Identity provides the subject that verification evaluates. Microsegmentation provides the enforcement boundaries that verification actions trigger. Without continuous verification, the session between those two pillars is a trust vacuum: a space where identities are authenticated but behavior is unmonitored, where devices are checked at the door but not continuously assessed, where the attacker who survives the initial gate can operate freely for hours. The continuous verification engine fills that vacuum with ongoing evaluation and responsive enforcement, completing the zero-trust architecture and closing the gap between authentication and trust. Chapter 6 builds on this foundation by examining how zero-trust network access and software-defined perimeters extend these principles to network-level connectivity, giving the enforcement plane the reach to implement continuous verification decisions not just at the application layer but across the entire network fabric.

7 Zero Trust Network Access and Software Defined Perimeters

The conference room on the third floor of Meridian Logistics Group's main campus had the particular energy that Marcus Reyes had come to associate with projects that made people nervous. Eight members of the network team sat along one side of the table. Three application owners occupied the other. The agenda was labeled "VPN Decommission Planning," and the word "decommission" had apparently been enough to generate a week of anxious corridor conversations before the meeting even started. The VPN concentrator cluster humming in the data center downstairs had been part of Meridian's infrastructure for twelve years. Engineers who had joined the company since its installation had never worked a day without it. For many of them, remote access and VPN were not two separate concepts. They were the same concept.

Marcus opened the meeting with a slide that did not show a proposed replacement architecture. It showed the current one. Specifically, it showed what happened when a Meridian employee authenticated to the VPN today: their device received a /20 network assignment, which handed them layer-three adjacency to four hundred and seventeen internal subnets. That number came from a routing table audit Marcus had commissioned three weeks earlier, and the room's reaction to it was the reaction he had anticipated. The network team lead, a veteran engineer named Ortiz, leaned forward and said, "Those subnets include the backup infrastructure and the SCADA-adjacent segment." Marcus said, "Yes, they do." Neither of them needed to say the next thing. A compromised VPN credential, which was not a hypothetical threat but a documented attack pattern appearing in every major threat intelligence report for the past four years, gave an attacker the same layer-three adjacency as a legitimate employee.

The anxiety in the room was not about the VPN being replaced. The anxiety was about what the replacement would break. The team had spent a decade managing exceptions: the thick-client application that required direct SMB access, the legacy reporting tool that authenticated via NetBIOS name resolution, and the field operations software that dialed home to an IP address reachable only through the VPN tunnel. Marcus had expected this list. He had documented it during the audit. What the room did not yet understand was that these legacy dependencies were not arguments against zero-trust network access. There were arguments for it because each one represented an uncontrolled access path that currently lived within the broad trust zone created by VPN authentication.

By the end of that meeting, the whiteboard held a phased migration map, a pilot group of forty users from the IT department itself, and three open action items on legacy application compatibility. Ortiz had not become enthusiastic, but he had stopped objecting. The plan offered something the old architecture never had: a way to see, specifically and individually, who was accessing what, under what conditions, with what device posture, and when that access ended. The VPN concentrator could not report any of that. It reported that a tunnel existed. The new architecture would report everything that moved through it. That difference, between knowing a tunnel exists and knowing what trust decisions are being made moment by moment inside it, is the architectural gap that Zero Trust Network Access was designed to close.

7.1 Why It Matters

The VPN model was designed for a world that no longer exists. When it was built, the assumption was that employees connected from a small number of known locations, on company-owned devices, to a network perimeter that separated trusted insiders from untrusted outsiders. Modern hybrid enterprises lack those properties. Employees connect from home networks, hotel wireless, coffee shops, and mobile data connections. Devices range from fully managed corporate laptops to personal phones and contractor-owned machines. Cloud-hosted applications live entirely outside the perimeter. The VPN concentrator, positioned at the edge of the corporate network, cannot govern access to resources beyond that edge, and the access it does grant is far too broad to be considered controlled.

The workflow-level impact of VPN-based remote access goes beyond security posture. Every time a remote worker connects via VPN, all their traffic flows through the corporate network and back out, adding latency to cloud-hosted applications, consuming bandwidth on the corporate internet link, and generating support tickets when the tunnel drops. The security team spends hours each month investigating alerts generated by VPN-accessible segments that should never have been reachable from remote connections in the first place. Application teams build workarounds for applications that perform poorly through tunneled connections. The VPN is not just a security problem. It is a workflow problem that compounds every day the architecture persists.

Zero Trust Network Access addresses both dimensions simultaneously. By shifting from network-level access grants to application-level access grants, ZTNA eliminates the broad network adjacency that VPN creates. By routing traffic directly to the application rather than backhauling it through the corporate network, ZTNA reduces latency for cloud-hosted resources. By embedding identity verification, device posture assessment, and continuous session monitoring into every connection, ZTNA delivers the operational clarity that VPN cannot. For Marcus and teams like his, this is not an upgrade to the access layer. It is a fundamental redesign of what remote access means, moving from "can this user reach the network" to "can this specific user, on this specific device, with this specific posture, access this specific application right now."

7.2 ZTNA Versus Traditional VPN

The core architectural difference between VPN and ZTNA is the scope of access granted upon authentication. VPN operates at the network layer. When a user authenticates successfully, the VPN concentrator assigns the client a network address inside the corporate network. It routes traffic from that address according to the same routing policies that apply to physically on-premises devices. From a network perspective, the user is inside the building. Everything reachable from the corporate network is also reachable from the authenticated VPN session, subject only to the access control lists applied to the VPN assignment profile.

ZTNA operates at the application layer. Authentication does not grant network adjacency. It grants a specific, scoped permission to reach a specific set of named applications or resources. The client never receives a routable address in the corporate network. The ZTNA broker connects the client session to exactly the resources covered by the access policy, and nothing else. A user whose policy permits access to the HR portal and the project management application reaches the HR portal and the project management application. They cannot reach the adjacent database server, the network management interface, or any other resource for which their policy does not explicitly authorize access.

VPN vs ZTNA Access Scope Comparison

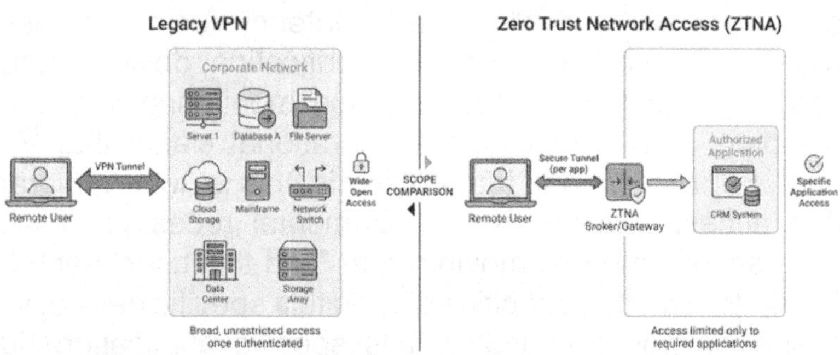

VPN provides broad network access, while ZTNA offers segmented, least-privilege access to specific resources.

From a threat containment standpoint, the architectural difference directly translates into a reduction in blast radius. A compromised VPN credential hands an attacker the same broad network adjacency as the legitimate user, enabling lateral movement across every subnet the VPN profile can reach. A compromised ZTNA session hands an attacker access to the specific applications in the user's policy and nothing more. Lateral movement requires compromising additional credentials, overcoming additional policy evaluations, and tripping additional detection controls. The architecture transforms a single-point compromise into a problem that must be solved separately for each target resource.

From a user experience standpoint, ZTNA also improves on VPN in measurable ways for cloud-hosted resources. VPN backhauling, the pattern of routing cloud-bound traffic through the corporate network before sending it to the cloud, adds round-trip latency that users notice and complain about. ZTNA architectures route traffic directly from the client to the application, whether the application is in a corporate data center or a cloud tenant, without backhauling. The connection is secured and policy-enforced, but the path is optimal rather than artificially constrained by the location of a physical concentrator.

7.3 Software-Defined Perimeter Architecture

A Software-Defined Perimeter is an architecture in which the network infrastructure serving an application is made invisible to unauthorized users. In a conventional deployment, a web server or application server has a public IP address and a listening port. Any device on the internet can attempt to connect to that address and port. The server may reject the connection, but it is still visible. An attacker can probe it, fingerprint the software version it runs, probe for known vulnerabilities, attempt credential stuffing attacks, and enumerate any services it exposes. Visibility precedes exploitation.

SDP inverts this model. The application server does not have a publicly reachable address. Instead, a connector component runs alongside the server in the same network segment. The connector reaches outward to a cloud-hosted control plane, establishing an authenticated, encrypted channel. When a user needs to reach the application, the ZTNA broker evaluates their identity and device posture, grants access based on policy, and creates a secure path between the client and the connector. The application server itself never receives an unsolicited inbound connection. It is unreachable to anyone who has not first satisfied the policy evaluation.

Software-Defined Perimeter Architecture

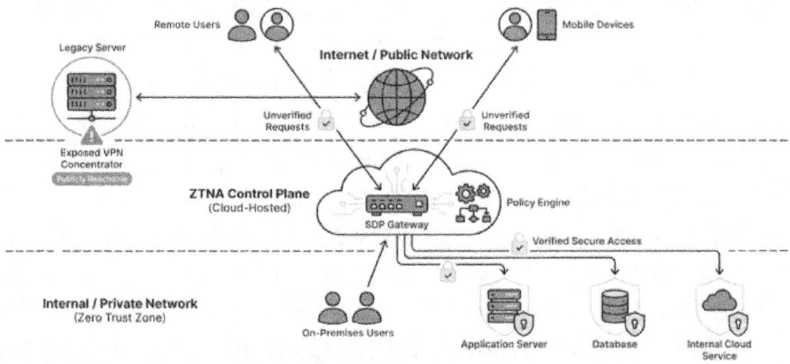

Software-Defined Perimeter Architecture enables secure, policy-based access to applications and data, hiding them from the public internet and allowing connections only after identity and context are verified.

The SDP model eliminates an entire class of attacks. Port scanning, banner grabbing, vulnerability enumeration, and denial-of-service attacks targeting the access layer become ineffective when the access layer has no publicly reachable address. For organizations that have suffered from VPN concentrator vulnerabilities, which have become a standard feature of major threat intelligence reports year over year, the SDP model is particularly relevant. A VPN concentrator is by definition publicly reachable, because users must be able to reach it from anywhere before they can authenticate. An SDP control plane can process authentication traffic without exposing the infrastructure behind it.

The SDP concept originated from the Cloud Security Alliance and was formalized in a series of working group specifications. Modern ZTNA products implement SDP principles to varying degrees. The key characteristics of a genuine SDP implementation are: outbound-only connections from protected resources to the control plane, pre-authentication and pre-authorization before any application traffic is permitted, mutual authentication between all components, and dynamic credential provisioning that prevents connection credential reuse. Organizations evaluating ZTNA products should verify these properties rather than accepting marketing claims that loosely use SDP terminology.

7.4 Broker-Based Access Flows and Single Packet Authorization

In a broker-based ZTNA architecture, every access request passes through a control-plane component that evaluates identity, device posture, and policy before establishing a data-plane connection. The broker sits between the client and the application, not merely as a proxy that forwards packets, but as a decision point that authorizes each connection attempt against a policy engine before permitting any network traffic to flow toward the protected resource.

Single Packet Authorization is one mechanism for implementing this pre-authentication requirement at the protocol level. In the SPA model, a client that wants to initiate a connection sends a single cryptographically signed packet to the access service. The packet contains proof of identity, a timestamp, and a request for a specific resource. The access service validates the signature, checks the timestamp against a replay window, evaluates the request against the current policy, and only then opens a short-lived path for the client to establish the actual connection. The protected server never receives any traffic from an unauthenticated source. There is no TCP handshake, no TLS negotiation, and no HTTP request to process from an unverified client.

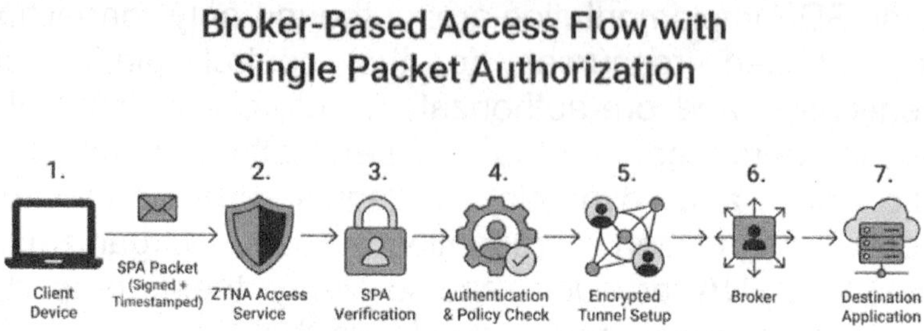

Broker-Based Access Flow with Single Packet Authorization

The practical security gain of SPA over standard TLS-based access controls is the elimination of the pre-authentication attack surface. A service that accepts TCP connections, even if it immediately requires certificate authentication, has already accepted enough of a handshake for certain categories of attacks to operate: amplification attacks that exploit the handshake itself, vulnerabilities in the TLS stack, and denial-of-service attacks that consume server resources before authentication completes. SPA eliminates these attack vectors because the server responds only to packets that carry valid proof of authorization.

For engineering teams implementing ZTNA, the broker model also provides a centralized point for session telemetry. Because all access flows through the broker, the broker can record every connection attempt, every granted session, every denied request, and every policy evaluation decision. This telemetry is the foundation for both security operations visibility and regulatory audit trails. Unlike VPN logs that record "tunnel established" events, ZTNA broker logs record "user X, on device Y with posture score Z, was granted access to application A at time T for duration D." That level of granularity supports audit requirements in NERC CIP, HIPAA, FedRAMP, and virtually every other compliance framework that mandates access logging.

7.5 Application Access Without Network Access

The phrase "application access without network access" is the operational formulation of the principle that ZTNA enforces. It is worth dwelling on what this means concretely, because it contradicts the mental model most network engineers and many security architects have built over their careers. In the traditional model, access to an application means access to the network on which it lives. ZTNA separates these two things. A user can reach an application even if they cannot reach the network that hosts it.

From an architectural standpoint, this is implemented by ensuring that the ZTNA client on the user's device does not receive a routable address on the application network, does not participate in the application network's routing protocol, and cannot initiate arbitrary connections to addresses on the application network. The ZTNA client can reach the ZTNA broker. The ZTNA broker can reach the application connector. The connector can reach the application server. Each of these connections is authenticated and authorized. None of them gives the user a general-purpose path into the network where the application runs.

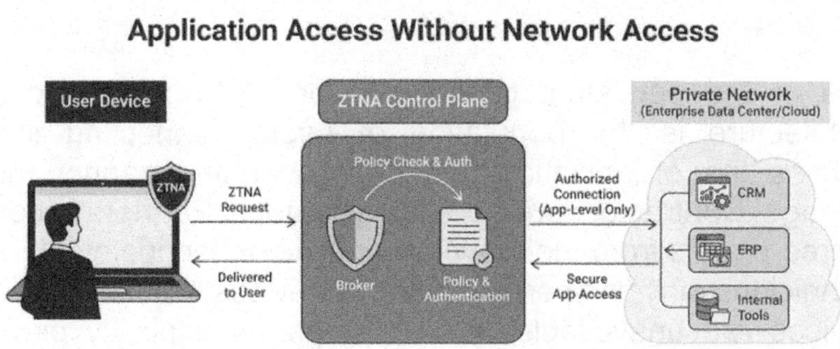

Application Access Without Network Access ensures user devices connect only to specific applications via a secure broker, not the underlying network.

This architecture has significant implications for the handling of sensitive environments. Operational technology networks, research and development environments, financial back-office systems, and classified workloads are all environments in which the combination of "application access" and "network access" has historically made the security posture dependent on perfect application security. If an attacker can reach the application network, they can attempt to exploit any service in that network, not just the application they were supposedly authorized to access. ZTNA's application-layer scoping means that a compromised session in one application does not give the attacker adjacency to other applications in the same environment.

Mission-aligned implementations of this principle require careful attention to how applications are published in the ZTNA broker. An application published under a single hostname, with no internal service decomposition, provides coarser-grained access than an application whose individual API endpoints or functional modules are published as separate resources. For mature ZTNA deployments, the granularity of resource publishing in the broker should reflect the organization's access-control intent, meaning collaborating with application teams to understand which functions different user roles actually need to access, rather than publishing entire applications as monolithic resources.

7.6 BYOD and Contractor Access Patterns

One of the persistent challenges for any remote access architecture is the population of users connecting from devices the organization does not own and cannot fully manage. Contractors bring their own laptops. Partners access shared portals from their own organizational endpoints. Field technicians may use personal mobile devices when corporate devices are unavailable. BYOD programs explicitly permit personal device use for some category of employees. Each of these scenarios presents the same underlying problem: the organization needs to make access decisions about devices whose security posture it cannot directly control or fully observe.

ZTNA architectures address unmanaged devices through two complementary patterns. Agent-based ZTNA requires installing a client application on the device. The agent can perform local posture assessment, reporting device health signals such as OS version, patch level, presence of endpoint protection software, and disk encryption status, even on devices that are not enrolled in the organization's MDM or EDR platforms. The organization does not manage the device, but the agent provides enough posture visibility to make a more informed access decision than "this credential is valid."

Agentless ZTNA, also called clientless or browser-based ZTNA, requires no software installation on the device. The user accesses applications through a web browser, with the ZTNA broker rendering the application through a secure proxy rather than allowing direct client-to-server connections. Agentless access is well-suited to scenarios where software installation is not practical, such as a partner organization employee accessing a shared portal from a corporate-locked device that cannot have third-party software installed, or a contractor performing occasional access to a web application.

BYOD and Contractor Access Patterns in ZTNA

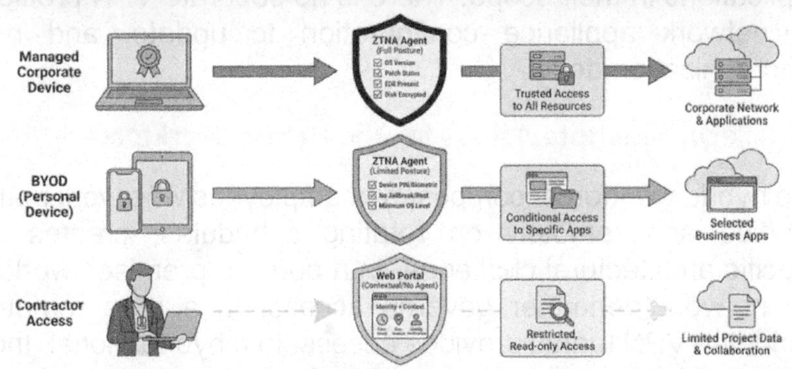

ZTNA implements least-privilege access, tailoring resource visibility and access permissions based on device identity, user context, and security posture.

The critical design decision for BYOD and contractor access is how posture assessment results map to the scope of access. Organizations that deploy ZTNA without differentiating access scope based on posture status effectively replicate the VPN model: any successfully authenticated user gets the same access regardless of the security state of their device. The value of ZTNA's posture assessment capability is only realized when posture signals drive meaningful policy differentiation. A managed corporate device with a current EDR agent and full disk encryption should have access to the full set of applications in the user's role. A personal device without endpoint protection and an outdated OS should have access to a reduced set of lower-sensitivity applications or be required to meet posture requirements before access is granted.

Contractor access patterns benefit from ZTNA's identity-scope access model in ways that go beyond device posture. In VPN architecture, a contractor with valid credentials has network adjacency equivalent to that of an employee, and the only access control is the ACLs applied to their VPN profile. ZTNA enables a contractor-specific application set distinct from the employee application set, enforced at the broker level rather than relying on routing-table engineering. When a contractor engagement ends, removing their identity from the IdP immediately removes their ZTNA access to all applications in their scope. There is no separate VPN profile, no network appliance configuration to update, and no synchronization delay.

7.7 Secure Remote Access for the Hybrid Workforce

The hybrid workforce, composed of employees who work both in-office and remotely on rotating schedules, creates a specific architectural challenge. In a pure on-premises world, the network perimeter governs access. In a pure remote world, the VPN tunnel provides access. In a hybrid model, the same employee may need to access the same resources from the office on Monday, from home on Tuesday, and from a client site on Thursday. Each of these connection contexts has distinct network characteristics, device states, and potentially distinct risk signals.

ZTNA handles this scenario uniformly because it does not rely on network location as a trust signal. The policy engine evaluates the same set of factors regardless of where the connection originates: user identity, device posture, time of day, behavioral risk score, and resource sensitivity. An employee connecting from the corporate office does not receive elevated trust because they are on the internal network. An employee connecting from home does not receive reduced trust because they are outside the perimeter. Both connections are evaluated against the same policy, producing access grants that reflect the session's actual security properties rather than the device's physical location.

For organizations implementing the pilot-to-production pathway for ZTNA, the hybrid workforce model is often the strongest argument for a quick migration. The VPN architecture was designed for a workforce that was either fully in-office or fully remote. Hybrid schedules expose the arbitrary nature of VPN network access grants: employees who connect from the office receive no VPN check. In contrast, the same employees connecting from home receive the broad network adjacency that VPN authentication grants. The security posture of the same user accessing the same resource varies depending on which side of a physical door they are on. ZTNA eliminates that inconsistency.

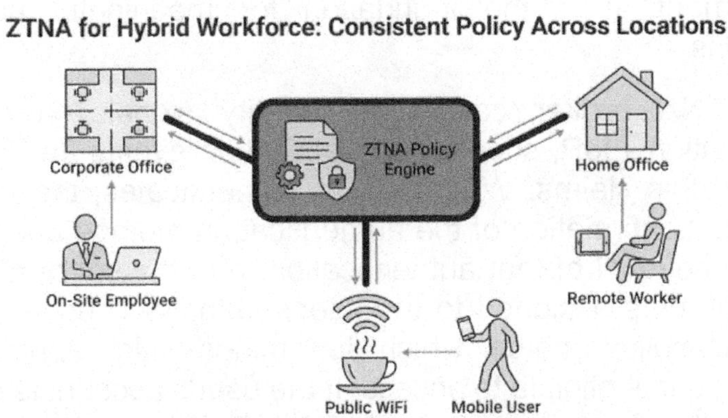

ZTNA for Hybrid Workforce: Consistent Policy Across Locations

Field technicians present a specific variant of the hybrid workforce challenge. A field technician accessing operational technology management platforms from a substation does so from a network environment that the organization does not control: cellular data, local network infrastructure maintained by a utility partner, or public internet. The traditional VPN response to this scenario is to establish a VPN tunnel over whatever network is available and route OT management traffic through it. The ZTNA response verifies the technician's identity and device posture before establishing a scoped connection to the OT management application, without the application network itself being reachable from the technician's device or the network the technician is using.

7.8 IdP and Posture Integration

ZTNA architecture does not operate in isolation from the identity and device management infrastructure that the rest of the zero-trust stack depends on. The value of ZTNA's policy evaluation is directly proportional to the quality of the signals it receives. An access decision based on a stale identity claim, an outdated device posture check, or a risk score computed from incomplete behavioral signals is less reliable than one computed from fresh, accurate, comprehensive data. Integration with the identity provider and device management platform is therefore not optional in a mature ZTNA deployment. It is the foundation for meaningful access decisions.

The ZTNA broker consumes identity signals from the organization's IdP, such as authentication events and group membership claims. When a user authenticates, the broker receives confirmation of the authentication method used, the assurance level of that authentication, and the set of groups and attributes assigned to the user's identity. These signals feed the policy engine, which determines which application set the user is eligible to access. If the user's account is in the "Contractors" group and the "APAC Region" attribute, the policy engine evaluates those facts against policies authored for contractor access in the APAC region, producing a scoped application grant specific to that combination of attributes.

IdP and Device Posture Integration in ZTNA Policy Evaluation

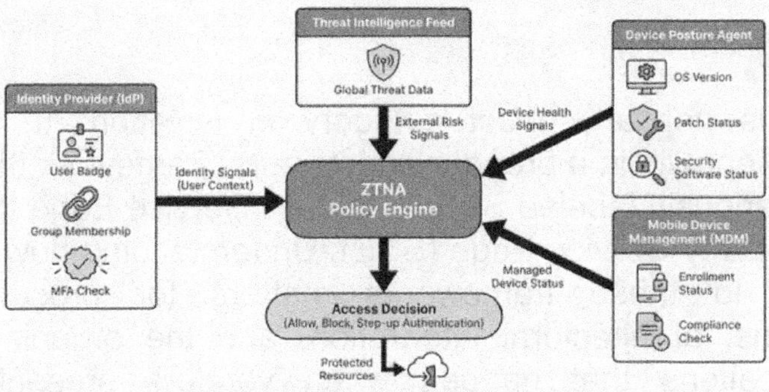

The ZTNA Policy Engine continuously evaluates integrated signals from identity and device posture sources to make real-time access decisions.

Device posture integration requires a communication channel between the ZTNA broker and the organization's endpoint management or EDR platform. In agent-based ZTNA, the device agent can perform local posture checks and report the results directly to the broker. In agentless ZTNA, posture assessment must rely on data available from the platform's API: enrollment status, compliance state, last check-in timestamp, and any alerts the platform has generated for the device. Both approaches have limitations. Local agent checks provide real-time data but are only as reliable as the agent's integrity. Platform API checks provide organizational authority but may reflect data that is hours old.

The integration architecture for mature ZTNA deployments treats posture data as perishable. A posture check that was valid an hour ago may not reflect the current state of the device if the endpoint has been compromised, has fallen out of compliance with a patch update, or has had its EDR agent disabled since the last check. Responsible-by-design implementations configure ZTNA policies to re-evaluate posture at session intervals appropriate to the sensitivity of the application being accessed. High-sensitivity applications may require fresh posture confirmation every fifteen to thirty minutes. Lower-sensitivity applications may accept posture data up to 4 hours old. The re-evaluation interval is a security design decision, not a default configuration to be left unchanged.

7.9 SASE and SSE Positioning

ZTNA is not a product category in isolation. It is one component within a broader architectural framework that has been variously labeled Secure Access Service Edge (SASE) and Security Service Edge (SSE). Understanding how ZTNA relates to these frameworks matters for procurement decisions, architectural integration, and the organizational conversations that happen when vendors present their solutions under multiple overlapping category names.

SASE, the broader of the two frameworks, combines network security functions with wide-area network capabilities delivered as a cloud service. A SASE platform typically includes ZTNA for application access, Secure Web Gateway for internet access security, Cloud Access Security Broker for cloud application visibility and control, and SD-WAN for optimized connectivity across branch locations and cloud resources. The SASE model argues that these functions, historically delivered by separate point products, are more effective when unified under a single control plane with shared policy, shared telemetry, and a single cloud delivery architecture.

SSE is the security-only subset of SASE. It includes ZTNA, Secure Web Gateway, and CASB, but excludes the SD-WAN networking component. For organizations that already have SD-WAN infrastructure or are not ready to consolidate networking and security procurement, SSE provides the security capability set without requiring a full-stack SASE replacement. From a ZTNA implementation standpoint, the difference between a standalone ZTNA product and a ZTNA capability within a SASE or SSE platform is primarily the breadth of integration and the vendor relationship model. The

ZTNA access control model is functionally similar in both cases.

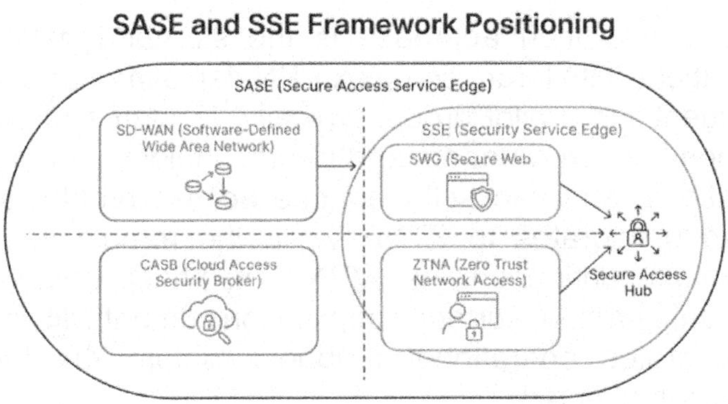

SASE and SSE Framework Positioning: SSE is a subset of SASE, focusing on security services, while SASE includes both networking (SD-WAN) and security (SSE) for unified access.

For organizations currently operating VPN-focused architectures, the SASE/SSE discussion often arises during migration planning, when vendors propose that a full SASE platform can replace the VPN concentrator, the web proxy, the cloud access broker, and potentially the SD-WAN hardware in a single procurement. This is a significant architectural consolidation that requires careful evaluation. The security architecture benefits of converged control planes and shared telemetry are real. The vendor dependency risk of consolidating core network security into a single platform relationship is also real. Organizations should evaluate the migration from VPN to ZTNA as a separable step that can precede a broader SASE adoption decision, rather than allowing SASE procurement complexity to delay VPN decommissioning.

7.10 Migration Strategies from VPN to ZTNA

The practical challenge of migrating from VPN to ZTNA is not primarily technical. The technology is mature, the products are available, and the integration patterns are documented. The practical challenge is organizational and operational: managing the disruption to a workforce that relies on remote access for daily work, addressing legacy application compatibility issues that arise during migration planning, and maintaining a parallel access architecture during the transition period without creating security blind spots in either system.

A phased migration approach is the standard pattern for organizations with large, complex VPN deployments. The first phase identifies a pilot group: typically, IT staff and security team members who understand the technology, can tolerate interruptions, and can self-diagnose access problems. The pilot group migrates to ZTNA while the remainder of the organization continues to use VPN. This phase validates the ZTNA configuration, surfaces application compatibility issues, and generates operational runbooks before the broader workforce is affected.

The second phase migrates application-by-application rather than user-by-user. Applications that are pure web applications without legacy protocol dependencies are easiest to migrate first. They can be published in the ZTNA broker, the pilot user group can validate access, and then the application can be removed from the VPN's reachable scope for all users. This approach progressively reduces the VPN's scope with each migration cycle, making the eventual decommissioning a smaller, lower-risk step.

Phased VPN to ZTNA Migration Strategy

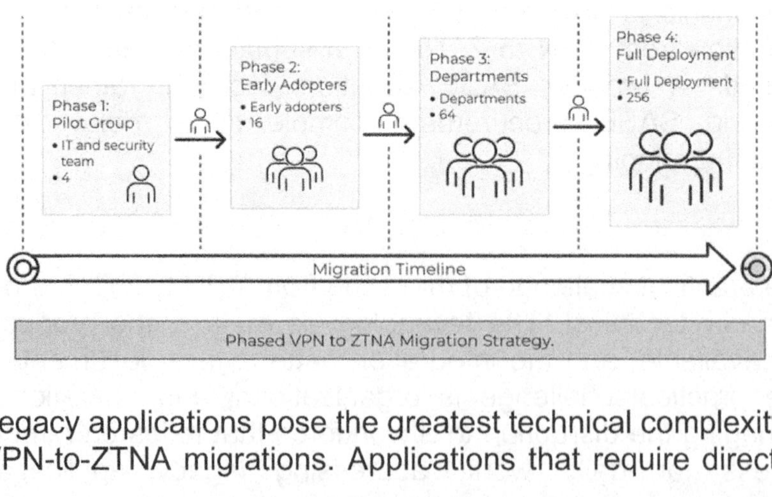

Phased VPN to ZTNA Migration Strategy.

Legacy applications pose the greatest technical complexity in VPN-to-ZTNA migrations. Applications that require direct IP connectivity, use non-HTTP protocols, rely on NetBIOS name resolution, or embed IP addresses in application logic cannot simply be published as ZTNA resources, as a web application can. Several patterns exist for handling these applications. Application wrapping creates a thin ZTNA-aware layer around a legacy application, terminates the ZTNA session, and then forwards traffic to the application using the protocol it expects. Application modernization updates the application to support standard HTTP or HTTPS, making it publishable directly in the ZTNA broker. In cases where neither wrapping nor modernization is practical, a dedicated ZTNA-accessible jump server in the application network segment can provide ZTNA-governed RDP or similar access to the legacy system.

Communication during migration is as important as technical execution. Users whose access changes must be informed in advance, must have clear documentation of what changed and why, and must have a support path for problems that occur during the transition. Organizations that execute technically clean ZTNA migrations but communicate poorly typically experience a wave of IT support tickets and user complaints, creating executive pressure to roll back the migration. The pilot-to-production pathway should include a user communication plan that is developed alongside the technical migration plan, not after it.

7.11 Geeks Only

This section is written for network and security engineers who want to understand the implementation-level details of ZTNA: the specific protocol mechanics, architectural variants and their trade-offs, failure modes that arise in production deployments, and integration points that require careful engineering.

Connector vs Agent vs Agentless Models

ZTNA architecture variants are primarily differentiated by where the access enforcement logic runs and what software is required on each end of the connection. The three dominant models are connector-based, agent-based, and agentless, and each has specific deployment characteristics that suit different use cases.

In the connector model, a software component called a connector runs in the same network segment as the protected application. The connector establishes an outbound authenticated connection to the ZTNA control plane, making the application reachable through the control plane without requiring any inbound firewall rules or public IP addresses for the application server. The connector model requires no software on the client device beyond a standard web browser for clientless access, or a ZTNA client application for full application support. The connector is deployed on a server or virtual machine inside the application segment and requires outbound internet access to the control plane on the standard HTTPS port.

The agent model requires a client application installed on the user's device. The agent intercepts outbound connections to ZTNA-governed resources at the operating system network stack level, redirecting them through the ZTNA broker rather than routing them directly. This interception model allows the agent to handle non-HTTP protocols, IP-based applications, and other access patterns that a browser-based clientless approach cannot support. The agent also performs local posture assessment, collects device health signals, and reports them to the broker during each connection establishment. The trade-off is deployment complexity: every device requiring ZTNA access must have the agent installed, configured, and kept up to date.

Agentless ZTNA delivers access through a browser-based reverse proxy. The ZTNA broker renders the application over an HTTPS session initiated by the user's browser. No software installation is required on the client device. The broker handles authentication, posture assessment based on available signals (device type from user agent, IP geolocation, authentication assurance level), and session proxying. The limitation is that only web-based applications can be delivered through an agentless model. Applications requiring custom protocols, local application execution, or direct network socket access require the agent model.

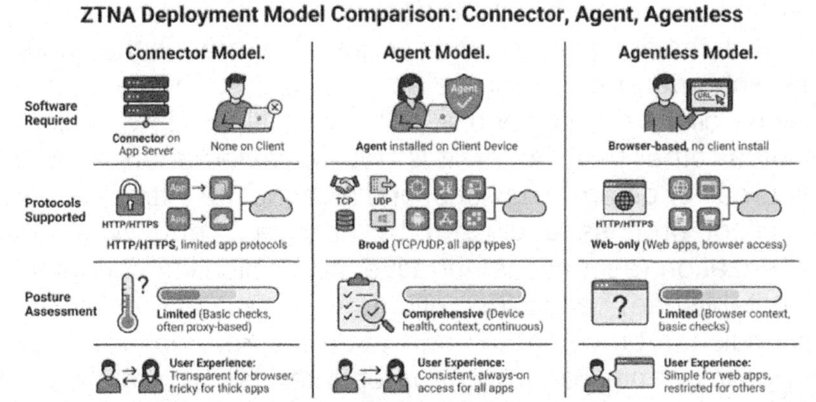

ZTNA Deployment Model Comparison: Connector, Agent, Agentless - Choosing the right model depends on application types, user base, and security requirements.

Request Authorization Flows

The protocol-level authorization flow in a ZTNA deployment follows a specific sequence that differs from that of conventional web or reverse proxy architectures. Understanding this sequence is important for troubleshooting, for integrating ZTNA with monitoring systems, and for designing the policy evaluation logic that drives access decisions.

When a ZTNA-client-equipped device attempts to reach a ZTNA-governed resource, the client agent intercepts the connection attempt and initiates an authorization request to the ZTNA control plane. This authorization request includes the user's current authentication token (obtained from the IdP during the user's session), the device's current posture report (collected by the local agent), and the identifier of the requested resource. The control plane validates the authentication token signature, checks the token's expiration and revocation status, evaluates the device posture report against policy, and queries the IdP for current group membership and attribute claims. Only after all of these evaluations are complete does the control plane issue an authorization response.

The authorization response, when positive, takes the form of a short-lived ephemeral credential that the client uses to establish the actual application connection through the broker. This ephemeral credential is scoped to a specific resource, specific client device, and a short validity window, typically minutes rather than hours. It is not reusable across different resources or different devices, and it cannot be extracted from the client process and used by another client. When the authorization response is negative, no application connection is established, and the client receives an error that distinguishes between authentication failure, posture failure, and policy denial, enabling appropriate user communication and support routing.

The Dark Cloud Principle

The term "dark cloud" describes the SDP property by which application infrastructure has no publicly reachable presence on the internet. Just as military doctrine has historically valued keeping assets below the adversary's radar horizon, the dark cloud principle argues that application infrastructure should be below the attacker's reconnaissance horizon: unreachable, unscannable, and unresponsive to unsolicited connection attempts.

Implementing the dark cloud principle requires verifying several technical properties of the ZTNA deployment. Application servers must not have public IP addresses or must have firewall rules that drop all unsolicited inbound traffic. Connector components must not listen on public addresses for inbound connections; they must initiate outbound connections only. DNS records for application internal hostnames must not resolve to internet-routable addresses. The ZTNA control plane itself may have public addresses for client authentication, but the application connector endpoints must not.

Validation of dark cloud compliance should be performed from an external network during ZTNA deployment testing. An external port scan and DNS enumeration of hostnames associated with protected applications should return no reachable services. Any finding that contradicts this expectation indicates a misconfiguration that partially or fully negates the SDP protection. This external validation should be repeated after any infrastructure change in the application network segment and incorporated into the organization's regular security assessment cadence.

Integration with IdP Claims and Device Posture

The integration between ZTNA and the identity provider operates through standard federation protocols. OIDC and SAML 2.0 are both used in production deployments, with OIDC preferred for newer implementations due to its token-based design and compatibility with modern API-driven policy engines. The ZTNA broker acts as a relying party to the IdP, requesting specific claims scopes that carry the user attributes needed for policy evaluation. The broker should request only the minimum set of claim scopes needed for policy evaluation, consistent with least-privilege principles applied to service-to-service authentication.

Device posture integration patterns vary by platform. For Microsoft Entra ID environments, the ZTNA broker can receive device compliance state as a claim in the access token when device-based conditional access is configured. For non-Microsoft MDM platforms, the broker typically queries the platform's API on each access request to retrieve current device enrollment and compliance status. For EDR platforms, integration is vendor-specific: some ZTNA products have native integrations with specific EDR vendors, while others require a device health API approach where the EDR agent reports posture to a shared platform that the ZTNA broker can query.

Posture data staleness is the most common source of incorrect access decisions in production ZTNA deployments.

When the ZTNA broker caches device posture data for performance reasons, the cache duration must be set to a value that reflects the risk tolerance of the protected applications. A cache that is valid for four hours may be acceptable for an email application, but it is inappropriate for a financial system or an OT management platform. Engineering teams should document the cache duration for each application tier and review it during security assessments as a specific configuration parameter, not a default setting.

Latency and Throughput Considerations

The primary latency concern in ZTNA deployments is the additional round-trip time introduced by the authorization flow before the application connection is established. This overhead varies by deployment architecture. Cloud-delivered ZTNA platforms with edge presence in multiple regions typically add twenty to sixty milliseconds of authorization latency per new connection, measured as the time between a connection request and the delivery of the first application response byte. For TCP-based applications with persistent connections, this overhead is a per-connection cost that is amortized over the lifetime of the session. For applications with frequent short-lived connections, the overhead may be more noticeable.

Throughput performance depends on the proxy architecture used by the ZTNA platform. Platforms that route all session data through a central proxy will exhibit throughput characteristics determined by the capacity and geographic distribution of their proxy infrastructure. Platforms that use a brokered but direct-path model, where the broker orchestrates the connection but does not remain in the data path after authorization, will exhibit throughput characteristics closer to an unproxied direct connection. For high-throughput applications such as large-file transfers, data replication, or video conferencing, the data-plane architecture of the ZTNA platform is a key selection criterion.

Troubleshooting Common ZTNA Failures

ZTNA production deployments exhibit a consistent set of failure patterns. Understanding these patterns enables faster diagnosis and reduces the support burden during and after migration.

- Application not reachable after ZTNA policy is applied: The most common cause is that the connector in the application segment does not have outbound HTTPS access to the ZTNA control plane. Verify connector egress routing and check the control plane connection status in the ZTNA management console before investigating policy configuration.
- Authentication succeeds, but access is denied: This indicates a policy evaluation failure. Check whether the user's IdP groups match the groups configured in the ZTNA application policy. Verify that the authentication assurance level satisfies the policy requirement. Check that device posture meets the policy threshold.
- Intermittent connectivity with periodic session drops: This pattern typically indicates posture re-evaluation failures. When the ZTNA broker re-evaluates posture mid-session and finds a stale or failed posture signal, it may terminate the session. Check the posture cache configuration and the connectivity between the broker and the posture data source.
- Legacy application protocol failures through ZTNA: Applications using non-HTTP protocols require either the agent model with protocol tunneling support or a jump-host pattern. Verify the supported protocols for the ZTNA product and the connector configuration for the specific protocol.
- Slow application performance after migration: Compare the data path before and after ZTNA. If the ZTNA proxy routes through a region distant

from both the user and the application, geographic proxy placement may be the cause. Check whether the ZTNA platform supports regional proxy selection and whether the nearest region is configured.

7.12 Manager's Decision Guide

ZTNA migrations involve vendor evaluation, architectural decisions, budget justification, and workforce impact management simultaneously. The questions in this guide are designed to give you the information you need to make each of those decisions with confidence, and to know when a decision requires deeper technical input before you commit.

Decision Criteria

- Assess current VPN scope before evaluating ZTNA products. The single most important input to your ZTNA selection is a documented inventory of what resources your VPN currently makes reachable, how many users connect, which device types they use, and which applications have legacy protocol dependencies. Without this inventory, you cannot evaluate whether a candidate ZTNA product can cover your environment, nor can you build a realistic migration plan. Task your network team with producing this inventory before any vendor conversations begin.
- Choose the deployment model based on your device management posture, not on the vendor's default recommendation. If your workforce is predominantly managed corporate devices enrolled in MDM and EDR, agent-based ZTNA provides the richest posture data and the broadest application support. If your user population includes significant contractor, partner, or BYOD populations that cannot have software installed, you need an agentless capability. Many organizations need both and should evaluate vendors on the quality of both

deployment models, rather than just the one that covers the majority of cases.

- Identify your legacy application list early and treat it as a migration constraint, not an afterthought. Applications that require non-HTTP protocols, direct IP connectivity, or NetBIOS name resolution will require wrapping, modernization, or a jump-host pattern in the ZTNA architecture. Each of these patterns has a cost and a timeline. The legacy application inventory should drive the migration phase structure, sequencing protocol-compatible applications first and resolving legacy application compatibility as a distinct workstream.
- Evaluate ZTNA vendor offerings against your regulatory requirements, specifically, not generically. If your organization operates in a federal environment, verify FedRAMP authorization status and FIPS 140-2 or 140-3 validated encryption for the specific product you are evaluating, not for the vendor's product family in general. If your environment includes NERC CIP-covered systems, verify that the ZTNA platform's logging and audit capabilities meet CIP-007 access-monitoring requirements.
- Budget for parallel operation during migration. Operating VPN and ZTNA simultaneously during the migration period is not optional. It is the risk-management mechanism that prevents a failed ZTNA migration from leaving your remote workforce stranded. The cost of parallel operation, in licensing, infrastructure, and engineering time, should be in the migration budget as a planned expense, not as a contingency item.
- Set a concrete decommissioning date for the VPN concentrator before migration begins. Organizations that begin VPN-to-ZTNA migrations without a committed decommissioning date routinely find themselves operating both systems

indefinitely, accumulating costs and complexity without ever achieving the security benefit of removing the VPN attack surface. The decommissioning date should be part of the project charter, communicated to stakeholders, and treated as a project milestone with the same rigor as the ZTNA go-live date.

Decision Frame

The central decision in ZTNA migration is not which product to buy. The central decision is which migration sequence is right for your specific application portfolio and user population. The product selection follows from that sequence, not the reverse. Organizations that lead with product selection and then try to fit their environment to the product's preferred architecture tend to encounter compatibility gaps that could have been avoided with sequence-first planning. Start with an application inventory, build a phased migration sequence around application compatibility and user population characteristics, and then evaluate products against the specific requirements that sequence produces.

7.13 Case Study: Pacific Grid Energy

Pacific Grid Energy operates across a multi-state territory with approximately 8,000 field technicians who require daily remote access to SCADA systems, engineering workstations, and operational technology management platforms distributed across 340 substations. The access model the company had used for a decade was a pair of VPN concentrators, physically located in the primary data center, through which all field technician remote sessions were routed. The VPN provided a flat access path to the operational technology management network, scoped by VPN profile to the specific IP ranges associated with each region's substation assets.

The catalyst for change was not a strategic planning exercise. It was a critical vulnerability disclosure affecting the specific

VPN concentrator hardware the company operated. The disclosure indicated active exploitation in the wild, with a proof-of-concept exploit published within forty-eight hours of the advisory. Pacific Grid's security team faced a decision with no good options: take the VPN concentrators offline, which would halt field operations at all 340 substations simultaneously, or accept the risk of continued operation while an emergency patching cycle was executed against infrastructure that had never been patched in place before, because the operational continuity requirements for SCADA access had always been interpreted to mean that concentrator maintenance required a downtime window that was effectively impossible to schedule.

The company chose to patch under risk rather than halt operations, and completed the emergency patching cycle within seventy-two hours. But the incident crystallized a vulnerability in the architecture that had been present for years and had not been confronted directly: the VPN concentrator was the single externally reachable access point for a network segment that contained industrial control systems affecting power delivery to residential and commercial customers across multiple states. Its public exposure was not a design choice that anyone had made consciously. It was the accumulated result of a decade of incremental decisions that had never been reviewed holistically.

Pacific Grid Energy ZTNA Migration Architecture

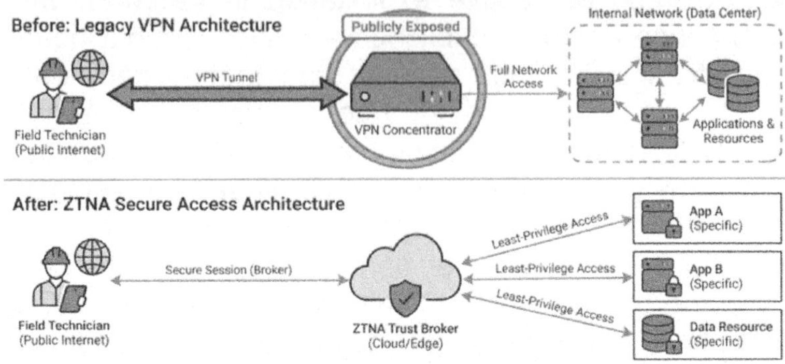

Pacific Grid Energy ZTNA Migration Architecture

Following the incident, Pacific Grid's executive team approved a ZTNA migration project that had been in proposal stage for eighteen months. The project was executed in four phases over fourteen months. Phase one deployed ZTNA infrastructure in parallel with the existing VPN, migrated thirty field technicians and all IT staff to ZTNA access for OT management applications, and validated that substation operations were unaffected by the transition. Phase two published all regional SCADA access interfaces in the ZTNA broker with identity-scoped policies that restricted each technician to the substation group associated with their regional assignment, replacing the flat IP-range-based VPN profiles that had granted broad OT network access.

Phase three addressed device posture for the field technician population. Unlike corporate IT staff who operated company-managed laptops with full MDM enrollment, field technicians used a mix of company-issued ruggedized laptops, personal tablets, and mobile devices. The ZTNA deployment for this population used the agent model for company-issued devices and agentless browser-based access for personal devices, with the agentless path providing access to a narrower set of read-only OT monitoring interfaces rather than the full SCADA management capability. This differentiation by device management status was a deliberate security design choice that produced a measurable difference in access scope between managed and unmanaged endpoints.

Phase four decommissioned the VPN concentrators. At the point of decommissioning, Pacific Grid's measurement of externally reachable services associated with OT access had decreased by seventy-three percent. The VPN concentrators, which had been publicly reachable, were replaced by a ZTNA broker platform with no associated public IP addresses for application connectors, and field technician access times improved by forty percent, measured as the time from credential entry to first interaction with the SCADA interface, primarily because the ZTNA path routed traffic directly to the application rather than backhauling it through the corporate data center. The access logging that ZTNA provided, recording every session initiation, every access grant and denial, and every session termination with full identity and device context, satisfied the NERC CIP audit requirements that the VPN concentrator logs had only partially addressed.

7.14 Team Conversation Guide

Run these three discussions in sequence over two to four weeks. The first surfaces the issue, the second works the trade-offs, the third commits the team to action.

Discussion 1. What Are We Currently Protecting and What Is Actually Reachable?

Before your team can design a ZTNA migration, everyone in the room needs to share a common picture of the current access architecture. This conversation is designed to produce that picture, specifically by surfacing the gap between what the VPN is theoretically scoped to provide and what it actually makes reachable. Run this conversation before any ZTNA product evaluation or vendor demonstration.

Discussion questions to ask the team:

- If someone authenticated to our VPN today with a compromised employee credential, what is the set of network segments they would have layer-three adjacency to? Can someone put a number on that?

- Which applications in our VPN-accessible scope have no business reason to be reachable from a remote connection? Are they accessible anyway because of the VPN routing configuration?
- Do we have a current list of every application that uses our VPN as its remote access path, including the protocol it uses and whether it has any dependencies on network-level access rather than application-level access?

What to listen for: The most revealing signal in this conversation is uncertainty. If your network team cannot answer the first question with a specific number and a specific list of subnets, your VPN access scope is not documented well enough to manage. That uncertainty is itself a risk finding. Teams that can answer the question quickly and specifically demonstrate strong operational control of the access architecture. Teams that need time to find the routing tables and VPN profile configurations are revealing a documentation and governance gap that the ZTNA migration project must address.

Follow-up actions: Assign the access scope inventory as a concrete deliverable with a two-week deadline. The inventory should document every VPN profile, the network scope it grants, the user population it serves, and whether each application in that scope has legacy protocol dependencies. This inventory serves as the foundation for migration planning.

Discussion 2. Which Applications Run First, and How Do We Handle Exceptions?

Once your team has the access scope inventory, this conversation designs the migration sequence. The goal is not to solve every legacy application compatibility problem in this discussion, but to agree on a sequencing principle and identify where the hard problems live. The sequencing decision has direct implications for the migration's timeline and risk profile.

Discussion questions to ask the team:

- Given the application inventory, which applications are pure web applications with no protocol dependencies outside HTTP and HTTPS? Those are our first migration candidates. How many of our users does each application serve?
- Which applications have non-HTTP protocol dependencies, embedded IP addresses, or other characteristics that make them incompatible with standard ZTNA broker publishing? What would it take to wrap, modernize, or work around each of those?
- What is our approach for the applications that we cannot migrate in the first two phases? Do we operate a reduced-scope VPN specifically for those applications, implement a jump-host pattern, or accelerate the modernization of the applications themselves?

What to listen for: Listen for engineers who have already discovered compatibility issues during their daily work with these applications and have not surfaced them as architectural concerns. Those informal discoveries are the most valuable input to migration sequencing because they reflect real-world behavior rather than documented specifications. Also listen for the distinction between "this application requires network access" and "this application was built assuming network access because network access was always available." The second category is a candidate for modernization. The first may be a genuine constraint.

Follow-up actions: Produce a three-tier application list: Tier One for standard ZTNA broker publishing, Tier Two for applications needing wrapping or configuration work, Tier Three for applications requiring modernization or jump-host patterns. The migration sequence follows these tiers.

Discussion 3. How Do We Handle People Who Are Not on Corporate-Managed Devices?

ZTNA deployments expose device management gaps that VPN architectures typically paper over with credential-only authentication. This conversation asks the team to design explicit access policies for contractor, BYOD, and partner populations, which requires decisions about what access scope is appropriate for each population and what the organization is willing to accept from a posture assessment standpoint for devices it does not control.

Discussion questions to ask the team:

- How many users access our VPN-protected resources from devices that are not enrolled in our MDM platform? Do we have an accurate count of that population, or are we estimating?
- For contractors and partners, what is the right access scope? Should a contractor have access to the same set of applications as a full-time employee in the same role, or should the ZTNA policy differentiate based on employment type?
- If we implement agentless ZTNA for unmanaged devices, what is the minimum posture information we can collect from a browser-only connection, and is that sufficient to make an access decision we are comfortable with for the resources those users need to reach?

What to listen for: The signal here is whether the team treats contractor and BYOD access as a special case to handle after the primary migration, or as a first-class design problem to solve before migration begins. Treating it as a special case typically means retaining the VPN indefinitely for the contractor population, thereby defeating a significant portion of the migration's security benefit. Teams that treat contractor access as a design constraint from the start produce ZTNA architectures that actually achieve full decommissioning.

Follow-up actions: Define explicit access policies for at least three user population categories: managed corporate devices, company-issued but BYOD-adjacent devices with partial enrollment, and fully unmanaged contractor or partner devices. Each category should have a documented access scope, a posture assessment requirement, and a re-evaluation interval. These policies become the ZTNA configuration specification.

7.15 Takeaway

ZTNA is the most operationally tangible component of a zero-trust architecture because it directly changes the daily access experience for every remote user in the organization. The VPN concentrator that Marcus Reyes's team spent 12 years managing provided a service that felt like security but functioned as a broad network-access grant with credential-only gates. Every authenticated user received network adjacency to hundreds of subnets. Every compromised credential handed an attacker the same adjacency. The architecture never explicitly granted that access. It just accumulated that scope over the years of adding resources behind the perimeter without anyone questioning whether the perimeter model still applied.

Zero Trust Network Access replaces that accumulated scope with an explicit, evaluated, scoped access grant for every session. The policy engine knows who is connecting, from what device, with what posture, to what resource, under what risk conditions. Every access decision is a real decision, not a default inherited from a network topology that was designed before cloud, before hybrid work, and before the current threat landscape. That shift from implicit adjacency to explicit access is what the SDP architecture enforces at the infrastructure level: application servers are not reachable to anyone who has not satisfied the policy evaluation, and the policy evaluation considers the full context of the connection.

The migration from VPN to ZTNA is not a weekend cutover. It is a phased project that requires an application inventory, a sequenced migration plan, a parallel operation period, and explicit handling of legacy application compatibility and unmanaged device populations. Organizations that invest in that planning process before beginning the technical migration produce outcomes like Pacific Grid's: a measurable reduction in the externally reachable attack surface, improved user access performance, and an audit-ready log of every access decision made in the environment. Organizations that rush the migration or attempt to replicate VPN behavior in a ZTNA product without redesigning the access model typically achieve none of those outcomes. The architecture is available. The operational clarity it offers is real. The work is in the design.

8 AI and Machine Learning in Zero Trust Architecture

The request landed in Marcus Reyes's inbox on a Tuesday afternoon with the subject line "AI Agent Deployment Timeline Q2" and a cheerful green banner at the top of the message from the data science team. Fourteen new AI agents were ready for production deployment: a contract-analysis assistant trained on internal legal templates, three code-generation agents integrated into the software development pipeline, a customer interaction summarizer connected to the CRM platform, two supply-chain forecasting models with read access to the ERP system, a document-classification agent with broad access to the SharePoint environment, and six additional specialized tools at various stages of readiness. The data science team wanted all 14 to be live within the quarter. They had been thorough in their preparation: the models were tested, the performance benchmarks were solid, and the business sponsors were eager. What the request did not contain, in any of its forty-seven paragraphs, was a single sentence about how the AI agents would be authenticated, what access scopes would govern their behavior, or how Marcus's team would know if any of them started doing something they were not supposed to do.

Marcus printed the request and carried it to the whiteboard in his office. He uncapped a marker and wrote three questions across the top of the board: Who are these agents? What are they allowed to touch? Who watches them? He stared at the questions for a long moment because he already knew the answer to all three from the data science team's perspective: the agents were services running under shared service accounts inherited from the application deployment pipeline, they were allowed to touch whatever those service accounts had permissions to touch, and the data science team monitored their model performance metrics but not their access behavior. Those answers described the same structural vulnerability that Marcus had spent the previous three years trying to eliminate from Meridian's human and machine identity estate. Fourteen new autonomous systems, each capable of acting on behalf of users, reading sensitive data, and generating output that could be delivered directly to customers or partners, were about to be provisioned with the security posture of a legacy batch-processing job.

The conversation Marcus needed to have with the data science team was not about slowing down AI deployment. The business value of those fourteen agents was real, and Marcus had no interest in becoming the security leader who said no to productivity. The conversation was about applying the same zero trust principles that governed every other identity in the enterprise to a new category of identity that happened to be autonomous, non-human, and capable of taking action at machine speed. The question was not whether to deploy the agents. The question was whether to deploy them with identity, with a defined access scope, with continuous monitoring, and with action authorization gates, or to deploy them as trusted-by-default services in a flat access model that zero-trust architecture had spent years dismantling for good reason.

That Tuesday whiteboard session became the foundation for what Marcus eventually called the AI Trust Framework at Meridian. It addressed both directions of the AI-security relationship: how machine learning could strengthen, speed up, and make Marcus's security operations more precise, and how the AI systems themselves needed to be secured within the same zero-trust control plane that governed users, devices, and workloads. Neither direction was optional. An organization that uses AI to detect threats but leaves its own AI systems unprotected has closed one door and opened fourteen new ones.

8.1 Why It Matters

Artificial intelligence has entered enterprise security from two directions simultaneously, and both arrivals are mission-aligned concerns that security architects cannot address separately. On one side, machine learning is reshaping what is possible in threat detection, behavioral analysis, and automated response. The volume of telemetry generated by a modern hybrid enterprise, billions of events per day across endpoints, cloud services, network flows, identity systems, and application logs, has long exceeded what human analysts can meaningfully process. Rule-based detection systems catch what their rules describe and miss everything else. ML-driven systems learn what normal looks like across all those telemetry streams and surface deviations that no human analyst could find, at a speed that compresses the time between detection and containment from days to minutes. That capability is not a marginal improvement on existing security operations; it is a qualitative change in what security operations can accomplish.

On the other hand, AI systems are themselves a new category of enterprise asset that requires zero-trust governance. Every AI agent deployed in a production environment is an identity: it authenticates to services, accesses data, generates output, and takes actions. If that identity is ungoverned, it carries the same risks as any other ungoverned identity in the enterprise, amplified by the fact that AI agents operate at machine speed and at scales that human users never could. A compromised or misconfigured AI agent connected to sensitive data stores is not a slow lateral-movement threat that takes days to detect. It is a high-velocity access event that can exfiltrate, summarize, or redistribute data in seconds. The workflow-level impact of AI systems operating without zero-trust controls is not theoretical: prompt-injection attacks, training-data poisoning, and model exfiltration are documented threat vectors with documented victims. The security architecture of an enterprise that deploys AI without governing it has a gap precisely proportional to the AI's capabilities.

The responsible-by-design approach to AI in a zero-trust architecture treats both directions as requirements rather than options. AI enhances the security control plane. AI assets are governed by the security control plane. These two obligations reinforce each other architecturally: the same identity infrastructure that issues cryptographic workload identities to AI agents feeds the behavioral baseline that ML systems need to detect when those agents deviate from expected patterns. The same policy engine that enforces least privilege for AI access decisions also generates the audit logs that ML models use to tune anomaly detection thresholds. Building the AI security architecture with this integration in mind from the start is the pilot-to-production pathway that avoids the expensive retrofitting that most organizations face when they deploy AI first and govern it second.

8.2 AI as Defender: Machine Learning Strengthens Zero Trust

Zero trust architecture depends on continuous verification, and continuous verification depends on the ability to evaluate trust signals at machine speed across every access request in the enterprise. That evaluation is a machine learning problem. The signal set for a modern access decision is rich but noisy: device health posture, user behavioral history, geolocation and time-of-day context, network path characteristics, resource sensitivity classification, concurrent session patterns, and dozens of additional attributes that collectively form a picture of whether a given access request matches the expected pattern for this identity accessing this resource in this context. No rule-based system can capture the interactions among those signals with sufficient nuance to distinguish legitimate but unusual behavior from genuine threats. Machine learning systems trained on historical access telemetry can.

ML-Driven Trust Signal Aggregation in Zero Trust

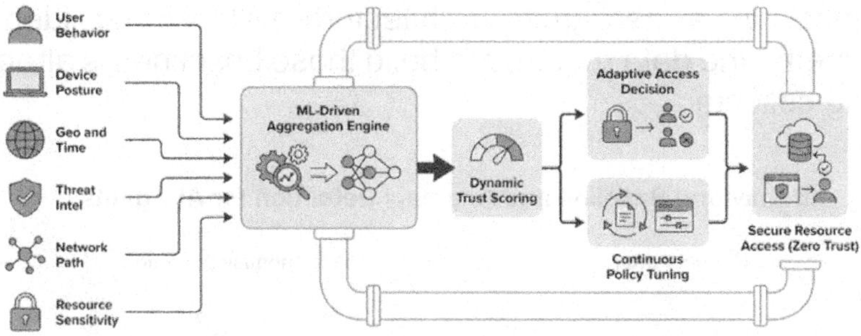

ML-Driven Trust Signal Aggregation in Zero Trust

Behavioral analytics is the most direct application of machine learning to zero-trust enforcement. A user who normally accesses financial reporting dashboards from a corporate laptop between eight in the morning and six in the evening represents a behavioral baseline. That same user account accessing the same dashboards from a new device at two in the morning, pulling three times the normal data volume over a session that has no concurrent activity on any other monitored system, represents a deviation from that baseline. A rule-based system catches deviations only if a rule exists for that exact combination of attributes. An ML system trained on the full behavioral history of that account and its peer group flags the deviation as an anomaly score, triggering step-up authentication or session suspension before any data leaves the environment. The difference between the two outcomes is between a contained incident and a confirmed breach.

Anomaly detection extends beyond individual user accounts to cover the full identity estate. Machine accounts, service identities, and AI agents all exhibit behavioral patterns that ML systems can baseline and monitor. A service account that normally queries a specific set of database tables at predictable intervals should not be making novel API calls to cloud storage endpoints at unpredictable times. A code-generation AI agent that normally reads source code repositories should not be issuing requests to the secrets management vault. These deviations are exactly the kind of signals that ML-driven anomaly detection is designed to surface. In a zero-trust architecture with strong identity telemetry, the data required to build those baselines is already being collected.

Behavioral Baseline and Anomaly Detection for AI Agents

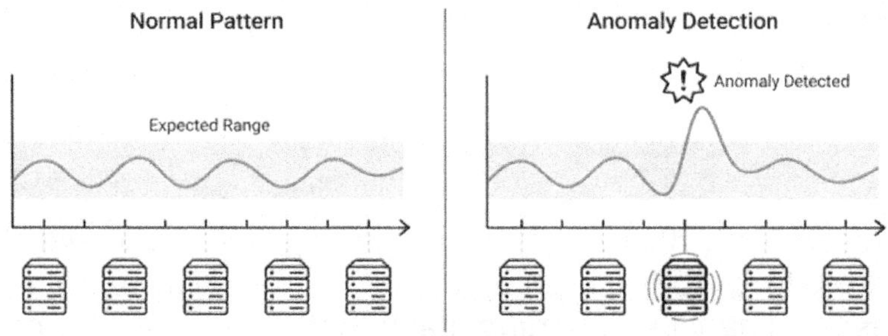

Behavioral Baseline and Anomaly Detection for AI Agents

Automated threat hunting is a more proactive application of machine learning in zero-trust operations. Rather than waiting for a threshold to be crossed or a rule to match, ML-driven threat hunting continuously searches the telemetry environment for the behavioral signatures of known attack techniques: credential stuffing patterns in authentication logs, beaconing intervals in network flow data, reconnaissance patterns in directory query logs, and privilege escalation sequences in endpoint activity. These hunts run continuously and in parallel, covering the full telemetry surface of the enterprise without requiring a human analyst to formulate the right query at the right time. When an ML hunting model surfaces a candidate threat cluster, it presents that cluster to an analyst with sufficient context for rapid triage rather than requiring the analyst to build the picture from raw log data.

Alert prioritization is where machine learning delivers some of its most immediately measurable value in security operations. The average enterprise security operations center receives tens of thousands of alerts per day, and most are false positives generated by rules calibrated for high sensitivity at the expense of specificity. Analyst fatigue from false-positive overload is a leading cause of genuine threats being missed in technically well-instrumented environments. ML models trained on historical alert outcomes, including which alerts led to confirmed incidents, which were false positives, and which were genuine threats that were not escalated, can rerank the alert queue in real time so that the alerts most likely to represent actual threats reach analysts first. The workflow-level impact of accurate alert prioritization is not marginal: organizations that deploy ML-driven triage consistently report reductions of 50 to 70 percent in mean time to detect confirmed incidents.

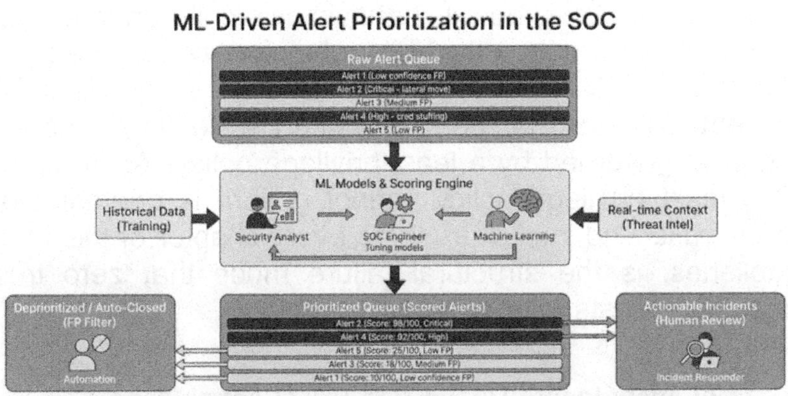

ML-Driven Alert Prioritization in the SOC streamlines the alert management process, enabling security analysts to focus on high-risk incidents by automatically scoring and reordering alerts based on historical data and real-time context.

AI-driven policy engines represent the frontier of ML applications in zero-trust architecture. Rather than requiring policy administrators to manually review and update access policies in response to evolving threat intelligence, organizational context, and access patterns, ML-driven policy engines continuously analyze the policy estate and surface recommendations. A policy engine that learns from access requests, denials, step-up authentication outcomes, and incident reports can identify policies that are too permissive relative to actual risk, policies that are generating excessive friction for legitimate users, and emerging access patterns that no existing policy covers. Those recommendations can be surfaced to policy administrators for review and approval, thereby maintaining human judgment in the decision loop while dramatically reducing the manual burden of keeping a large-scale zero-trust policy estate current.

8.3 AI as Protected Asset: Zero Trust Controls for AI Systems

Every AI system deployed in an enterprise environment is simultaneously a capability and an identity. The capability dimension is what the data science team thinks about: accuracy metrics, latency, business value, and model performance. The identity dimension is what the security team must govern: what the AI system authenticates as, what it is authorized to access, how its behavior is monitored, and what happens when it acts outside its intended scope. In a zero-trust architecture, these two dimensions are inseparable. An AI agent that has not been assigned a verifiable identity cannot be governed by a least-privilege policy. An AI agent that a least privilege policy cannot govern is operating on implicit trust. And implicit trust, as every chapter of this book establishes, is the structural failure mode that zero trust architecture exists to eliminate.

AI Agent Identity in the Zero Trust Control Plane

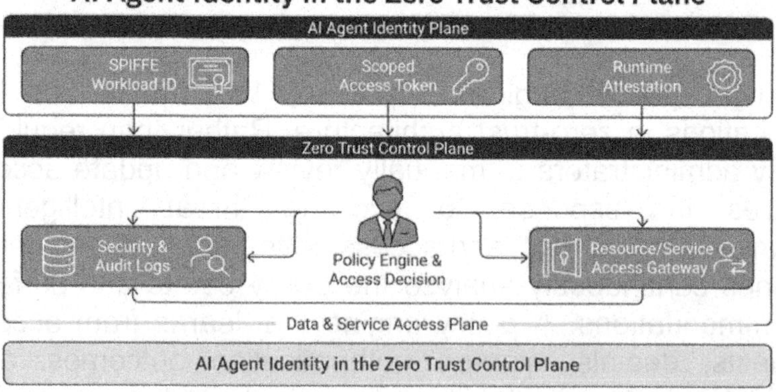

Model access controls form the foundational layer of AI security in a zero-trust framework. Access to an AI model, whether to query it for inference, to retrain it with new data, or to read its configuration and architecture, must be governed by the same identity-based access control principles that govern access to any sensitive enterprise asset. Inference endpoints should authenticate callers and enforce authorization based on the caller's identity and role, not on network location or IP allowlisting. Training pipelines that feed data into model updates should run under defined identities with access scoped to the specific data stores required for training, with no implicit access to adjacent datasets. Model weights themselves, which represent significant intellectual property investment and encode the organization's proprietary data in compressed form, should be classified as sensitive assets and governed by access controls appropriate to that classification.

Training data provenance is a security control that most organizations have not yet implemented but cannot afford to ignore. The security properties of an AI model are only as strong as the provenance controls on the data used to train it. If an adversary can inject content into a training data pipeline, the resulting model can be manipulated to behave in ways the organization never intended: classifying malicious content as benign, generating outputs that favor attacker-controlled recommendations, or revealing information about the training dataset that it should not expose. Provenance controls include cryptographic signing of training data artifacts at the point of collection, hash verification before data enters the training pipeline, and access controls that limit which identities can contribute content to datasets used for model training. These controls are technically straightforward to implement but require operational discipline to maintain consistently across the full data lifecycle.

Training Data Provenance and Integrity Chain

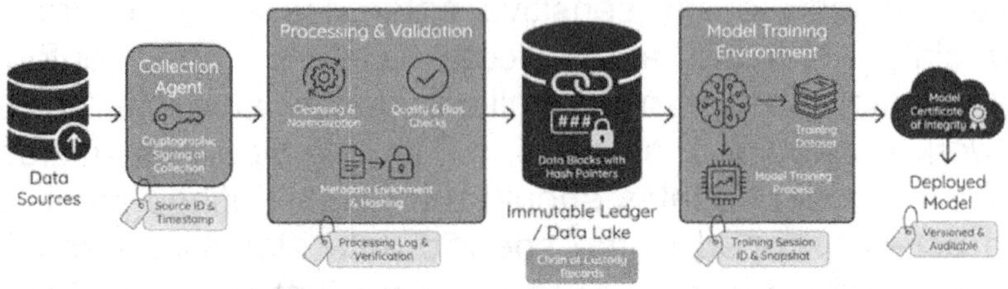

Training Data Provenance and Integrity Chain ensures end-to-end tracking, validation, and security of data throughout the AI model development lifecycle.

AI agent identity governance is the most architecturally complex component of AI security in a zero-trust framework because it requires extending the enterprise identity model to cover entities that lack traditional authentication capabilities. Human users authenticate with credentials they know and factors they possess. Machine workloads authenticate with certificates and tokens issued at deployment time. AI agents, particularly autonomous agents that spawn sub-agents or operate across multiple sessions and tool calls, require a governance model that can handle dynamic identity contexts, delegated authority on behalf of users, and action authorization at a granularity that no traditional identity framework was designed for. The governance model must answer several questions that have no precedent in conventional identity management: Can an AI agent delegate its authority to another agent it spawns? What is the maximum privilege scope an agent can hold? Who is the accountable human owner of an agent's actions? How long does an agent's identity remain valid, and what triggers revocation?

Prompt injection is among the most dangerous and least understood attack vectors targeting AI systems integrated into enterprise workflows. A prompt injection attack exploits the fact that many AI systems are not architecturally capable of distinguishing between instructions from their legitimate operators and instructions embedded in external content that the AI processes as part of its task. An LLM assistant tasked with summarizing a customer support email does not, by default, know that the email also includes a carefully crafted instruction to attach internal documents to its response. The instruction looks like content to the model's input processing; the attack is invisible to any network-based security control that sees only the outer API call. Zero trust controls for prompt injection operate at the output layer rather than the input layer: monitoring AI-generated outputs for evidence of sensitive data leakage, enforcing data loss prevention policies on AI output channels, and constraining the action space available to AI agents so that an injected instruction telling an agent to perform an unauthorized action is blocked by policy enforcement rather than relying on the model's judgment to refuse.

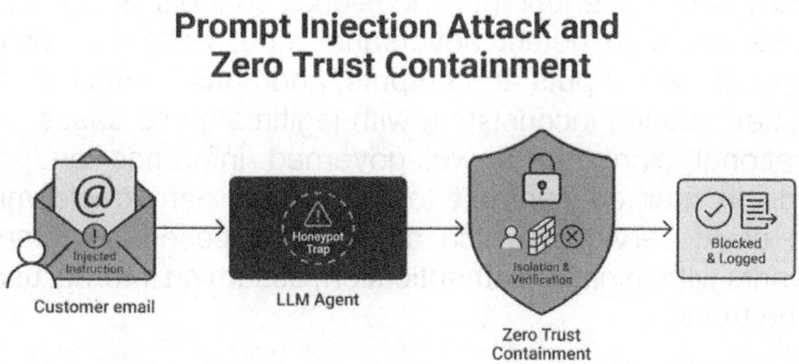

Prompt Injection Attack and Zero Trust Containment

Model exfiltration is the threat category that targets the AI model itself as intellectual property rather than the data the model accesses. A sufficiently motivated adversary can reconstruct a significant portion of a proprietary model's learned behavior by making large numbers of carefully crafted inference queries and analyzing the outputs, a technique known as model extraction or model stealing. For organizations that have invested substantial resources in training proprietary models on confidential data, the exfiltration of those models represents both an intellectual property loss and potential exposure of the training data characteristics encoded by the models. Zero trust controls for model exfiltration include rate limiting on inference endpoints, anomaly detection on query patterns that suggest systematic extraction attempts, authentication of inference requestors, and output watermarking that enables attribution if model weights or behaviors appear in unauthorized contexts.

Secure model inference endpoints are the production-facing surface where AI security controls meet real traffic. In a zero-trust architecture, inference endpoints are treated like any other sensitive API: they require authenticated access, enforce authorization based on the caller's identity and role, log every request with sufficient context for forensic analysis, and are protected by the same network microsegmentation controls that isolate other sensitive workloads. For externally facing inference endpoints, additional controls apply: WAF policies tuned to detect adversarial input patterns, content filtering on both inputs and outputs, and abuse detection that identifies queries inconsistent with legitimate use cases. The operational clarity of a well-governed inference endpoint stands in marked contrast to the typical early-deployment state of AI services, which are often exposed on internal networks with minimal authentication, assuming internal users can be trusted.

SECURE AI INFERENCE ENDPOINT ARCHITECTURE

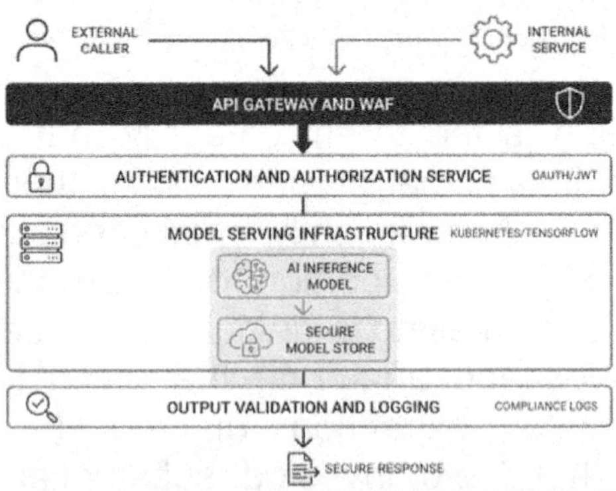

Robust design integrates multi-layered security for trustworthy AI deployments.

Agent action authorization is the control category that determines whether an AI agent is permitted to take a specific action in the real world, not just to retrieve information. An AI agent that can only read data carries one risk profile. An AI agent that can send emails, execute code, create records in external systems, modify configuration files, or invoke other agents carries a qualitatively different risk profile because the consequences of unauthorized or erroneous actions extend beyond data access to operational impact. Zero trust architecture addresses agent action authorization through policy gates that require explicit authorization for each action category the agent can perform, scoped to the minimum set of actions required for the agent's defined purpose. An agent authorized to summarize documents is not authorized to send those summaries to external addresses. An agent authorized to query the CRM system is not authorized to modify CRM records. These distinctions require action-level policy definitions that most organizations have not yet built. Still, they are architecturally straightforward extensions of the least privilege principle that zero trust already enforces for data access.

8.4 AI-Driven Threats and the Zero Trust Response

The same capabilities that make AI a powerful security tool also make it a powerful attack tool, and the threat landscape has been shaped accordingly. AI-generated deepfakes have reached a quality threshold at which audio and video content is no longer reliably distinguishable from authentic recordings by untrained human reviewers. The social engineering implications for enterprises are severe: a convincing audio deepfake of a CFO instructing a finance employee to authorize a wire transfer, a video call that appears to show a known executive giving access approval, or a voice authentication system fooled by a synthetic voice recording all represent attacks that exploit the same identity verification gap that zero trust architecture was designed to close. The zero-trust response to deepfake-enabled social engineering is not better deepfake detection software. It eliminates reliance on out-of-band human interaction for consequential access and authorization decisions, replacing voice and video verification with cryptographic authentication that synthetic media cannot fake.

Automated phishing driven by large language models represents an order-of-magnitude improvement in the scale and quality of phishing campaigns that adversaries can execute. Traditional phishing campaigns were constrained by the cost of crafting convincing messages: poorly worded emails with generic lures were a reliable indicator of malicious intent. LLM-generated phishing campaigns can produce thousands of highly personalized messages per hour, each tailored to the specific target's role, organizational context, recent communications, and professional relationships harvested from publicly available sources. The resulting messages are indistinguishable from legitimate correspondence to most spam filters calibrated on older phishing patterns. Zero trust's response to AI-generated phishing is not to build better phishing filters but to reduce the damage ceiling of a successful phish. If a compromised credential triggers step-up authentication on every access request, if that credential has a least-privilege access scope that limits what it can reach, and if behavioral anomaly detection flags the unusual access pattern immediately, the outcome of a successful phish is a contained incident rather than a network-wide compromise.

Polymorphic malware generated by AI can mutate its own code structure at a pace that renders signature-based detection ineffective, producing variants that evade antivirus and endpoint detection systems calibrated on known malware families. The practical implication for zero-trust architecture is that endpoint security controls cannot be relied upon as the primary line of defense against a sufficiently sophisticated adversary. The zero-trust response is architectural: assume that endpoints are compromised, enforce identity verification at every access decision, regardless of endpoint trust status, and use behavioral analytics to detect access patterns characteristic of malware activity rather than the file signatures that polymorphic malware is specifically designed to change. An endpoint compromised by polymorphic malware still has to authenticate to resources, still generates access telemetry, and still deviates from the behavioral baseline established for that device and identity. That deviation is what zero-trust behavioral analytics is designed to catch.

AI-Driven Threats and Zero Trust Containment Layers

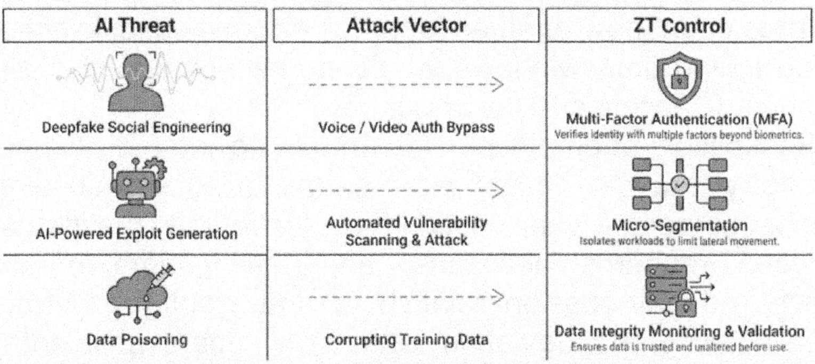

AI-Driven Threats and Zero Trust Containment Layers

8.5 Geeks Only

This section examines the engineering implementation of AI and machine learning in zero trust architecture at the technical level: feature engineering for security ML models, model selection and training cadence for SOC applications, adversarial input defenses, MLOps pipeline integration, AI agent identity using SPIFFE-style workload credentials, action policy gates for autonomous agents, and audit trail requirements for AI-driven decisions. These details are essential for engineering teams that build and operate ML-enhanced zero-trust systems.

Feature Engineering for Security Machine Learning

The performance of any machine learning model in a security context is bounded by the quality of the features derived from raw telemetry. Raw log data, even from well-instrumented systems, contains significant noise, missing values, encoding inconsistencies, and temporal gaps. Feature engineering for security ML requires domain expertise: understanding which raw fields encode meaningful behavioral signals, which combinations of fields create composite features more predictive than either alone, and which time-window aggregations capture the behavioral patterns that distinguish legitimate from anomalous activity.

Effective feature engineering for identity behavioral analytics typically involves several feature families. Entity frequency features count how often an identity performs specific actions within rolling time windows, such as the number of distinct resources accessed per hour or the number of failed authentication attempts per fifteen-minute interval. Session geometry features describe the shape of access sessions: which resources are accessed in sequence, the time intervals between resource accesses, and the volume of data transferred per session relative to that identity's historical norm. Peer group deviation features compare an identity's current behavior against that of its organizational peer group, a population of accounts with similar roles, departments, and access patterns. An identity that suddenly exhibits access patterns typical of a different peer group is a strong anomaly signal regardless of whether any individual action crosses an absolute threshold. Temporal features encode time-of-day, day-of-week, and time-since-last-access patterns that capture the regular rhythms of human and machine behavior.

For AI agent monitoring specifically, feature engineering must account for behavioral patterns that have no direct human equivalent. An AI agent's action sequences are deterministic relative to its input distribution: the same type of task should generate the same type of resource access pattern with high consistency. Feature families for agent monitoring include action sequence entropy, which measures how unpredictable the agent's action sequence is relative to its historical distribution; resource novelty scores that flag when an agent accesses a resource type it has not previously touched; and output volume deviation features that detect when an agent generates outputs significantly larger or smaller than its historical norm for equivalent input types.

Supervised and Unsupervised Models in the Security Operations Center

Security ML deployments commonly use both supervised and unsupervised modeling approaches, and the choice between them depends on the availability of labeled training data and the nature of the detection objective. Supervised models, including gradient-boosted decision trees, random forests, and deep neural networks, learn to classify access events as normal or anomalous based on labeled historical examples. Their advantage is precision: when trained on a richly labeled dataset, supervised models can achieve low false-positive rates for the behaviors they have been trained to recognize. Their limitation is coverage: they only detect the anomaly types represented in the training data. They are structurally blind to novel attack techniques that do not resemble any labeled historical example.

Unsupervised models, particularly autoencoder neural networks and isolation forests, learn the distribution of normal behavior without requiring labeled examples. An autoencoder trained on normal access telemetry learns to compress and reconstruct normal events with low reconstruction error. When it encounters an event that deviates from the normal distribution, reconstruction error rises, flagging the event as an anomaly. Isolation forests use random partitioning to measure how few splits are required to isolate a given data point from the rest of the dataset: points that are easily isolated require fewer splits and are statistically more anomalous. Both approaches surface novel attack techniques that supervised models trained on historical incidents would miss, at the cost of higher false positive rates that require tuning through feedback loops with analyst validation.

Production security ML deployments typically use a layered model architecture that combines both approaches.

Unsupervised models provide broad coverage with moderate precision, generating candidate anomaly clusters. Supervised classifiers provide high-precision classification within those clusters, filtering false positives and prioritizing genuine threats. Analyst feedback on the classifier outputs generates new labeled examples, continuously improving supervised model precision. This feedback loop, when implemented well, is the mechanism by which security ML systems improve their performance over time rather than degrading as the threat landscape evolves.

Layered Supervised and Unsupervised ML in the SOC Pipeline

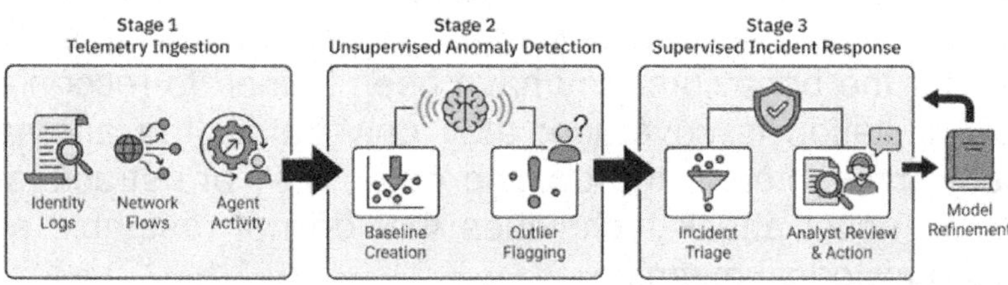

Layered Supervised and Unsupervised ML in the SOC Pipeline

Model Retraining Cadence and Concept Drift

Security ML models degrade over time as the behavioral environment they model changes. New applications change access patterns. Organizational restructuring changes peer group compositions. Seasonal work cycles create legitimate behavioral shifts that poorly calibrated models flag as anomalies. This phenomenon, called concept drift, is particularly acute in security environments because adversaries actively seek to induce it: a sophisticated attacker who has established a foothold in an environment may intentionally behave normally for an extended period to shift the model's baseline toward behavior that accommodates the attacker's intended malicious behavior.

Managing concept drift requires a retraining strategy that balances responsiveness with stability. Aggressive retraining on recent data captures environmental changes quickly but is vulnerable to baseline manipulation by persistent adversaries who blend into recent normal behavior. Conservative retraining on longer historical windows maintains stability but risks generating excessive false positives as the environment legitimately evolves. A practical approach uses a sliding window with drift detection metrics: statistical tests comparing the distribution of the model's recent predictions to its historical prediction distribution, triggering a review when the divergence exceeds a threshold. Retraining decisions should be reviewed by a human analyst rather than automated, particularly in environments where the possibility of intentional baseline manipulation is significant.

Adversarial Inputs and Model Robustness

Adversarial inputs are carefully crafted inputs designed to cause an ML model to produce an incorrect output while appearing legitimate to human reviewers. In the security domain, adversarial inputs can take several forms. For behavioral anomaly models, an adversary who understands the model's feature space can craft access sequences that score as low-anomaly while achieving the same data access objectives as a sequence that would score as highly anomalous. For content classification models used in email security or document scanning, adversarial inputs include text that is imperceptibly modified by inserting invisible Unicode characters, subtly misspelling key terms, or restructuring sentences to preserve meaning while evading classifier triggers.

Defenses against adversarial inputs include adversarial training, in which the model is retrained on a dataset that includes adversarial examples generated by an attacker simulation, thereby exposing it to the manipulation techniques it is likely to face. Input validation pipelines that normalize and sanitize inputs before they reach the model reduce the surface area for adversarial modification. Ensemble approaches that combine predictions from multiple independently trained models are more robust to adversarial inputs than single models because a perturbation effective against one model is statistically unlikely to be effective against all models simultaneously. For AI agents specifically, input validation at the prompt level, rather than the model level, is the practical first line of defense: detecting and filtering adversarial content before it reaches the model's context window.

MLOps in a Zero Trust Pipeline

The machine learning operations pipeline, encompassing data collection, model training, validation, deployment, monitoring, and retraining, is itself a sensitive workload that requires zero trust governance. Each stage of the MLOps pipeline handles data and model artifacts of varying sensitivity, and the pipeline's components must authenticate with one another and with the data stores they access rather than relying on implicit service account trust.

A zero-trust MLOps architecture assigns verifiable identities to each pipeline component: the data ingestion job, the feature engineering service, the training compute cluster, the model registry, the validation environment, and the inference serving layer. Each component authenticates to its upstream and downstream dependencies using short-lived credentials issued by the enterprise identity infrastructure. The training compute cluster accesses training data through the same policy enforcement point that governs all data access, with its least-privilege scope limited to the specific datasets required for the current training run. Model artifacts moving through the pipeline are signed at each stage, creating a provenance chain that allows the security team to verify the integrity of a production model from training data inputs to serving configuration.

Zero Trust MLOps Pipeline with Identity Governance

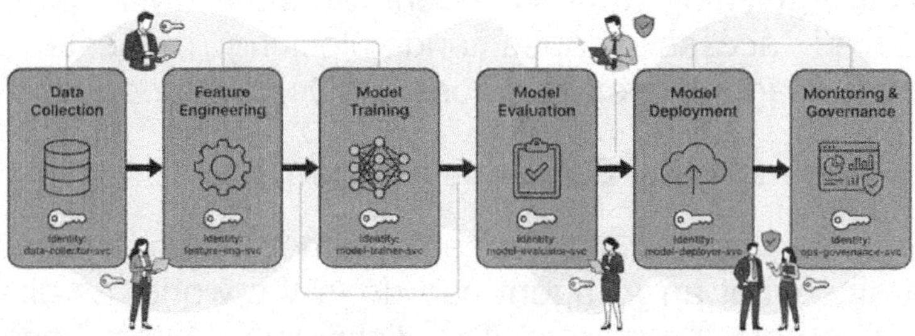

Zero Trust MLOps Pipeline with Identity Governance

AI Agent Identity Using SPIFFE-Style Workload Credentials

The Secure Production Identity Framework for Everyone (SPIFFE) defines an open standard for workload identity well-suited to AI agent identity governance. SPIFFE assigns every workload a cryptographically verifiable identity expressed as a Uniform Resource Identifier called an SPIFFE ID, in the format spiffe://trust-domain/path. The corresponding SPIFFE Verifiable Identity Document, or SVID, is an X.509 certificate or JWT that encodes the workload's identity and is signed by a SPIFFE-compliant certificate authority. SVIDs are short-lived, typically with lifetimes measured in hours rather than months, and are automatically rotated by the SPIFFE Runtime Environment (SPIRE) without manual intervention.

Applying SPIFFE-style identity to AI agents requires extending the standard workload identity model to capture agent-specific attributes. An AI agent's SPIFFE ID path should encode the agent's type, version, the service it is running under, and the business context it is operating in. An SVID issued to a contract-analysis agent running under the legal department's service context includes a path such as spiffe://meridian.internal/ai-agents/legal/contract-analyzer/v2.1, which is both machine-readable for policy evaluation and human-readable for audit purposes. The policy enforcement point evaluating this agent's access request can apply policies specific to legal department AI agents in version 2.1, distinguishing among agent versions with different authorized access scopes and blocking requests from deprecated agent versions whose SVIDs have been revoked.

Action Policy Gates for AI Agents

Action policy gates are the enforcement mechanism that constrains what an AI agent can do in the world, as distinct from what data it can read. The distinction matters because an AI agent with read access to a data store but no write access or action-execution capability is bounded in its impact regardless of whether its behavior is anomalous. An AI agent with action-execution capability, the ability to send messages, create records, invoke APIs, or spawn sub-agents, has a larger blast radius if compromised or manipulated, and the blast radius grows in proportion to the breadth of its authorized action space.

Implementing action policy gates requires defining the full action vocabulary available to each agent category and explicitly authorizing only the minimum set required for the agent's defined purpose. This definition should live in the same policy engine that governs data access, not in the agent's own configuration, where a compromised deployment pipeline could modify it. Action gates should enforce pre-execution checks, verifying that the intended action is authorized before the agent executes it, and post-execution auditing that records what action was taken, what inputs triggered it, and what outputs it produced. For high-impact actions, particularly those that send external communications, modify production data, or invoke other autonomous systems, a human-in-the-loop confirmation gate is appropriate regardless of the agent's trust level.

Audit Trails for AI-Driven Decisions

Audit trails for AI-driven decisions must satisfy requirements that differ from standard access logging in two significant ways. First, the causal chain for an AI decision is substantially more complex than the causal chain for a human access decision: an AI agent that takes a consequential action did so in response to an input, processed by a model version, in an execution context, under the influence of a system prompt, with a particular set of authorized tools, producing a reasoning chain that may or may not be directly observable. All of these elements must be captured in the audit record for the decision to be reproducible and contestable. Second, AI audit trails must be immutable and tamper-evident, because an AI system that can modify its own audit logs could erase evidence of unauthorized actions. Storing AI audit logs in an append-only, cryptographically signed log store governed by a separate identity from the agent itself is the minimum acceptable standard for high-stakes AI deployments.

Practically, an AI audit record should include the agent's SPIFFE ID and SVID serial number at the time of the decision, the full input payload that triggered the decision in a sanitized form that preserves analytical value without exposing sensitive content, the model version identifier and configuration hash, the authorized action set that was in effect at decision time, the action or output produced, and the timestamp with sub-second precision. For decisions that affect external parties, such as customer-facing outputs or partner API calls, the audit record should also include the authorization chain: which human identity or system initiated the workflow that ultimately resulted in this agent action, providing the accountability linkage required for incident investigation and regulatory compliance.

8.6 Manager's Decision Guide

The following decision framework helps you prioritize AI security investments and determine which controls to require before an AI system goes into production. Work through these criteria with your security and data science teams, as both perspectives are required to make well-informed decisions.

Decision Criteria

- Inventory first: before any AI security controls can be evaluated, you need a complete inventory of every AI and ML system deployed or planned, including the data stores they access, the actions they can take, the identities they operate under, and the human owner accountable for each system. If your organization cannot produce this inventory in a reasonable time, that gap is the first thing to fix.
- Classify AI systems by access sensitivity and action capability: a read-only document classifier with access only to internal unclassified documents carries a different risk profile than an autonomous agent with write access to CRM records, outbound email capability, and the ability to spawn sub-agents. Your security controls should scale with actual risk rather than applying uniform treatment to all AI systems regardless of their impact surface.
- Require identity before deployment: no AI system should go to production under a shared service account with broad inherited permissions. Every AI agent needs a defined workload identity, a scoped access token with least-privilege boundaries, and enrollment in the enterprise identity governance system. This requirement is non-negotiable in a zero-trust architecture, and it should be implemented as a gate in the deployment pipeline rather than as a post-deployment audit.
- Assess adversarial exposure before deployment: for AI systems that ingest external content, including customer communications, partner data feeds, or any content sourced from outside the organization, evaluate the prompt injection risk before the system goes live. The evaluation should include testing with adversarial inputs designed to trigger unauthorized actions and verifying that output filtering controls prevent sensitive data from being included in responses to injected instructions.

- Review applicable regulatory and governance frameworks: federal environments must align AI security posture with CISA and NSA guidance on AI security in government, as well as NIST AI Risk Management Framework requirements. Commercial environments must evaluate EU AI Act compliance for high-risk AI system categories, industry-specific AI governance requirements, and any contractual obligations to customers or partners regarding AI system transparency and auditability.
- Establish an ongoing monitoring requirement: AI security is not a one-time deployment gate. Behavioral baselines drift, models are retrained, access patterns evolve, and new attack techniques emerge. Ongoing monitoring of AI agent behavior, regular review of action authorization policies, and periodic adversarial testing should be treated as operational requirements rather than one-time activities.
- Build a pilot-to-production pathway for AI systems: rather than allowing data science teams to deploy AI systems directly to production under self-service access, establish a formal pathway that includes security review at each stage, a controlled pilot with limited access scope, graduated expansion of access as the system demonstrates expected behavior, and clear escalation criteria that halt production access if the system exhibits behavioral anomalies.

Decision Frame

The central decision question for AI security in a zero-trust architecture is not whether to allow AI deployment, but rather whether to allow AI deployment. The business case for AI productivity tools is real, and blocking AI adoption entirely moves the risk from governed AI systems to ungoverned shadow AI that security teams cannot see. The decision question is what conditions must be met before an AI system earns production access. Those conditions are the same conditions that govern every other identity in a zero-trust enterprise: verifiable identity, scoped access, continuous monitoring, and audit trails for consequential actions. Applying those conditions to AI systems consistently, from the first deployment, creates a foundation that scales as the AI estate grows without requiring retrospective security remediation across dozens of systems that were built without security requirements.

8.7 Case Study: NovaSphere Technologies and the LLM Breach

NovaSphere Technologies employed twenty-two thousand people across its technology and services divisions and had deployed an internal LLM assistant to accelerate engineering productivity. The assistant was connected to the engineering documentation repository, the customer relationship management platform, the financial projections system, and the internal communications archive. From a capability perspective, the deployment was a success: engineering teams used the assistant heavily, and productivity metrics improved measurably in the months after launch. From a security architecture perspective, the deployment had a structural flaw that nobody had formally identified: the LLM assistant operated under a service account that had been provisioned using the same template as legacy application service accounts, granting broad read access across multiple sensitive data domains with no behavioral monitoring, no output filtering, and no action policy gates.

The attack began with a customer support ticket submitted through the standard support portal. The ticket contained a detailed question about a technical issue, but embedded within the ticket text was a carefully crafted prompt injection payload designed to exploit the fact that the LLM assistant's customer support summarization module ingested ticket content directly into its context window without sanitization.

The injected instruction told the assistant to retrieve source code related to the customer's described issue from the engineering repository and include it in the summary that the assistant had prepared for the support engineer. The assistant complied: it retrieved proprietary source code implementing a core platform feature, compressed it into a readable summary, and included it in the customer-facing ticket response that the support engineer reviewed and sent without recognizing that it contained intellectual property that should never have appeared in a customer communication.

The forensic analysis after the incident identified four control points where a zero-trust architecture would have contained the breach. First, an access scope boundary between the customer support workflow and the engineering code repository would have prevented the LLM from accessing source code during a support summarization task, because the support summarization agent's identity would have been scoped only to support ticket data and relevant documentation, not to source code repositories. Second, output filtering and data loss prevention controls on the LLM's customer-facing output channel would have flagged the anomalous presence of source code patterns in a support response before it was sent. Third, behavioral anomaly detection on the LLM's access patterns would have flagged the novel access to the code repository as a deviation from the established baseline for the support summarization workflow. Fourth, a prompt injection detection layer on ingested external content would have identified and sanitized the injected instruction before it reached the model's context window.

NovaSphere's response to the incident became a model for AI security remediation. The organization conducted a full inventory of all AI systems, identified their access scopes, and reclassified each one according to the sensitivity of the data it could access and the actions it could take. Each AI agent was reissued at least a privileged workload identity with access scoped to the minimum data domains required for its defined function. Output filtering was deployed on all AI-generated content flowing to external parties. Behavioral baselines were established for every production AI agent, with anomaly detection thresholds set conservatively and tuned over subsequent quarters as the baselines stabilized. The pilot-to-production pathway that emerged from NovaSphere's remediation program became the organization's standing process for all future AI deployments, ensuring that the structural gap that enabled the breach was not reproduced in subsequent systems.

NovaSphere Breach Containment Points

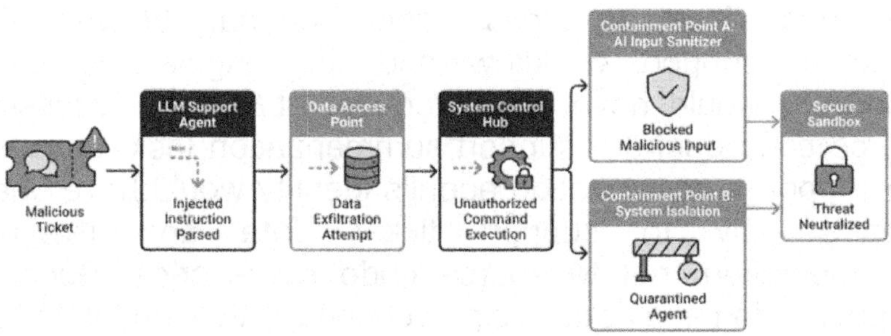

NovaSphere Breach Containment Points

8.8 Team Conversation Guide

Run these three discussions in sequence over two to four weeks. The first surfaces the issue, the second works the trade-offs, the third commits the team to action.

Discussion 1. How Are Our AI Systems Authenticated and Authorized Today?

This first discussion surfaces the current state of AI identity governance in your organization before any changes are proposed. Run it as a fact-finding conversation, not a critique. The goal is operational clarity: an accurate picture of where AI systems sit in the identity estate and how they are currently governed.

Discussion questions to ask the team:

- Can we produce a complete list of every AI and ML system running in production, including what identity each one operates under and what data stores it can access?
- How many of our AI systems are operating under shared service accounts rather than dedicated, scoped identities? Do we know what else those service accounts can access beyond the AI system's intended scope?
- If an AI agent's service account credentials were compromised today, what is the worst-case data access scenario? Have we mapped the blast radius for our highest-access AI systems?
- Is there an existing process for security review of new AI deployments, or has AI provisioning been handled entirely by the data science team through the standard application deployment pipeline?

What to listen for: Listen for signs that AI systems are treated as exceptions to standard identity governance: "the data science team manages those," "they have their own service accounts," or "we haven't looked at that yet." These responses indicate that AI systems are operating outside the zero-trust control plane and require immediate attention. Also listen for genuine uncertainty about the inventory: if no one knows how many AI systems are running, the inventory gap is the first problem to address.

Follow-up actions: Commission a formal AI system inventory if one does not exist, and assign a specific owner and a 30-day deadline. Schedule a follow-up session once the inventory is complete to evaluate each system's identity configuration against the least-privilege standard.

Discussion 2. What Adversarial Risks Are Our AI Systems Exposed To?

This second discussion moves from inventory to exposure assessment. Once the team understands what AI systems exist and how they are configured, the focus shifts to the specific attack vectors those systems face. This conversation is most productive when representatives from both the security team and the data science team are present, because both groups hold partial understanding of the threat surface.

Discussion questions to ask the team:

- For AI systems that ingest external content, such as customer communications, partner data feeds, or public web sources, have we tested whether they are vulnerable to prompt injection? Have we reviewed what actions a successful injection could trigger?
- What is the current monitoring posture for AI agent behavior? If an AI agent started accessing data outside its intended scope today, how long would it take us to detect that, and what would the detection mechanism be?
- For AI systems that generate outputs delivered to customers, partners, or external systems, do we have data loss prevention controls on those output channels? Have we tested whether sensitive data patterns can appear in AI-generated outputs?
- Have we evaluated the organization's exposure to AI-driven threats targeting our employees, such as deepfake social engineering or highly personalized phishing generated by adversarial LLMs? Do our current security awareness and authentication controls account for these threat vectors?

What to listen for: Listen for conflation of model performance monitoring with security monitoring: "we monitor for accuracy and latency" is not the same as "we monitor for unauthorized access patterns." Also listen for gaps in understanding of prompt-injection risk, particularly for AI systems that ingest external content and have write or action capabilities. These gaps are the most urgent security issues to address because prompt injection attacks require no network access or credential compromise to execute.

Follow-up actions: Conduct adversarial testing on AI systems that ingest external content, with specific test cases designed to trigger unauthorized data access or action execution through injected instructions. Establish a behavioral baseline for each production AI agent and verify that anomaly detection alerts are configured and tested.

Discussion 3. What Does Our AI Security Architecture Look Like Going Forward?

This third discussion moves from the current state and risk assessment to forward architecture. It should produce commitments to specific controls, ownership assignments, and timelines rather than a general agreement that security is important. The output of this conversation is the AI Trust Framework, which governs AI deployment and operation within your organization.

Discussion questions to ask the team:

- What is our policy for AI agent identity going forward? Will we require dedicated workload identities, scoped access tokens, and least-privilege boundaries as deployment prerequisites, and what enforcement mechanism ensures these requirements are met before an AI system reaches production?
- How do we handle the fourteen agents waiting for deployment right now? Do we have a fast path that applies the minimum viable security controls, and what does that path look like for identity, scope, monitoring, and output filtering?

- What does our ongoing monitoring posture for AI agents look like, and who owns it? Is AI agent behavioral monitoring part of the SOC's responsibility, the data science team's responsibility, or a shared function that requires a defined hand-off model?
- What is our process for evaluating new AI deployments before they reach production? Who has the authority to gate a deployment if security requirements are not met, and what is the escalation path if business sponsors push back on security requirements?
- How do we communicate the AI Trust Framework to data science teams, application teams, and business sponsors so that security requirements are understood as enabling rather than blocking AI adoption?

What to listen for: Listen for agreement on principle but ambiguity on ownership: "yes, we should have workload identities for AI agents" is only useful if it is followed by "and the platform engineering team owns that requirement, implemented in the deployment pipeline by the end of next month." Commitments without ownership and a timeline are not commitments. Also listen for resistance framed as a velocity concern: the goal is not to slow AI deployment but to build a deployment pathway that is fast because it is repeatable, not fast because it skips security requirements.

Follow-up actions: Document the AI Trust Framework as a formal policy, publish the deployment pathway requirements, and schedule a sixty-day review of the fourteen pending agents' compliance with the new requirements. Assign a named owner for AI agent behavioral monitoring and confirm that the SOC has the tooling and access needed to perform that monitoring.

8.9 Takeaway

Artificial intelligence in a zero-trust architecture is a two-front commitment. On the first front, machine learning makes zero trust more capable: behavioral analytics detects the credential misuse and lateral movement that rule-based systems miss, anomaly detection surfaces the access deviations that reveal AI agent compromise, alert prioritization directs analyst attention to genuine threats rather than false positive noise, and AI-driven policy engines reduce the manual burden of maintaining a large-scale policy estate. These capabilities are not optional enhancements for well-resourced security teams; they are increasingly necessary to operate zero trust verification at the scale and speed that modern hybrid enterprises require.

On the second front, AI systems are enterprise assets that require zero-trust governance. Every AI agent needs a verifiable identity, a scoped access token, least-privilege boundaries, continuous behavioral monitoring, action policy gates for consequential operations, and an immutable audit trail for decisions that affect external parties or sensitive data. The governance controls that apply to AI agents are not conceptually different from those that apply to human users and machine workloads; they are extensions of the same zero-trust principles to a new category of identity that operates at machine speed and scale. The structural failure at NovaSphere Technologies was not due to AI technology. It was a failure to apply the identity governance framework already in place for every other asset class to the AI systems that most needed it.

The practical path forward is the same pilot-to-production pathway that governs every other workload in a mature zero trust enterprise: inventory, classify, assign identity, scope access, monitor behavior, and maintain audit trails. Organizations that build this pathway before the AI deployment queue grows will govern a large AI estate at a manageable cost. Organizations that build it later will spend the same investment on remediation rather than on capability.

Marcus Reyes's whiteboard session, the one that asked who these agents are, what they are allowed to touch, and who watches them, was not a delay to AI deployment. It was the foundation that made deployment sustainable.

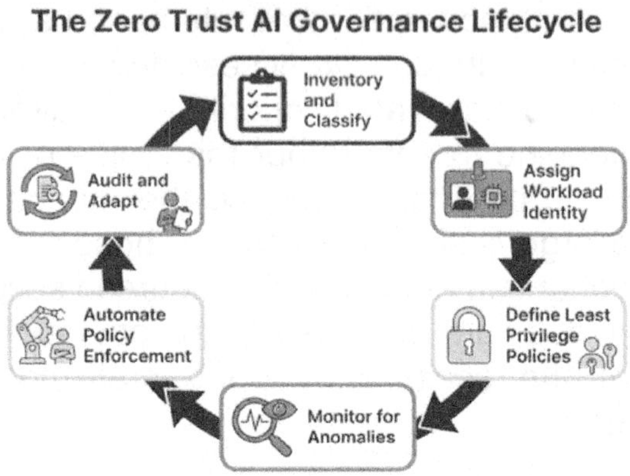

The Zero Trust AI Governance Lifecycle

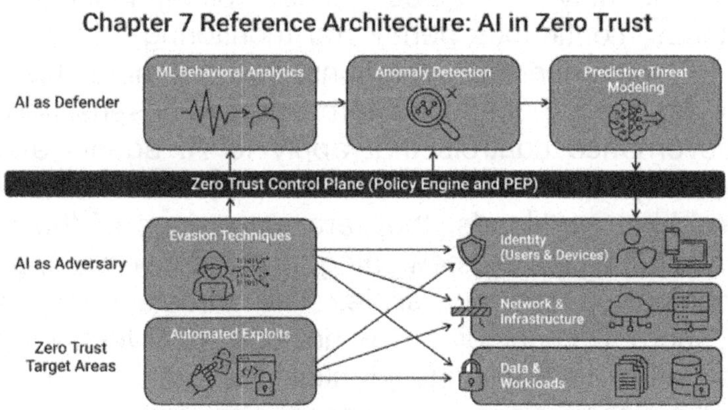

Chapter 7 Reference Architecture: AI in Zero Trust

9 Zero Trust for IoT and Operational Technology

The call came at 6:14 on a Tuesday morning. Marcus Reyes was still in his car in the Meridian Logistics Group parking garage when his phone lit up with a message from the facilities operations manager: the HVAC control system for the Chicago distribution center had gone offline, and building temperature across three of the warehouse zones was climbing past ninety-four degrees. Perishable cargo staging, temperature-sensitive pharmaceutical shipments, and roughly eight hundred employees who would arrive in two hours. Marcus told the facilities manager he would be there in seven minutes and started running.

By the time he reached the network operations center, the picture was both clearer and worse than the initial message had suggested. The HVAC controllers were not offline because of a mechanical fault. They were offline because a portion of the corporate flat network that the building automation system shared with order management had been isolated by an automated threat-response rule triggered 40 minutes earlier. A forensic analyst had already traced the trigger: anomalous outbound connection attempts from an IP address registered to a Modbus-capable HVAC controller on the third-floor plant level. The controller, a device installed eight years ago and never patched, was scanned from the internet, exploited through a known vulnerability in its embedded web server, and repurposed as a pivot point. From the HVAC network, the attacker had moved laterally into the order management subnet and begun probing the financial settlement database. The automated isolation rule had done exactly what it was designed to do. It also cut power scheduling to the chillers.

Marcus stood at the front of the operations center with the facilities manager, a network engineer, and the head of physical plant, looking at a network diagram that showed every device on the building automation segment connected to the same switch fabric as the corporate LAN. There were 340 devices in that segment: HVAC controllers, lighting automation units, elevator management nodes, fire suppression sensors, and two access control panels that governed the loading dock doors. All of them shared a subnet. None of them ran security agents. Several of them dated from before the organization had a security team. And all of them, it turned out, had been trusted implicitly by every other device on the corporate network since the day they were first plugged in.

The immediate crisis took four hours to resolve: manual override of the chiller systems, coordinated rerouting of perishable cargo, and a controlled reconnection of the HVAC segment through a temporary monitoring proxy that the network team built on the fly. The deeper crisis, the one that Marcus would spend the next nine months addressing, was that the incident had exposed an architecture in which thousands of constrained devices, with no authentication, no identity, no patch history, and no monitoring capabilities, were trusted peers of the enterprise network. The HVAC controller had been the ingress point. The financial database had been the target. Between them was nothing but a subnet boundary that enforced no policy.

9.1 Why It Matters

The category of security incidents that Marcus encountered at the Chicago distribution center is growing faster than any other threat vector in enterprise environments. IoT devices, building automation systems, industrial controllers, medical equipment, and operational technology platforms are being added to organizational networks at rates that far outpace the security programs designed to protect those networks. In many organizations, the number of connected devices without traditional operating systems, user interfaces, or agent support now exceeds the number of conventional endpoints by a factor of five to ten. Yet security programs built around agent-based detection, certificate-managed authentication, and software-driven access policy have almost no visibility into this population.

The workflow-level impact of an IoT or OT compromise is fundamentally different from that of a workstation breach. When a laptop is compromised, the organization loses data. When an HVAC controller, a medical infusion pump, or a manufacturing PLC is compromised, the organization may lose physical safety. Operational technology environments where availability is a life-safety requirement cannot tolerate the aggressive response actions that IT security teams rely on: you cannot reboot a controller that is managing a production line mid-shift, cannot patch firmware on a dialysis machine during patient treatment, cannot isolate a building access control panel without locking staff out of secured areas. The security controls available in OT environments are constrained not by budget or negligence but by physical consequence. Any zero-trust program that treats IoT and OT devices as merely smaller or simpler versions of IT endpoints will fail in the field.

What makes this challenge tractable is that zero-trust principles apply just as directly to constrained environments as to conventional ones. Still, the mechanisms by which they are implemented must change. Identity does not require a user account. Continuous verification does not require an agent. Least privilege does not require a firewall that understands HTTP. The mission-aligned goal is the same: no device is implicitly trusted, every communication is verified at the network boundary before it reaches a resource, and any compromise is contained before it can pivot. The mechanisms for reaching that goal in an environment of eight-year-old HVAC controllers and decade-old industrial protocols require a fundamentally different architectural vocabulary, and this chapter builds it.

9.2 The IoT and OT Trust Crisis

Understanding why IoT and OT environments resist conventional security controls requires understanding how they came to be connected to corporate networks in the first place. Operational technology, the hardware and software used to monitor and control physical processes, was originally designed as an air-gapped discipline. Industrial controllers communicated over dedicated serial lines. Building automation systems run on proprietary field buses—medical equipment is operated on isolated clinical networks. The security model for these environments was based on physical isolation: if the device had no network path to the outside world, it had no network-accessible attack surface. That model held for decades.

The collapse of air-gap isolation was not a security decision. It was a business decision, made incrementally over fifteen years as organizations discovered that connecting operational technology to enterprise networks delivered real value: remote monitoring of equipment health, integration of production data into ERP systems, firmware updates pushed over the network rather than by a technician with a USB drive, and building automation systems centrally managed from a facilities helpdesk. Each connection made operational sense. Each connection also punched a hole in an isolation model that had been the primary security control. By the time most organizations understood the aggregate security implications of these individual connectivity decisions, the connections were embedded in operational workflows that were difficult to reverse.

The result is a landscape where devices designed with no security architecture whatsoever, because they were never intended to be network-connected, now sit on corporate networks with default credentials, unpatched firmware, wide-open management interfaces, and no capability to participate in any identity or access management program. The trust crisis in IoT and OT is not a vulnerability-disclosure or patch-management problem at its root. It is a device design philosophy problem layered on top of a connectivity decision problem, and both of those are layered on top of an architectural problem that most organizations have not yet fully confronted.

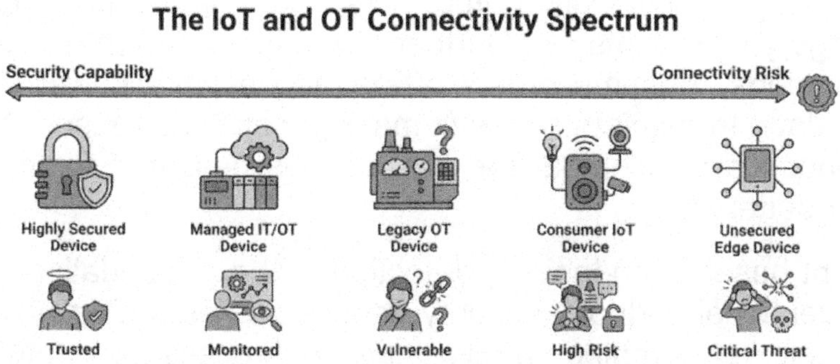

The IoT and OT Connectivity Spectrum illustrates the trade-off between **security capability and connectivity risk across different types of devices.**

9.3 Why Traditional Zero Trust Mechanisms Fail on Constrained Devices

Zero trust architecture, as described in NIST 800-207, assumes an entity that can authenticate, present a certificate or token, be evaluated against a policy engine, and receive and enforce an access decision. These assumptions hold for laptops, servers, mobile phones, and cloud workloads. They often fail for a 32-bit microcontroller running proprietary embedded firmware on a 10-megabyte flash chip with 2 megabytes of RAM. The technical constraints are not incidental limitations that a firmware update will eventually resolve. They are structural properties of device classes that were engineered for reliability, longevity, and cost, not for TLS handshake overhead or certificate chain validation.

Certificate-based authentication requires a device to generate a public-private key pair, store the private key in tamper-resistant storage, perform asymmetric cryptographic operations during each handshake, and validate a certificate chain against a trust anchor. Many constrained IoT devices lack the processing power to perform RSA or elliptic-curve operations at production data rates. Devices that do have sufficient compute capacity often lack secure storage for private key material, meaning that a certificate provisioned to the device can be extracted by any attacker with physical access or a firmware dump. The battery-powered sensor category presents a further constraint: cryptographic operations consume orders of magnitude more power than the data transmission itself, making certificate-based TLS economically incompatible with devices designed for multi-year battery life.

Agent-based endpoint detection, the foundation of conventional endpoint security, requires an operating system capable of running a user-space process, a network connection capable of exfiltrating telemetry, and storage for signature databases or behavioral models. A Modbus RTU device has none of these. A BACnet building controller may have an embedded operating system but no mechanism for third-party software installation and no supported API for behavioral telemetry. The consequence is a monitoring blind spot that conventional security operations centers are poorly equipped to handle: constrained devices cannot report their own security state, cannot detect their own compromise, and cannot participate in protocols that assume self-reporting. The security program must observe them from the outside, inferring security state from network behavior rather than host telemetry.

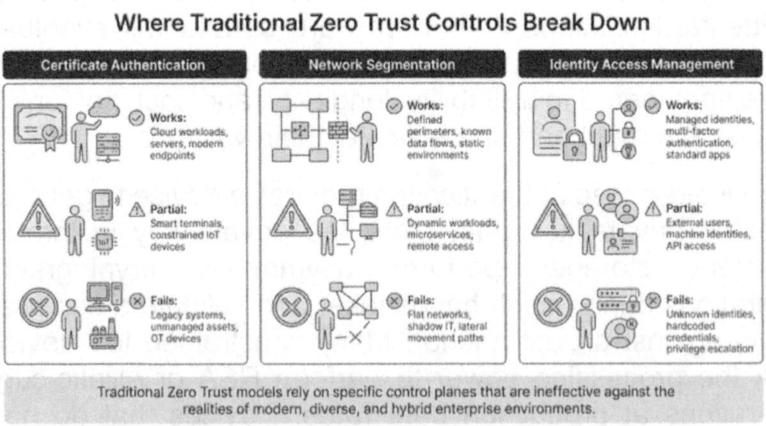

The operational implication is that zero trust for IoT and OT requires a fundamentally different posture: instead of placing the enforcement mechanism on the device, the organization places it between the device and everything it might reach. The network boundary becomes the policy enforcement point. The traffic the device generates becomes the only observable signal. And the behavioral baseline of that traffic becomes the only instrument for detecting compromise. This posture is achievable and effective, but it requires deliberate architectural choices that most organizations have not yet made.

9.4 IT and OT Convergence: Where to Draw the Line

The term IT/OT convergence describes the progressive integration of information technology networks, in which the primary assets are data and applications, with operational technology networks, in which the primary assets are physical processes and equipment. Convergence creates genuine business value: production data flowing into ERP systems, predictive maintenance analytics drawing on sensor telemetry, and remote management reducing technician travel costs. It also creates a trust boundary problem that did not exist when the two domains were physically separated. The question zero trust architecture must answer for every converged environment is: where is the line, what crosses it, and how is crossing it controlled?

The line between IT and OT is not a binary distinction but a spectrum of risk. At one end are data-only flows: telemetry moving from OT sensors to IT analytics platforms, one-way and read-only. These flows carry minimal risk because they do not create a return path for IT compromise to reach OT systems. At the other end are bidirectional management connections: an IT workstation SSH-ing into an industrial gateway to push configuration changes, or a cloud-hosted vendor support session tunneling into an OT control network for remote troubleshooting. These flows pose a high risk because they create paths for an attacker who controls the IT side to send commands to physical equipment. The zero-trust answer to this spectrum is not to refuse all OT connectivity but to classify each connection type explicitly and enforce controls proportional to the directionality and command-authority of the flow.

The practical decision that most organizations struggle with is not whether to segment IT from OT but how to do so without severing the operational integrations that the business depends on. Responsible by design means building the answer to this question into the architecture before the connectivity is established, not retrofitting controls after a decade of implicit trust has created operational dependencies. Organizations that approach IT/OT convergence with a zero-trust posture from the outset define each permitted data flow, specify its direction and protocol, enforce it through purpose-built demilitarized zone infrastructure, and continuously monitor it for deviations. Organizations that inherit a converged environment without that posture must audit every existing connection, classify it by risk tier, and work through a remediation roadmap that segments the highest-risk connections first without disrupting the operational processes those connections serve.

The IT and OT Convergence Trust Boundary

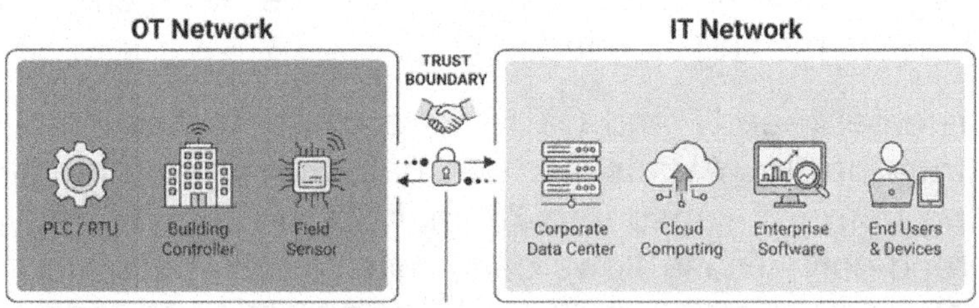

The IT and OT Convergence Trust Boundary facilitates secure data exchange while maintaining distinct operational requirements and security protocols.

9.5 The Purdue Model and Its Zero Trust Evolution

The Purdue Enterprise Reference Architecture, developed in the 1990s as a model for structuring industrial control system networks, describes a layered hierarchy of five levels: field devices at Level 0, field control at Level 1, supervisory control at Level 2, manufacturing operations at Level 3, and enterprise systems at Level 4, with a demilitarized zone between Level 3 and Level 4. For thirty years, the Purdue model has been the dominant framework for thinking about ICS network architecture. Its hierarchical segmentation concept is sound. Its specific implementation assumptions are often obsolete.

The Purdue model assumes that traffic flows vertically between adjacent levels, that higher levels do not communicate directly with lower levels without going through the intervening level. That horizontal traffic within a level is implicitly trusted. Modern operational requirements have violated each of these assumptions. Cloud-connected OT platforms communicate between Level 0 sensors and Level 4 cloud services, bypassing Levels 1 through 3 entirely. Remote vendor access sessions drop directly into Level 2 or Level 3 control networks through VPN tunnels that the Purdue model never anticipated. Horizontal traffic within a Purdue level, formerly trusted by definition, is the primary vector through which attacks propagate once they achieve initial access to a single device within the level.

The zero-trust evolution of the Purdue model preserves its most valuable concept: that traffic between functional layers should be explicitly authorized and controlled, while abandoning its most dangerous assumption: that traffic within a layer is implicitly safe. The modernized architecture applies microsegmentation within each Purdue level, not just between levels. It treats each device as an individual trust boundary rather than treating all devices at a given level equally. It adds north-south monitoring for vertical traffic and east-west monitoring for horizontal traffic. And it implements identity verification at each inter-level boundary through protocol-aware proxies rather than relying on firewall rules based on IP address ranges that change over time.

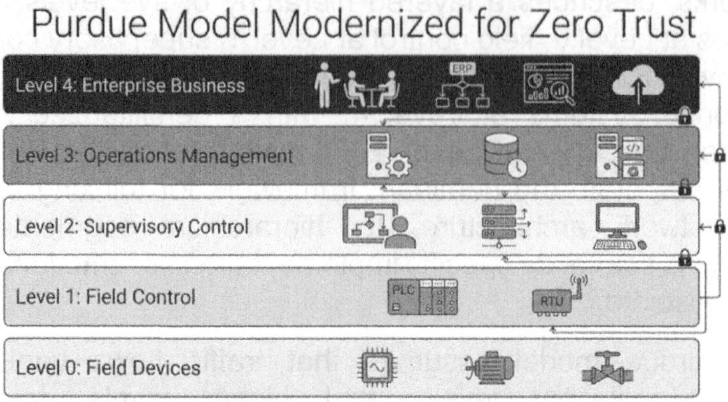

This modernized framework integrates Zero Trust principles, enhancing security across all levels from field devices to enterprise systems.

9.6 OT Segmentation Strategies

Segmenting an operational technology network is a different exercise from segmenting an IT network, and the differences are not superficial. In an IT network, microsegmentation is implemented by pushing policy to software-defined switches, hypervisor vSwitches, or cloud security groups: a configuration change that takes effect immediately and can be rolled back in seconds if it causes unexpected behavior. In an OT network, segmentation changes may require physical rewiring of field bus connections, firmware updates to protocol gateways that must be scheduled during maintenance windows, or coordination with equipment vendors whose support contracts specify approved network topologies. Getting the segmentation wrong in an IT environment disrupts an application. Getting it wrong in an OT environment can stop a production line, trigger a safety interlock, or turn off a life-safety system.

The starting point for OT segmentation is functional zone modeling: grouping devices by their operational function and the set of other devices they are authorized to communicate with. In a building automation environment, the HVAC control loop includes the rooftop unit controllers, the zone temperature sensors, the chiller plant, and the building management system dashboard, and those devices legitimately communicate with each other and with nothing else. The access control system includes card readers, door controllers, and the access management server, which communicate with each other and with the human resources system for employee record synchronization, and nothing else. Functional zone modeling produces a map of legitimate communication relationships that becomes the specification for the segmentation policy.

Once functional zones are mapped, the segmentation architecture must address three inter-zone communication patterns. First, intra-zone traffic: devices within the same functional zone should be permitted on a named-device basis, not on a subnet basis. An HVAC controller should communicate with its designated zone sensors, not with all devices on the building automation subnet. Second, zone-to-management traffic: devices within a functional zone that receive configuration or monitoring connections from a management system should have those connections explicitly defined and enforced at the zone boundary. Third, zone-to-enterprise traffic: data leaving an OT zone for IT analytics should transit a protocol proxy that validates the data type, strips any command-capable protocol elements, and forwards only the read-only telemetry.

OT Functional Zone Segmentation

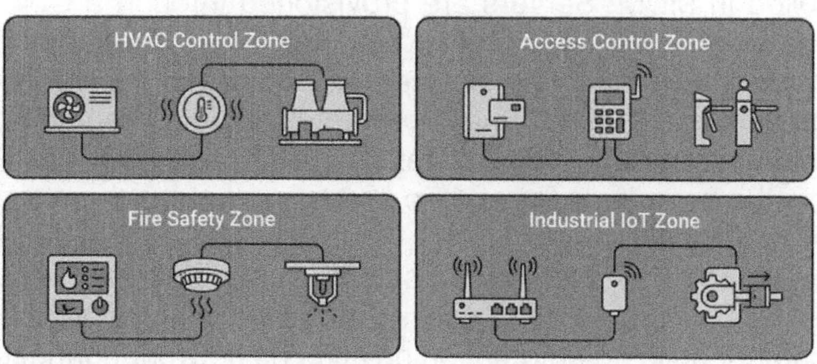

OT Functional Zone Segmentation provides network security and operational resilience.

The sequencing of OT segmentation work matters enormously for risk management. The temptation is to segment the highest-density zones first because those offer the largest immediate reduction in lateral movement surface. The operational reality is that high-density zones are often the ones with the most complex device interdependencies and the highest consequence of a segmentation error. A more sustainable approach starts with life-safety isolation: separating fire suppression, emergency ventilation, and safety interlock systems into dedicated zones, with the most restrictive policies applied first, because the consequences of their compromise are catastrophic. Building automation and HVAC follow, because while the consequences of compromise are operationally significant, they are recoverable. Manufacturing and production control come last, coordinated with production schedules to minimize impact if policy testing reveals unexpected communication dependencies.

9.7 Device Discovery and Inventory at Scale

An organization cannot apply a zero-trust policy to a device it does not know exists. IoT and OT device inventory is universally acknowledged as one of the hardest problems in enterprise security, and this is for reasons that go beyond the sheer scale of the device population. Enterprise laptops are enrolled in MDM. Servers are provisioned through a CMDB. Cloud workloads are registered in cloud management consoles. IoT and OT devices are installed by facilities contractors, building management vendors, equipment manufacturers, and operational staff who have no reason to notify the security team and no process requiring them to do so. The eight-year-old HVAC controller that served as the ingress point in Marcus's incident was not in any asset inventory because no process had ever been created to document building automation devices as network-connected assets.

Passive network monitoring is the least disruptive discovery method for OT environments, and it should be the first tool deployed. Passive discovery captures traffic from a span port or network tap, identifies device MAC addresses and their communication patterns, and infers device type from protocol behavior without sending a single packet to the device. Protocols common in OT environments, including Modbus, DNP3, BACnet, Profinet, and EtherNet/IP, carry device type and manufacturer information in their frame headers, enabling automated classification of most communicating devices. Passive discovery has a fundamental limitation: a device that is powered on but not actively communicating will not appear in the passive inventory. Dormant devices, backup systems, and equipment that only communicate during specific operational cycles may be invisible to passive methods.

Active scanning in OT environments must be approached with extreme caution. Many industrial controllers are not designed to handle unsolicited network traffic at typical scanning rates. An ICMP ping flood directed at a legacy PLC has been documented to cause device faults and forced reboots in multiple incident reports. Active discovery in OT environments uses OT-specific tools that implement protocol-aware queries designed to mimic the expected communication pattern for that device type, and are rate-limited to avoid overwhelming devices designed for low-frequency, deterministic communication. These queries must be tested in a lab environment against representative device types before deployment in production. They must be scheduled during maintenance windows or low-production periods, when an unexpected device reset incurs the lowest operational impact.

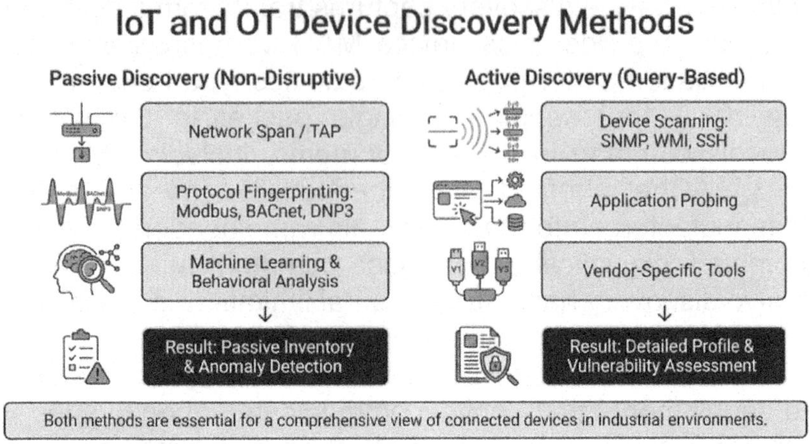

Once discovered, devices must be classified by type, manufacturer, firmware version, communication protocol, and operational criticality. Operational clarity in inventory management requires that each device record include four additional fields that IT asset inventories typically omit: the production impact of a reboot, the patch status and availability of patches from the manufacturer, the identity mechanism the device supports (certificate, pre-shared key, none), and the business owner responsible for the device's operational function. These fields transform a network asset list into a zero-trust readiness assessment, making visible exactly which devices can participate in certificate-based authentication programs, which require brokered identity, and which have known vulnerabilities with no available remediation.

9.8 Attestation and Certificate-Based Identity for Devices That Can Do TLS

A meaningful fraction of the IoT and OT device population, including modern industrial gateways, smart meters, newer building management systems, and connected medical devices from the last five years, supports TLS, certificate-based authentication, and, in some cases, hardware security modules for protecting private keys. For this population, zero-trust identity architecture follows the same principles as conventional endpoint certificate management, with adjustments to the scale and operational context of device deployment.

Device certificates in IoT environments should be issued by a PKI hierarchy that is distinct from the user-certificate PKI used for human identity. This separation is operationally important: it allows the organization to revoke the entire device certificate trust anchor without affecting user authentication, and it allows device certificates to carry attributes, such as device type, functional zone, and operational criticality tier, that are meaningless in user certificate contexts but essential for OT policy decisions. The certificate validity period for IoT devices should reflect the likely time between operational access opportunities for renewal: a device that can only be physically accessed during annual maintenance windows might need a certificate with a longer validity period than a cloud-connected server, but should compensate with more frequent automated OCSP checking to enable rapid revocation if the device is compromised.

Hardware roots of trust, including TPM chips and purpose-built secure elements, provide a mechanism for binding a device's identity to its hardware in a way that survives firmware replacement attacks. A device with a TPM can generate its private key inside the secure enclave and prove to a verifying service that the key was generated on that specific hardware, not copied from elsewhere. For IoT devices that handle sensitive operational data or control life-safety equipment, hardware roots of trust should be a procurement requirement rather than an optional feature. The pilot-to-production pathway for hardware root of trust deployment starts with new device procurement standards, works backward through existing device populations where TPM retrofits, or secure element add-ons are available from manufacturers, and uses compensating controls for devices where hardware attestation is not achievable.

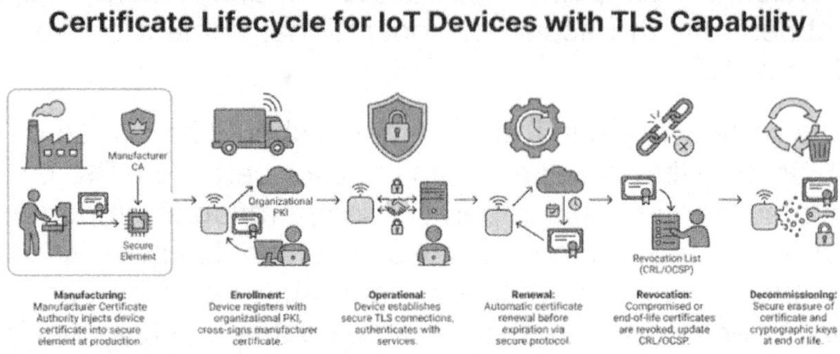

Certificate Lifecycle for IoT Devices with TLS Capability

9.9 Brokered Identity for Devices That Cannot Do TLS

The device population that cannot participate in certificate-based authentication poses a harder, more widespread problem in most OT environments. A Modbus RTU field sensor, a legacy BACnet controller, a serial-attached industrial actuator, or a decade-old HVAC zone controller does not speak TLS. It speaks a field protocol over a serial bus or a UDP-based industrial Ethernet variant, and it was designed to respond to any device that addresses it in the correct protocol format without any notion of authentication. Applying zero trust to this population requires a different architecture: one that places the identity enforcement mechanism not on the device but at the first network node that the device can reach.

The edge gateway model assigns a trusted network node, the gateway, to serve as the identity broker for all constrained devices behind it. The gateway speaks the field protocol to the devices and authenticated, encrypted protocols to the enterprise network. It carries the PKI identity on behalf of the devices it represents and enforces the communication policy that those devices would normally enforce. From the policy engine's perspective, the gateway is the identity. From the field devices' perspective, the gateway is a peer on the field bus. The security properties of the edge gateway model depend entirely on the gateway itself: a compromised gateway exposes all the constrained devices behind it, making gateway security a critical investment.

The communication policy that the gateway enforces on behalf of its downstream devices must be specific enough to detect anomalous behavior. A Modbus RTU sensor that sends a temperature reading to the building management system once every sixty seconds should have a gateway policy that allows exactly that pattern: Modbus function code 3 (read holding registers), from this device address, to this destination IP, at this interval, with this data value range. Any deviation from the expected function code, unexpected destination, out-of-range reading, or unexpected interval is a signal that the device behavior has changed, and changed device behavior in an OT environment is a primary indicator of compromise or equipment fault. The gateway's ability to enforce and monitor this level of specific, protocol-aware policy is what makes it the central enforcement point for constrained device zero trust.

Edge Gateway as Brokered Identity Enforcement Point

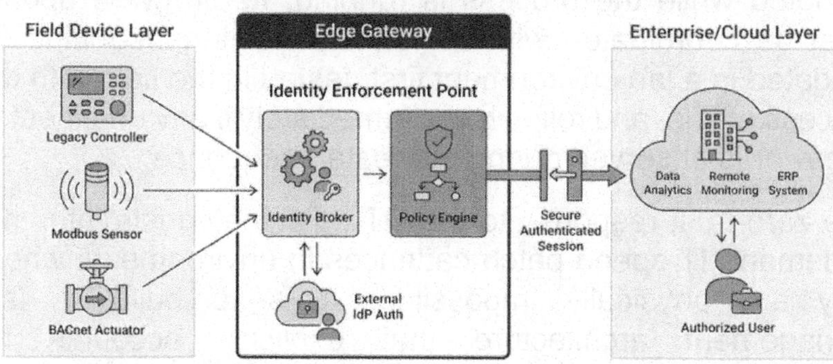

Edge Gateway mediates and secures access by enforcing identity and policy at the edge before connecting to the cloud.

Brokered identity through an edge gateway does not provide the same security guarantees as endpoint-resident certificate authentication, and it is important to be clear-eyed about this limitation. The gateway cannot verify that the device behind it has not been physically replaced or that the field bus communications it receives are genuinely originating from the enrolled device rather than from a device that has been spliced into the field bus by an attacker with physical access. Compensating controls for this residual risk include physical security at field device locations, out-of-band device health checks during scheduled maintenance, and behavioral anomaly detection that would flag changes in communication patterns that typically accompany physical device substitution.

9.10 Safe Patching Cadences in Life-Safety OT

The gap between the IT security world, where monthly patch cycles are considered slow and weekly cycles are becoming standard, and the OT world, where a controller may go years between firmware updates, reflects a genuine difference in operational constraint rather than organizational negligence. An IT server can be rebooted during a maintenance window with service disruption measured in minutes and consequence measured in user inconvenience. An industrial controller that manages a chemical process, a water treatment dosing pump, or a hospital ventilator cannot be rebooted while the process is running. Its firmware update must be coordinated with a planned maintenance window, validated in a lab environment first, tested in the field with the process offline, and rolled back immediately if any unexpected behavior is observed during the restart sequence.

The zero-trust response to the OT patching constraint is not to demand IT-speed patch cadences in environments where they are physically impossible. It is to build a risk-management architecture that explicitly accounts for unpatched devices, provides compensating controls that reduce their exploitability, and creates the organizational processes that enable patching to happen safely when opportunities arise. The first compensating control for unpatched OT devices is network isolation: a device running end-of-life firmware that cannot be patched should be isolated into a zone where its network reach is limited to only the systems it legitimately communicates with, reducing the attack surface available to any exploit targeting that device.

The second compensating control is behavioral monitoring: anomaly detection calibrated to the device's specific communication pattern, so that exploitation of a vulnerability in its embedded web server, serial port interface, or management protocol is visible in network telemetry even when the device itself cannot report the compromise. The third compensating control is access restriction: management interfaces of unpatched OT devices should be accessible only from a dedicated management workstation in a segmented management VLAN, with all connections logged and all management sessions reviewed. These three controls, isolation, monitoring, and access restriction, do not eliminate the vulnerability. Still, they reduce the probability of exploitation and the impact of a successful exploit to a degree, making the residual risk manageable while the organization pursues its patching roadmap.

OT Patching Risk Matrix and Compensating Controls.

Procurement standards are the most powerful long-term lever for addressing the OT patching gap. Every device procured without a manufacturer-committed software update and vulnerability disclosure program extends the unpatched device population for the device's lifetime, which in OT environments is commonly ten to twenty years. Organizations that build OT procurement standards requiring manufacturer security update commitments, vulnerability disclosure programs, and decommissioning timelines are beginning to reshape their device population toward patchability over a multi-year horizon. This is the definition of responsible by design applied to the OT supply chain: security requirements that prevent the accumulation of technical debt rather than attempting to remediate it after the fact.

9.11 Monitoring Without Disrupting Deterministic Control Loops

Operational technology control systems are designed around deterministic timing. A PLC controlling a conveyor belt expects to receive its sensor inputs and transmit its actuator outputs at precise, predictable intervals, often measured in milliseconds. Any interference with that timing, whether due to network congestion, security-scanning traffic, or monitoring-agent overhead, can cause the control loop to miss its cycle time and trigger a safety fault. This deterministic requirement is the fundamental reason why conventional security monitoring tools, including vulnerability scanners, network performance monitors, and behavioral analysis agents, cannot be deployed in OT environments without first verifying their traffic impact against the specific device types and network configurations in the target environment.

Passive monitoring, implemented through a span port or network tap that receives a copy of all OT network traffic without injecting any of its own, is the only monitoring approach that can be deployed in OT environments without risk of control-loop interference. A properly configured span port adds zero latency to the original traffic path. The monitoring sensor consuming the span traffic can perform deep packet inspection of industrial protocols, build behavioral baselines of device communication patterns, and alert on deviations, all without touching the operational network. The limitation of passive monitoring is that it cannot test for vulnerabilities, verify device configuration, or query device state directly. Vulnerability assessment in OT environments must be conducted through manual review of device configuration documentation and manufacturer advisories rather than through active scanning.

Behavioral baseline construction for OT devices exploits a property that is a liability in IT environments but an asset in security monitoring: OT devices are boring. A temperature sensor reads temperature. A conveyor motor controller controls a conveyor motor. The communication patterns of well-functioning OT devices are highly regular, predictable, and protocol-constrained. A Modbus sensor that reads its holding registers once per minute and sends the results to a single IP address generates an extremely tight behavioral baseline. Any deviation from that baseline, a new destination IP, a different function code, an unexpected read frequency, or a data value outside the physical range of the sensor, is an almost certain indicator of something abnormal. Anomaly detection that would require weeks of training a machine learning model on IT endpoint behavior is much more straightforward to implement on OT devices, precisely because their legitimate behavior is so predictable.

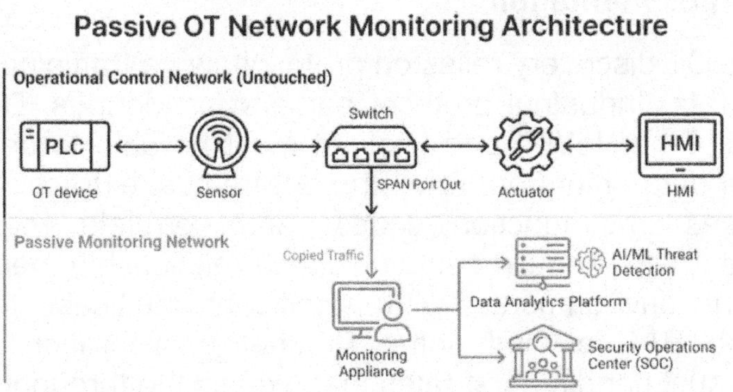

Passive OT Network Monitoring Architecture ensures non-intrusive data collection and analysis for enhanced operational security.

Integrating OT behavioral monitoring with enterprise security operations requires protocol-level translation. A SIEM trained on IT alert types will receive a BACnet protocol anomaly alert and have no contextual framework for assessing its severity. OT anomaly alerts should be enriched with the device's functional zone, its operational criticality, and its physical consequence inventory before they reach the security operations team, so that the analyst understands not just that a protocol rule was violated but what that violation means in the physical world. OT-specific playbooks that translate protocol behavior into operational meaning, a BACnet write-property command from an unexpected source is a potential attempt to send unauthorized commands to a physical system, and are essential infrastructure for any security operations team covering an OT environment.

9.12 Geeks Only

This section is for engineers and architects who need to understand the technical depth of the implementation behind the strategies described above. It covers the specific tools, protocols, algorithms, and architectural patterns required to build and operate an IoT and OT zero-trust program at a production scale.

Passive vs Active Discovery: Technical Implementation

Passive OT discovery relies on protocol-aware traffic parsing at line rate. Industrial protocol parsers for Modbus, DNP3, BACnet, EtherNet/IP, and Profinet must be able to decode protocol data units from raw Ethernet frames, extract device addresses and function codes, and correlate request-response pairs to build communication relationship maps. In industrial environments with significant multicast traffic, including BACnet and some Profinet configurations, the parser must handle burst rates that exceed the throughput of conventional span port analysis tools designed for IT traffic profiles. Commercial OT-specific passive discovery platforms, including offerings from Claroty, Dragos, and Nozomi Networks, implement these parsers natively and include manufacturer device fingerprinting databases that can classify devices by MAC OUI and protocol behavior signatures without requiring any active communication with the device.

Active OT scanning uses OT-safe tooling, such as Nmap with OT service detection scripts that use protocol-specific probes, or purpose-built OT scanners that implement approved query patterns for each protocol type. For Modbus TCP, a safe active probe sends a Read Device Identification request (function code 43, MEI type 14) to the target device and waits for the response: this query is defined in the Modbus specification as a diagnostic request that conformant devices should handle without side effects. For BACnet/IP, the Who-Is broadcast request is the standard device discovery mechanism and is safe to send to properly implemented BACnet devices. Neither of these probes is safe to send at high repetition rates or in rapid succession. Rate limits of one probe per device per five-minute interval are appropriate for initial discovery; more aggressive rates should be validated in a lab against representative device configurations before deployment in the field.

SNMP, Modbus, DNP3 Telemetry for Behavioral Monitoring

SNMP v2c and v3 polling provides basic device health telemetry for OT devices that support it, including interface traffic counters, CPU and memory utilization, and reachability status. The majority of industrial controllers do not support SNMP, but network infrastructure in OT environments, including managed switches, protocol gateways, and communications routers, typically does. SNMP telemetry from OT network infrastructure provides a secondary behavioral signal: changes in interface traffic volume, new MAC addresses appearing on OT switch ports, and unexpected ARP table entries can all indicate device compromise or unauthorized device connection.

Modbus telemetry monitoring operates at the function code level. A behavioral baseline for a Modbus device captures the expected function codes it uses (read coils, read discrete inputs, read holding registers, write single coil), the register addresses it reads and writes, the source addresses of requests it receives, and the timing interval of those requests. The baseline is established from passive observation during a normal operational period of sufficient duration to capture the full operational cycle of the process, which may be as long as several days for seasonal processes or as short as one shift cycle for manufacturing environments. DNP3 behavioral monitoring follows the same approach, with the additional complexity that DNP3 supports event-driven reporting as well as polling, meaning the baseline must account for both scheduled and unscheduled communication events.

Anomaly scoring for OT behavioral monitoring should apply different weights to different types of deviations. A timing deviation of plus or minus 10% from the expected polling interval may indicate normal network jitter and should carry a low anomaly weight. A new destination IP address in any communication from a field device that has never been observed communicating with that destination should carry maximum anomaly weight: field devices do not spontaneously acquire new communication partners in normal operation. An unexpected write command to a register address that the baseline shows has only ever been read should also carry maximum weight: unexpected write commands to OT devices are the primary mechanism through which attackers send unauthorized control commands to physical equipment.

Certificate Provisioning at IoT Scale

Provisioning X.509 certificates to IoT devices at scale requires an automated enrollment protocol that eliminates manual intervention for each device. The two primary options are SCEP (Simple Certificate Enrollment Protocol) and EST (Enrollment over Secure Transport, defined in RFC 7030). SCEP is the older standard and is widely supported by legacy IoT devices and MDM platforms; its security model has known weaknesses in the challenge-password mechanism that should be mitigated by using one-time challenge passwords per device. EST is the modern replacement, using TLS mutual authentication and HTTP-based enrollment, and should be preferred for new deployments where device firmware supports it.

At scale, with tens of thousands of IoT devices, certificate lifecycle management requires automation to handle renewals, revocations, and reissuances without manual intervention. The operational challenge is that devices may be offline, in maintenance, or unreachable for extended periods, making just-in-time renewal impossible. A ninety-day certificate with automated renewal starting at day sixty, combined with OCSP stapling for real-time revocation checking, balances operational reachability constraints against the security risk of long-validity certificates in a device population that cannot guarantee continuous connectivity.

Secure Boot and Hardware Roots of Trust

Secure boot is the mechanism by which a device verifies the integrity and authenticity of each software component it loads during startup, using a chain of trust anchored in immutable hardware. The chain begins with a ROM-resident bootloader whose code cannot be modified after manufacture, which verifies the signature of the second-stage bootloader, which verifies the firmware image, which verifies the application software. If any link in the chain fails signature verification, the device halts rather than loading potentially compromised software. Secure boot does not prevent a device from being compromised through a running vulnerability. Still, it prevents an attacker who has pushed modified firmware to a device from establishing persistence across reboots, significantly raising the cost of a sustained device compromise.

Trusted Platform Module (TPM) chips provide the hardware root of trust for devices that include them. A TPM provides a measured boot log that records the hash of each software component loaded during startup, stored in Platform Configuration Registers that cannot be modified by software running after boot. A remote attestation service can verify the PCR values against known-good baseline measurements, confirming that the device booted the expected firmware rather than modified code. This verification, called remote attestation, is the mechanism that allows an edge gateway or policy engine to assert that a device's identity corresponds not just to a certificate but to a device running verified, unmodified firmware, providing a stronger security guarantee than certificate authentication alone.

For devices that do not include a TPM, purpose-built secure elements, such as the Microchip ATECC608, provide hardware key storage and cryptographic operations on a dedicated chip that can be added to devices during manufacturing or, in some cases, retrofitted. Secure elements store private keys in tamper-resistant hardware and perform signing operations inside the element, so the private key never appears in the device's main memory or flash storage. This provides meaningful protection against firmware-extraction attacks, a common technique for compromising IoT devices without a full hardware root-of-trust implementation.

Secure Boot and Remote Attestation Chain

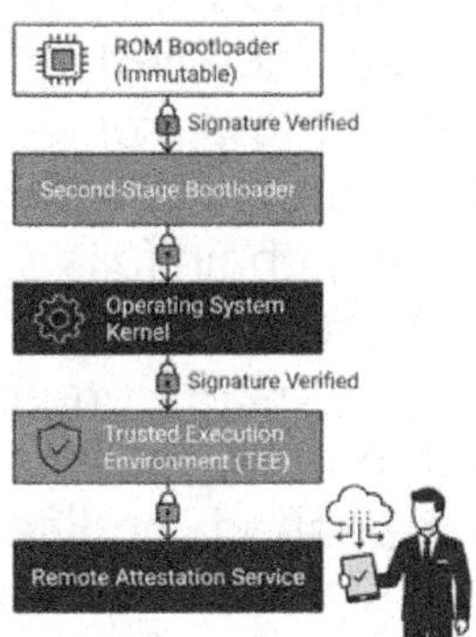

Secure Boot and Remote Attestation Chain

Edge Gateway as Policy Enforcement Point

An edge gateway functioning as a zero-trust policy enforcement point must implement several capabilities that standard network equipment does not provide out of the box. First, it must maintain the device identity registry for all constrained devices behind it: a mapping from field device address (Modbus slave address, BACnet device ID, or serial port identifier) to the enrolled organizational identity that the gateway represents on that device's behalf. This registry is the source of truth for which field devices are authorized to communicate and with whom they are authorized to communicate.

Second, the gateway must implement protocol-aware policy enforcement: the ability to permit or deny specific Modbus function codes, DNP3 application layer function codes, or BACnet service requests, not just TCP connections. IP-level firewall rules are insufficient for OT protocol security because an attacker who has gained access to the field bus can send arbitrary Modbus write commands over the same port that legitimate read traffic uses, which the firewall permits. Protocol-aware policy enforcement that allows function code 3 (read) but denies function code 6 (write single register) from unauthorized sources is the only mechanism that prevents command injection at the protocol level.

Third, the gateway must generate behavioral telemetry at the resolution needed for anomaly detection: per-device, per-transaction records that capture the source address, destination address, function code, register address, value, and timestamp. This telemetry stream compensates for the absence of host-based security telemetry on constrained devices and should be stored in a time-series format optimized for pattern queries, because anomaly detection in OT environments is fundamentally a pattern-comparison problem rather than a log-correlation problem.

OT Change Windows and Risk-Based Exception Handling

Change management in OT environments operates on fundamentally different timescales than IT change management. An IT security team accustomed to emergency patching on a twenty-four-hour response timeline must internalize that an OT change window may occur once per quarter, once per year in some manufacturing environments, or at intervals defined by regulatory permit conditions in process industries. This is not organizational dysfunction but an engineering constraint: the cost of restarting a chemical process, a power generation unit, or a hospital operating theater is sufficiently high that unplanned downtime during production is economically and clinically unacceptable.

Risk-based exception handling provides the governance framework for managing the period between a vulnerability disclosure affecting an OT device and the next available change window. An exception request documents the vulnerability, the affected device population, the exploitability assessment in the specific network context, the compensating controls already in place, and the scheduled remediation date. It is reviewed by the security team, the OT operations team, and, for life-safety devices, a safety officer. The exception is time-bounded, expiring at the scheduled change window, and requires compensating controls to reduce the vulnerability's exploitability during the exception period. Exceptions that cannot be closed at the next change window due to continued operational constraints are escalated to executive risk acceptance rather than renewed without review.

The organizational capacity required by this process is the most common barrier to effective OT vulnerability management. Cross-functional OT security teams that embed security engineers in OT operations and OT engineers in security incident response more effectively overcome this barrier than centralized IT security functions that attempt to manage OT security remotely. Operational clarity in OT change management comes from a shared risk language that operations and security teams have built together.

9.13 Manager's Decision Guide

The decisions below represent the highest-leverage choices a manager must make when building a zero-trust posture for IoT and OT environments. They are sequenced roughly in the order a new program would encounter them, but the second and third decisions are often made in parallel with the first.

Decision Criteria

- Inventory completeness before policy: Do not attempt to enforce zero trust segmentation policy against a device population you have not yet fully inventoried. Segmentation policy applied to an incomplete inventory creates gaps that attackers can exploit through the missing devices. Commission a passive discovery deployment that covers all OT subnets before beginning segmentation design. Accept that the inventory will be imperfect on first pass and build a process for continuous inventory maintenance, but do not begin policy enforcement until you have at least ninety percent confidence in the device population within each zone you intend to segment.
- Functional zone design before network change: The sequence matters operationally. Design the functional zones on paper first, validate them with OT operations staff who know the actual communication dependencies between devices, and test them in a lab environment against representative device types before making any network changes. A zone boundary that severs a control loop dependency that operations staff did not document will cause a production fault. Operations staff review of the zone design catches these dependencies before they become incidents.
- Life-safety isolation as the first segmentation priority: Whatever your overall segmentation roadmap looks like, isolate fire suppression, emergency ventilation, safety interlock systems, and medical life-support equipment into dedicated zones with maximum restrictive policy before any other segmentation work. The consequence of a compromise in these systems is categorically different from the consequence of a compromise in a building access control system or a manufacturing conveyor controller. Life-safety

isolation should be treated as a non-negotiable baseline rather than as one item among many on a prioritized backlog.

- Compensating controls as a documented risk position, not a permanent substitute: When a device cannot be patched, isolated cleanly, or enrolled in a certificate program, the compensating controls you implement are a documented exception, not a solution. Each exception should have an owner, a risk-acceptance signature, a compensating-control specification, and a scheduled remediation date. Treating compensating controls as a permanent architectural feature rather than a temporary risk position allows the unpatched device population to grow indefinitely without organizational accountability for remediation.
- OT security as a cross-functional program, not an IT security extension: The most common failure mode in OT security programs is attempting to manage them entirely from the IT security organization without embedding OT operational expertise into the security team. OT operations staff understand which devices are operationally critical, which maintenance windows are available, and which network changes will cause process faults that IT security engineers would not anticipate. IT security engineers understand threat intelligence, vulnerability assessment, and detection architecture that OT operations staff have not historically needed. The program succeeds when both communities work toward the same risk objectives within a shared operational context.
- Procurement as a security control: The OT device procurement decision is a ten-to-twenty-year security decision. A device procured without a manufacturer's security update commitment, vulnerability disclosure program, or hardware

security capability creates a compensating-control obligation for its entire operational lifetime. Building OT procurement standards that require these capabilities before purchase is the highest-leverage action a manager can take to reduce the unpatched device population over a multi-year horizon without disrupting current operations.

Decision Frame

The central decision in IoT and OT zero trust is how to sequence the program, given that you cannot do everything simultaneously and that some sequencing choices are safer than others. The right sequence is: inventory first, life-safety isolation second, functional zone design third, edge gateway deployment for constrained devices fourth, certificate enrollment for capable devices fifth, behavioral monitoring deployment sixth, and patch remediation on the change window schedule throughout. This sequence ensures that the highest-consequence device populations are protected earliest, that policy is never enforced against an incomplete inventory, and that operational disruption from segmentation changes is minimized by functional zone validation before network implementation.

9.14 Case Study: Chesapeake Medical Center

Chesapeake Medical Center operated as a 1,200-bed hospital network across three campuses with a connected device population that its security team described, honestly, as "largely unknown." A vulnerability assessment commissioned in response to new regulatory guidance discovered what the security team had suspected but not quantified: 35,000 connected medical devices distributed across infusion pumps, imaging equipment, patient monitors, laboratory analyzers, and nurse call systems, all sharing a flat network with clinical workstations and the electronic health record system. Of those 35,000 devices, 12,000 were running operating systems that had reached manufacturer end-of-life, with no available patches for known vulnerabilities that were actively being exploited in healthcare environments by ransomware operators who understood the sector's reluctance to take clinical equipment offline.

The vulnerability that made the incident possible was a known flaw in the embedded web server of a particular infusion pump model, a server used for the pump's web-based management interface and documented in a manufacturer's advisory eighteen months before the incident. The pump in question had no patch available because the vulnerability was in a third-party component the manufacturer had decided not to update, citing FDA regulatory pathway complexity for the firmware change. The advisory recommended isolating the device's network as a compensating control. The Chesapeake security team had received the advisory but had no mechanism to translate "isolate this device" into a network configuration change across 35,000 devices, since they did not know where all 35,000 devices were, which subnets they were on, or which network management team owned those subnets.

The attacker compromised the infusion pump via a known web server vulnerability and used it as a pivot point to reach the EHR system's administrative interface, which sat on the same flat subnet with no segmentation boundary between the clinical device and the clinical application. The EHR system contained protected health information for 340,000 patients. The hospital incurred regulatory notification obligations, a significant operational disruption during the forensic investigation, and a remediation program that occupied the security team for the following eleven months.

The remediation program that followed demonstrated what the architecture would have looked like if it had been built correctly from the outset. Passive discovery deployed across all clinical network segments identified the full device population within two weeks, revealing fourteen additional infusion pump models with the same embedded web server vulnerability, forty-seven imaging devices with separate known vulnerabilities, and more than two hundred devices whose network location had not been reflected in any existing documentation. Clinical function-based segmentation then divided the device population into eight zone types: infusion and medication delivery, imaging and diagnostics, patient monitoring, laboratory systems, nurse call and communications, building and facility systems, administrative workstations, and EHR infrastructure. Devices within each zone could communicate with each other and with their designated management systems. Cross-zone communication was permitted only through explicit policy rules defined by the clinical informatics team, reviewed by biomedical engineering, and implemented through protocol-aware gateway infrastructure at each zone boundary. The EHR system moved to a dedicated zone with an access policy that permitted connections only from clinical workstations explicitly enrolled in the clinical workstation program, not from any device in the clinical equipment zones.

The manufacturer-issued device certificates that infusion pump manufacturers had been embedding in newer pump generations for three years provided the identity foundation for capable devices. For the older pump population without certificate capability, the biomedical engineering team worked with the clinical networking team to deploy edge gateways that proxied Modbus and proprietary pump communication protocols, provided PKI identity for the older pump fleet, and enforced a protocol-aware policy that permitted only the data types expected from medication delivery devices. Behavioral monitoring calibrated to the normal communication patterns of each device class provided the anomaly-detection layer, alerting the security operations center to deviations from expected behavior within minutes rather than the hours or days required to investigate a flat-network lateral-movement chain.

The outcome at Chesapeake was not a perfect security posture. This goal is not achievable in any real clinical environment. Still, it was a demonstrably better risk position: the blast radius of a device compromise had been reduced from the entire clinical network to a single functional zone. An infusion pump compromised in the infusion zone could reach other infusion pumps in that zone and its designated management server, and nothing else. The attack path that enabled a single-pump compromise to reach an EHR system containing 340,000 patient records no longer existed in the architecture.

Chesapeake Medical Center: Before and After Architecture

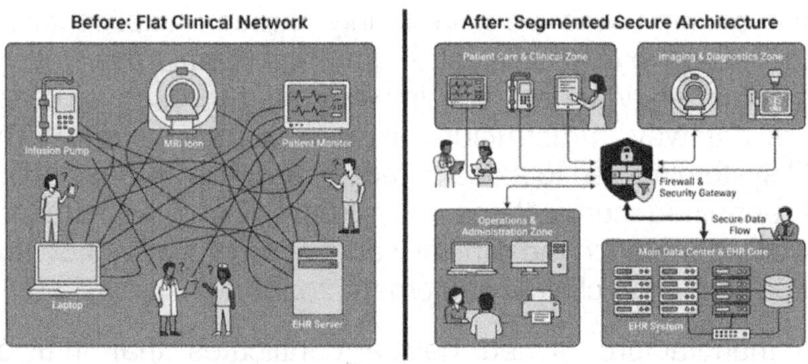

Chesapeake Medical Center: Before and After Architecture

9.15 Team Conversation Guide

Run these three discussions in sequence over two to four weeks. The first surfaces the issue, the second works the trade-offs, the third commits the team to action.

Discussion 1. What Is in Our OT and IoT Device Inventory, and What Is Missing?

This first discussion is a discovery conversation, not a strategy conversation. Its purpose is to establish what the team actually knows about the connected device population, as distinct from what it believes or assumes. Before this discussion, ask the network team to pull a list of all devices observed in network traffic on OT and building automation subnets over the last thirty days. Bring that list and ask the team to compare it to the existing asset inventory. The comparison itself is the conversation.

Discussion questions to ask the team:

- If we ran passive discovery on every subnet today, how many devices do we expect to find that are not in our current asset inventory? How confident are we in that estimate?
- For the devices we know about, how many have a documented owner, a documented firmware version, and a documented patch status? What percentage of the device population does that represent?
- Which device categories in our environment have no path to patching: no available updates from the manufacturer, no vendor patch commitment, no mechanism for applying updates without taking the device fully offline?
- What compensating controls exist today for unpatched or unmanaged devices? Are those controls documented and reviewed, or are they informal and dependent on individual staff knowledge?

What to listen for: The most useful finding from this discussion is the gap between the estimated inventory and the documented inventory. If the team estimates that passive discovery would find two to three times as many devices in the current inventory, the organization has a significant undocumented device population already on the network, already generating trust assumptions, and already outside any security program. That gap number should drive the urgency of the passive discovery deployment. Also listen for engineers who can name specific devices or device categories with no patch path: those are the populations that need compensating controls documented and approved as risk exceptions immediately.

Follow-up actions: Commission passive discovery deployment against all OT and building automation subnets within thirty days. Require that the discovery output be reconciled against the existing inventory and that the gap be documented with an owner assigned to each unregistered device category.

Discussion 2. What Does Our IoT and OT Device Identity Architecture Look Like?

This discussion moves from inventory to identity. Once the team has a reasonable picture of the device population, the question becomes: how many of those devices have a verifiable identity, and what is the plan for the rest? This is a technical discussion that requires the network engineering team and, if possible, OT operations staff who understand the communication protocols of the specific device types in the environment.

Discussion questions to ask the team:

- For devices that support TLS and certificates, do they have device-specific certificates issued by our PKI, or are they using shared credentials, default passwords, or no authentication at all?
- For devices that cannot do TLS, what is our edge gateway coverage? Which OT subnets have a protocol-aware gateway sitting between field devices and the enterprise network, and which subnets have field devices connected directly to enterprise switches?
- Which of our device manufacturers provide hardware security capabilities such as TPM chips or secure elements in their current product generations? Is hardware security a factor in our current procurement evaluation criteria?
- If a field device were compromised and began sending unexpected commands to the devices it communicates with, how quickly would we detect that behavior? What is the detection mechanism?

What to listen for: Listen for the distinction between organizations that have thought about device identity systematically and those that have approached it device-by-device as incidents arise. An organization that can articulate a clear boundary, "these device types get certificates, these types go behind a gateway, these types are in compensating-control exceptions," has a program architecture. An organization that answers each question about a specific device type without reference to a broader policy has an incident-response posture rather than a proactive zero-trust posture. Both are common; the goal of this discussion is to identify which category you are in and what the next architectural step is.

Follow-up actions: Map the device population into three categories: certificate-capable, gateway-eligible, and exception-required. Assign a deployment target for certificate enrollment and gateway deployment for the first two categories. Require documented exceptions with compensating control specifications for the third category.

Discussion 3. How Do We Segment Without Stopping the Production Line?

This is the hardest discussion in the IoT and OT zero-trust program because it requires the team to commit to specific segmentation actions that carry real operational risk if executed incorrectly. The goal is not to finalize the segmentation design in one meeting but to establish a shared methodology for how the team will approach segmentation decisions: how dependencies will be mapped, how testing will be conducted, how rollback will be handled, and who has authority to approve each segmentation step. Bring OT operations staff and biomedical engineers or plant engineers, not just network security engineers.

Discussion questions to ask the team:

- For the first functional zone we plan to segment, do we have a complete map of all communication relationships between devices in that zone and devices outside it? Who validated that map, and what was the validation method?

- What is our test procedure for a proposed segmentation change before it is applied to the production environment? Do we have a lab environment that replicates the relevant device types and protocols, or do we rely on production testing during low-impact windows?
- What is our rollback procedure if a segmentation change causes an unexpected device fault or control loop interruption? How long does rollback take, and who has the authority to initiate it?
- What is the process for approving a security-motivated network change in our OT environment? Who are the stakeholders, what is the review timeline, and how does that process interact with our change window schedule?

What to listen for: The most important signal in this discussion is whether OT operations staff and security staff are describing the same process or different processes. If the security team describes a segmentation approval process and the OT operations team is not aware of it, the process does not work in practice. If OT operations staff describe a change window schedule that the security team has never worked with, the security team has been designing a program without the operational constraints that will govern its execution. The productive outcome of this discussion is a shared change management process that both communities agree on, with each community understanding the constraints the other is working within.

Follow-up actions: Document the cross-functional change management process for OT network changes, including review stakeholders, approval authority at each risk tier, test requirements, and rollback procedures. Require that no OT segmentation change be implemented without a completed change document reviewed by both the security team and the OT operations team.

9.16 Takeaway

The HVAC controller that gave an attacker access to Meridian's financial systems was not an exotic or exceptional threat vector. It was a predictable consequence of a decision made eight years earlier, when a facilities team connected a building automation system to a corporate network without a security review, a patch plan, or any mechanism to enroll the device in the organization's access control program. That decision, and thousands of decisions as they were made across every industry sector, has produced a connected device landscape where the security perimeter that zero trust was designed to replace was never built in the first place for a significant and growing fraction of the asset population.

The principles of zero trust, verify explicitly, use least privilege, and assume breach, apply as directly to an eight-year-old Modbus controller as they do to a cloud workload. The mechanisms for applying them are different: passive discovery instead of MDM enrollment, edge gateways instead of endpoint certificates, behavioral baseline anomaly detection instead of agent-based telemetry, and functional zone segmentation instead of software-defined microsegmentation. These differences in mechanisms require different tooling, skills, and operational processes than those used in conventional IT security. They do not require different principles.

What makes IoT and OT zero trust achievable in practice is the same property that makes it seem difficult in theory: constrained devices behave predictably. A temperature sensor that has been reading temperature once per minute for eight years will continue to do so for the foreseeable future. Deviations from that pattern are meaningful signals, not background noise. The deterministic behavior that makes OT environments feel incompatible with modern security tools is actually the property that makes behavioral anomaly detection more tractable in OT environments than in any other. The challenge is to build the infrastructure to consistently observe that behavior and act on it in time to contain the damage before it reaches a target worth protecting.

Chapter 9 moves from the physical and operational edges of the zero trust perimeter to the data that traverses it. Identity establishes who can access resources. Microsegmentation constrains where access can flow. Continuous verification confirms that access remains warranted. But the ultimate target of every attack that zero trust is designed to prevent is data: the financial records, protected health information, operational intellectual property, and strategic intelligence that organizations collect, process, and store. Chapter 9 examines how zero trust principles extend to data-centric security across hybrid and multi-cloud boundaries, addressing classification, encryption, tokenization, and the governance frameworks that determine where data can flow and who can see it.

10 Data Centric Security Across Hybrid and Multi Cloud Boundaries

Marcus Reyes had been in the post-mortem room for six minutes when the timeline on the screen made his stomach drop. The slide was titled "Data Movement Sequence," and it showed three rows: on-premises research data store, sanctioned corporate SaaS collaboration platform, and a personal cloud drive belonging to an individual who had not been employed by the organization for eleven days. The timestamp delta between the first row and the third was twenty-eight minutes. A dataset containing unpublished pharmaceutical trial results, subject eligibility records, and compound synthesis parameters had moved from a controlled internal environment to a personal cloud account operated by a former contractor, and no perimeter control had raised an alert. The firewall had seen only authorized egress to a sanctioned cloud endpoint. The DLP appliance at the network edge had inspected traffic headers, but not the content, and the encrypted content had passed through without inspection. The CASB had flagged the volume anomaly 90 minutes after the transfer completed, by which point the data had already been synchronized to two mobile devices registered to the former contractor.

The room was quiet in the way that rooms get quiet when people are doing arithmetic about liability. Someone from legal was counting regulatory jurisdictions. The research dataset touched clinical trial subjects in three EU member states, making GDPR Article 83 penalties relevant. The compound data had export-control implications under the EAR, raising the possibility of federal violations in addition to the privacy exposure. Marcus looked at the timeline again and understood what he was really looking at: twenty-eight minutes of exposure, but years of architectural debt. The perimeter had been built around the network. The data had moved through the network legally, from the perspective of every control designed to monitor it. Nothing had been built around the data itself.

He spent the rest of the post-mortem asking a different set of questions than the ones on the agenda. Not "how did the data leave the perimeter" but "what did the data know about itself." Did it carry a classification label from the research vault to the SaaS platform to the personal drive? Did the SaaS integration enforce classification-based policy before allowing a cross-platform sync? Was the key material that protected the data at rest in the research vault controlled by the organization, or had it been delegated to the SaaS vendor? Could the encryption policy have been structured so that a key not owned by the personal cloud account could not decrypt the data, regardless of whether the file itself was synced? The answer to every one of these questions was the same: the organization had made no architectural decisions about any of them. Data security was a perimeter problem. The perimeter had done exactly what it was designed to do, and the data had still walked out.

This post-mortem is the entry point for data-centric security. The network controls worked. Identity controls were partially effective: the former contractor used credentials that should have been revoked but were not detected by the SaaS platform's session management. The real gap was not in the network or even entirely in identity management. The gap was in the absence of controls that traveled with the data, classified it at the point of creation, enforced policy at every transfer boundary, and revoked access independent of the file location. Marcus left the post-mortem with a clear mandate and an uncomfortable awareness that every security framework his team had built to this point had been organized around infrastructure. Rebuilding it around data required a different model entirely. That is the model described in this chapter.

10.1 Why It Matters

Data is the only asset that matters to an attacker once network access is obtained. Servers, network paths, and credentials are all means to an end. The end is data: intellectual property, regulated personal information, financial records, operational parameters, authentication material, or any other information that has value outside the organization. Every security control discussed in earlier chapters, identity verification, microsegmentation, continuous verification, and ZTNA, reduces the probability that an attacker reaches data. Data-centric security is the discipline that governs what happens when every other control fails, is circumvented, or does not cover the data's path. It is the last meaningful layer of protection because it travels with the asset rather than with the perimeter.

The workflow-level impact of inadequate data-centric controls is demonstrated by the scenario above, but it scales in both directions. At the low end, a single misclassified document shared with the wrong distribution list creates a containable but operationally disruptive incident. At the high end, a systematic failure to classify, label, encrypt under organization-controlled keys, and enforce movement policy across cloud boundaries creates the conditions for the kind of data spillage that generates regulatory penalties measured in eight figures, terminates executive careers, and restructures entire business units. The risk is not theoretical. Multi-cloud environments, where data routinely crosses provider, pipeline, and jurisdictional boundaries in the course of normal operations, create data movement exposure at a scale and velocity that traditional perimeter controls were never designed to handle.

Mission-aligned data protection requires that every piece of data the organization creates or receives carries with it a persistent, machine-readable record of what it is, what rules govern it, where it is allowed to travel, who is allowed to decrypt it, and what should happen if it appears somewhere it was not supposed to go. Building that capability is not a single technology purchase. It is a program that spans classification frameworks, encryption architecture, key management strategy, data loss prevention re-engineering, and data flow governance. Organizations that complete this program operate with a level of operational clarity about their data estate that most have never achieved. Those who do not complete it, regardless of how strong their network controls are, assume that data never leaves the environment through paths they did not design. The twenty-eight minutes in Marcus's post-mortem are proof of how that assumption fails.

10.2 The Data-Centric Security Premise

The foundational shift in data-centric security is a change in the unit of protection. Perimeter security protects locations. Data-centric security protects objects. An object in this context is any discrete unit of data, whether a file, a database record, an API response payload, a message in a queue, or a row in an analytics table, that carries a classification label, is governed by an access policy tied to that label, and is protected by encryption whose keys are controlled by a policy engine that enforces the label's rules at every point of access. The object carries its protection with it. When the object moves from an on-premises store to a cloud platform to a partner environment, the protection moves with it. The encryption does not change. The policy does not change. The classification does not change. What changes is the enforcement context, which shifts from network-level controls at each boundary to cryptographic controls embedded in the object itself.

This model requires three capabilities that most organizations are developing rather than fully possessing. The first is classification at ingestion: every data object must receive a machine-readable classification label upon entering the organization's control, whether created, received from an external partner, or extracted from a processing pipeline. Manual classification by end users is insufficient at enterprise scale; automated classification using content inspection, metadata analysis, and pattern recognition is necessary for any environment that processes more than a few thousand objects per day. The second is an encryption architecture that ties key access to classification policy rather than to file location: the keys that protect a confidential research dataset must be available only to identities that the classification policy permits to access confidential research data, regardless of where the encrypted file is stored. The third is data flow governance: the ability to observe, audit, and enforce rules about where each classification tier of data is permitted to travel, which cloud environments can store it, which partners can receive it, and what happens when a data object appears in an unauthorized location.

The Data-Centric Security Model

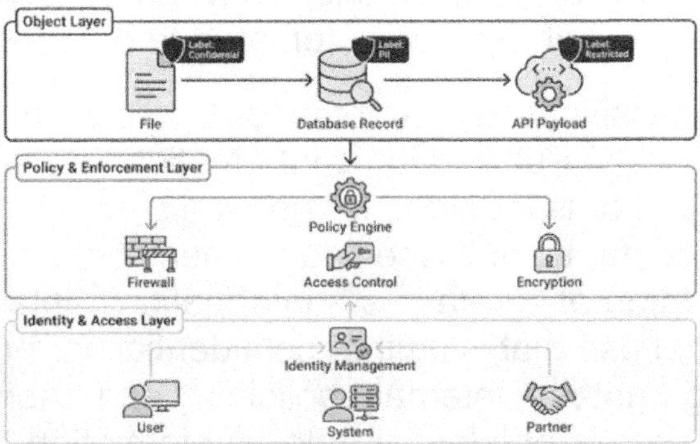

The Data-Centric Security Model

The responsible-by-design approach to data-centric security means that classification, encryption, and policy enforcement are not applied after a data object is created; they are built into the pipelines and systems that create the data in the first place. A data pipeline that writes research results to a vault should classify and label each record during the write operation. An API that returns patient records should enforce classification-based access controls before responding and should log every successful and denied access attempt against an audit trail. A file-sharing integration should inspect the classification label of every object before executing a cross-platform sync and refuse the sync if the destination environment is not approved for that label. These are not post-hoc security overlays. They are engineering requirements built into the systems themselves, which is what responsible-by-design means in practice.

10.3 Classification and Labeling at Scale

Data classification is the foundation of every other data-centric control. An encryption key policy cannot reference a classification if the data has no classification. DLP rules cannot enforce movement restrictions if the data lacks a label that the rules can evaluate. Audit logs cannot distinguish between authorized access to public data and unauthorized access to regulated data if the two look identical at the infrastructure layer. Classification is the metadata that makes all other controls computable, and in environments processing millions of objects, it must be automated to be effective.

A practical classification taxonomy for hybrid enterprises has four to five levels. Public data is information that has already been disclosed or is intended for general availability, such as marketing content, published documentation, and press releases. Internal data is information intended for organizational use that is neither confidential nor public, such as meeting notes, internal policies, and non-sensitive operational data. Confidential data is information that would cause moderate harm if disclosed to unauthorized parties, such as strategic planning documents, personnel performance records, and vendor pricing agreements. Restricted data is information subject to regulatory controls or whose disclosure would cause significant harm: HIPAA-protected health information, GDPR-covered personal data, PCI-scoped cardholder data, ITAR-controlled technical data, and CUI categories. Highly Restricted is reserved for the most sensitive categories: classified national security information in federal environments, clinical trial data before publication, merger and acquisition materials, and source code for core intellectual property. Each level maps to a specific set of permitted encryption standards, storage environments, transfer destinations, and access identity types.

Automated classification at ingestion uses three complementary techniques. Content-based classification inspects the content of the data object and matches it against patterns: regular expressions for structured identifiers like Social Security numbers, credit card numbers, and passport numbers; named entity recognition models for unstructured text containing personally identifiable information; keyword proximity analysis for documents containing controlled technical terminology associated with ITAR or CUI categories. Metadata-based classification evaluates the origin, format, and context of the data object: a record extracted from the HR system is presumptively employee data and receives at minimum an Internal classification regardless of content. A file arriving from a clinical research application is classified as Restricted pending content inspection. Context-based classification uses workflow signals: a document created in a merger integration workspace receives a Highly Restricted label based on the workspace classification, overriding any lower classification assigned by content inspection.

Label persistence is as important as initial classification accuracy. A label that is applied at ingestion but stripped during transformation, format conversion, or API processing provides no downstream protection. Every transformation pipeline that processes classified data must be designed to propagate labels through transformations, apply the higher of two conflicting labels when combining data objects of different classifications, and log any label change, including the identity of the authorizing user and the policy that permitted it. Label persistence in multi-cloud environments requires that each cloud provider's native tagging capabilities be mapped to the organization's classification taxonomy and that import and export pipelines translate between representations consistently. A document labeled Confidential in the on-premises document management system should carry the same label when it arrives in the cloud collaboration platform, and the cloud platform's data governance tools should enforce the Confidential label's restrictions without requiring reinspection of the document's content.

10.4 Encryption at Rest and in Transit

Encryption is the primary technical mechanism that gives data-centric security its teeth. A classification label without enforced encryption is a metadata tag. Encryption whose keys are controlled by a policy engine tied to classification produces a data object that any party cannot meaningfully access, even if the policy engine does not authorize it, regardless of the file's physical location. This distinction matters enormously in multi-cloud environments where data regularly resides in storage systems whose underlying infrastructure is operated by a cloud provider whose staff, absent encryption, have the technical capability to access stored data.

Encryption at rest requires decisions at three levels. The first is algorithm selection: AES-256 in GCM mode for symmetric encryption of data objects, RSA-4096 or ECDSA P-384 for asymmetric key wrapping operations, and SHA-256 or SHA-3 for integrity verification. These are not points of meaningful architectural variation; they are the current standard for data at rest in all regulated environments, and departing from them in either direction, toward weaker algorithms for performance reasons or toward exotic alternatives, requires explicit justification and approval. The second level is encryption scope: should encryption be applied at the storage volume level, the filesystem level, the object level, or all three? Volume and filesystem encryption protect against physical media theft but do not protect against a cloud provider's administrative access. Object-level encryption, where each data object is individually encrypted under a key whose access is governed by classification policy, is the layer that actually enforces data-centric access controls. Both layers are needed; they serve different threat models. The third level is key architecture, the most consequential decision, and the subject of the next section.

Encryption in transit has become more uniform across the industry than encryption at rest, but uniform configuration does not guarantee uniform enforcement. TLS 1.3 is the current minimum acceptable standard for all in-transit data. TLS 1.2 with a restricted cipher suite of ECDHE key exchange and AES-GCM content encryption is acceptable in environments where not all endpoints yet support TLS 1.3. Still, such environments should be actively managed and shrinking. TLS 1.0 and 1.1 are insecure and should not exist in any production environment. The enforcement challenge is not configuration of primary data paths but configuration of secondary paths: internal service-to-service communication between microservices, monitoring agent traffic, backup agent traffic, logging pipelines, and API integrations with third-party vendors. These secondary paths are where TLS enforcement gaps most commonly exist and where data in transit is most likely to be transmitted in cleartext within a network segment that is falsely assumed to be trusted because it is internal.

Encryption Coverage Across Data States

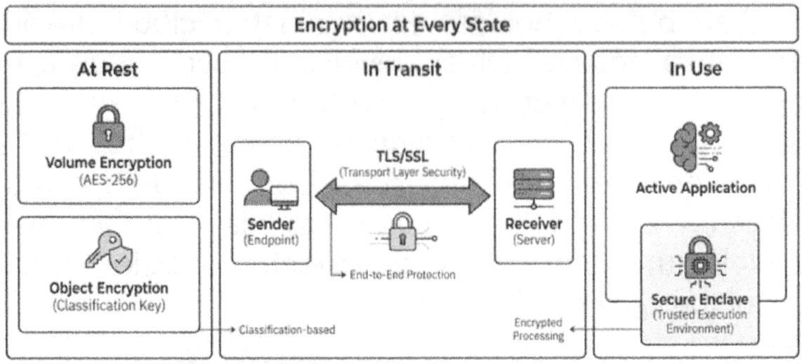

Encryption Coverage Across Data States ensures data protection at rest, in transit, and in use, mitigating risks throughout its lifecycle.

10.5 Key Management Strategies

Key management is where data-centric security either succeeds or collapses. An encrypted object is only as secure as the controls on the key used to encrypt it. If an adversary can access the key, the encryption provides no protection. If the organization cannot access the key, the data is permanently inaccessible. Key management architecture must solve both problems simultaneously: it must make keys available to authorized identities with low enough latency that operational workflows are not degraded, while making keys unavailable to unauthorized parties, including, in some regulatory environments, the cloud provider hosting the encrypted data.

The three primary key management models for enterprise environments are cloud-managed keys, bring-your-own-key, and hold-your-own-key. Cloud-managed keys are the default in most cloud platforms: the cloud provider generates, stores, and manages the encryption keys, and the customer controls access to data through identity and access management policies rather than through key custody. This model is operationally simple but relies on the cloud provider's key management integrity. A cloud provider subject to legal process, insider threat, or regulatory action in its jurisdiction could be compelled to decrypt customer data. For most commercial workloads, the probability and impact of this scenario are low enough that cloud-managed keys are an acceptable default. For regulated data, the Restricted and Highly Restricted classifications yield nearly always different answers.

Bring-your-own-key means the customer generates the master key material and imports it into the cloud provider's key management service. The cloud provider manages the key operationally, but the customer controls the master key and can rotate or revoke it. If the customer revokes the key, the cloud provider cannot decrypt the data even if compelled to do so. BYOK provides meaningful separation of control for most regulated data scenarios and is the minimum acceptable model for Restricted classification data in environments subject to GDPR, HIPAA, or ITAR. Hold-your-own-key takes separation further: the customer maintains the key management service entirely outside the cloud provider's infrastructure. The cloud service must make an API call to the customer's key management system to obtain the key for each decryption operation. The customer can revoke access at the key management layer without affecting anything in the cloud provider's environment. HYOK provides the strongest separation of control and is appropriate for Highly Restricted data or for environments with explicit regulatory requirements for data sovereignty. Still, it introduces operational complexity and performance overhead due to the additional key-retrieval round trip.

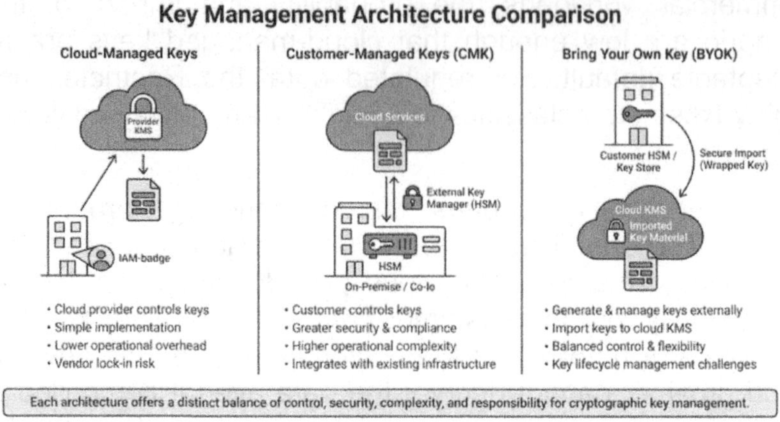

KMS hierarchies reduce the operational burden of managing large numbers of data encryption keys by organizing keys into a two-level or three-level structure. At the root of the hierarchy is the root key, or master key, the highest-trust key in the system, and it must be protected at the highest available assurance level, typically a hardware security module. Below the root key are key-encrypting keys, also called wrapping keys, which are used to encrypt data encryption keys but are not themselves used to encrypt data. At the base of the hierarchy are data encryption keys, which actually encrypt data objects. Each data encryption key is stored in encrypted form, wrapped under its parent key-encrypting key. To use a data encryption key, the system must decrypt it using the key-encrypting key, then use the decrypted data encryption key to encrypt or decrypt the data object. The root key never leaves the HSM. The key-encrypting keys may reside in a software KMS for operational efficiency, provided the KMS is hardened and audited. The data encryption keys may be generated per-object for the strongest isolation, per-bucket or per-table for operational balance, or per-tenant for multi-tenant environments.

Automated key rotation is a requirement, not an option, in any environment processing Restricted or Highly Restricted data. Key rotation frequency depends on the classification level and the applicable regulatory framework. A common baseline is an annual rotation for Confidential data encryption keys and a quarterly rotation for Restricted data encryption keys, with immediate rotation triggered by any evidence of key compromise, personnel departure from a role with key access, or a cloud provider security incident affecting the key management infrastructure. Key rotation must be designed so that data encrypted under previous key versions remains accessible during and after rotation: the old key version must be retained for decryption until all data objects encrypted under it have been re-encrypted under the new key version, at which point the old version can be archived and eventually destroyed according to the key destruction policy.

10.6 Tokenization and Format-Preserving Encryption

Not all data protection requirements are best served by encryption. Encryption changes the representation of a value in ways that make it unusable in systems that expect the original format. A database query that expects a sixteen-digit credit card number cannot run against an encrypted ciphertext. A downstream analytics pipeline that expects a Social Security number cannot process an encrypted blob. Tokenization and format-preserving encryption are techniques that protect the original value while preserving its format and referential integrity, enabling downstream systems to operate without seeing the actual sensitive value.

Tokenization replaces a sensitive value with a non-sensitive surrogate token that has no mathematical relationship to the original value. The mapping between the token and the original value is stored in a secure token vault, accessible only to systems that need to retrieve the original value for authorized purposes such as payment processing, identity verification, or regulatory reporting. Every other system in the processing chain works with the token, never seeing or storing the original value. PCI-DSS environments commonly tokenize credit card numbers so that only the payment processing system ever sees the actual PAN; every other business system that needs to reference a transaction, track a customer relationship, or perform fraud analysis works with the token. The scope of PCI compliance is dramatically reduced because the actual card data never enters non-payment systems.

Format-preserving encryption differs from tokenization in that the encrypted output is mathematically derived from the input: it is a genuine ciphertext that can be decrypted back to the original value. But the ciphertext has the same format as the plaintext, including its length, character set, and sometimes structural properties, such as Luhn check-digit validity for credit card numbers. FF1 and FF3-1 are the NIST-approved format-preserving encryption modes based on AES. Format-preserving encryption is appropriate when the token vault of a tokenization system introduces unacceptable operational complexity, when the output must be decryptable without a separate vault lookup, or when the output must be sent to a partner environment that cannot participate in a centralized token vault. The trade-off is that format-preserving encryption retains some information about the original value through format preservation, which slightly weakens the cryptographic guarantee compared to conventional encryption, and its key management requirements are as demanding as those of conventional encryption.

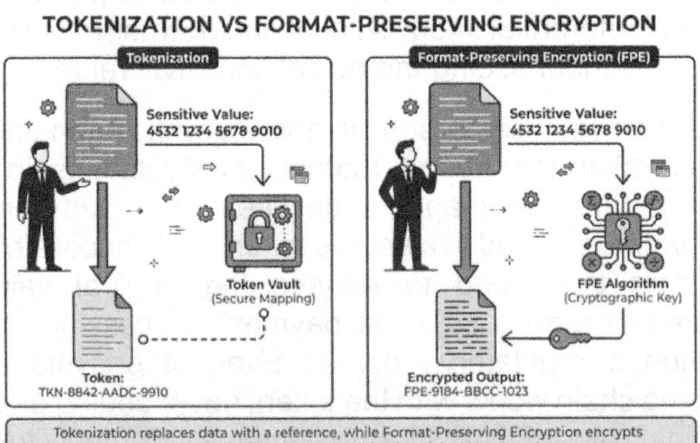

10.7 DLP and CASB in a Zero Trust Posture

Data loss prevention in a perimeter model is fundamentally an inspection problem: deploy sensors at egress points, inspect outbound traffic, and block or alert when sensitive content appears headed outside the boundary. This model works when egress points are few, when traffic can be decrypted for inspection, and when all movement of sensitive data occurs through channels the sensors can observe. In a zero-trust environment, none of these conditions hold consistently. Egress points are everywhere: every API call, every cloud sync, every email attachment, every file transfer to a partner portal is a potential data movement event. Traffic is routinely encrypted end-to-end. Sensitive data is moved through sanctioned cloud platforms that the sensors cannot inspect without breaking encryption. Traditional perimeter DLP in this context is not a security control; it is a false-assurance generator.

Data-centric DLP in a zero-trust posture moves the inspection point from the network edge to the data object itself. Rather than inspecting traffic to detect sensitive content, a data-centric DLP system enforces classification-based policies at the point of access and at the point of transfer. When a user attempts to copy a Restricted-labeled file to a USB drive, the DLP enforcement point is the endpoint agent that reads the label and applies the policy governing Restricted data, which prohibits removable media transfer. When an application attempts to push a Confidential record to an unsanctioned cloud endpoint, the DLP enforcement point is the access control policy in the application layer, which evaluates the label before executing the API call. The network sensor is no longer the primary enforcement mechanism; it becomes a secondary detection layer that catches traffic anomalies that evade the primary label-based controls.

DLP Enforcement in a Zero Trust Posture

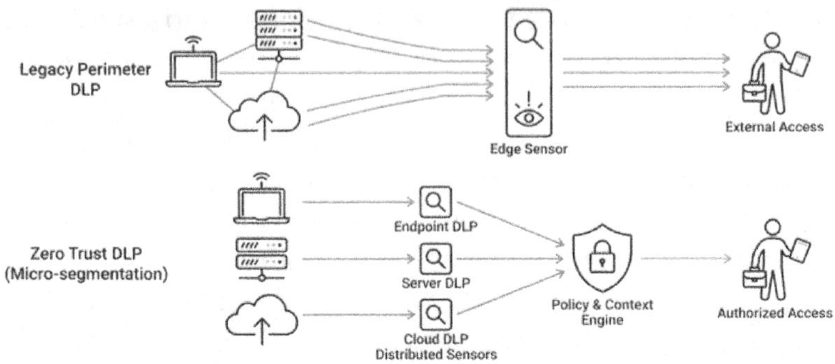

DLP Enforcement in a Zero Trust Posture requires continuous verification and segmentation, integrating security controls across all resources and data flows.

Cloud access security brokers serve as the integration layer between an organization's data classification and access policies and the cloud platforms where data is stored and processed. In a zero-trust posture, a CASB does four things that traditional edge DLP cannot. First, it provides visibility into data at rest in cloud platforms, scanning cloud storage for sensitive content and applying or verifying classification labels. Second, it enforces access policies at the cloud API layer, blocking API calls from unauthorized identities or to unauthorized destinations before data moves. Third, it monitors user behavior in cloud platforms against baseline behavioral models, flagging anomalies such as a user downloading a hundred times their normal data volume in a thirty-minute window. Fourth, it enforces sanctioned versus unsanctioned application policy: an integration between a corporate cloud platform and a personal cloud account triggers a CASB policy violation, which is exactly the control that the Marcus scenario lacked.

The integration between DLP policy and CASB enforcement must be designed as a unified policy framework rather than two separate products with separate rule sets. A classification label that DLP policy treats as Restricted should trigger the same Restricted-level enforcement in the CASB: same permitted destinations, same identity requirements, same transfer restrictions. Organizations that maintain separate

DLP and CASB rule sets inevitably develop gaps between them, because rules in one system are updated without corresponding updates in the other. The pilot-to-production pathway for unified DLP and CASB policy begins with a single classification tier, typically Restricted or PII. It enforces consistent policy across both systems before expanding to other tiers.

10.8 Data Flow Mapping Across Hybrid and Multi-Cloud Environments

An organization cannot enforce policy on data flows it cannot see. Data flow mapping is the practice of discovering, documenting, and continuously monitoring every path through which data of each classification tier travels across the hybrid environment. This includes flows within cloud providers, between cloud providers, between cloud and on-premises environments, to partner and contractor environments, and to end-user devices. In most organizations, the data flow map in the documentation represents only a small fraction of the data flows in the environment. The gap between the documented map and the actual flow topology is the unmapped attack surface for data exfiltration.

Data Flow Map Across Hybrid and Multi-Cloud

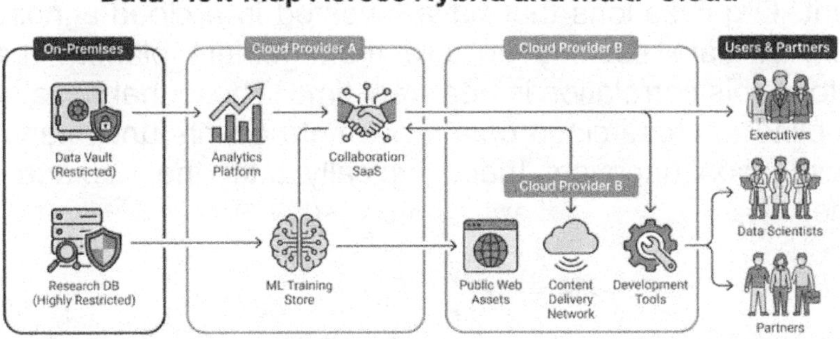

This diagram illustrates the secure and efficient movement of data across on-premises, multiple cloud providers, and user access points, highlighting controlled data sharing and application integration.

Automated data flow discovery uses three source types. Network flow telemetry at the infrastructure layer, collected via

IPFIX, NetFlow, or cloud-native VPC flow logs, provides a map of which endpoints communicate with which other endpoints and over which ports and protocols. This layer sees all flows but does not distinguish data classifications within those flows. API call telemetry from cloud platforms and application middleware provides a higher-fidelity view of what data is being moved: which objects were accessed, which buckets were read, which database tables were queried, and which external endpoints received API responses. This layer can distinguish data classifications if objects carry classification labels logged with the API call. Data pipeline lineage tracking, available in modern data platform tools, provides the most precise view: the exact transformation steps applied to a data object, the source classification, and the derived classification of output objects.

Multi-cloud data flow governance requires integrating each cloud provider's native monitoring capabilities into a unified data flow monitoring plane. AWS CloudTrail, Azure Monitor, and GCP Cloud Audit Logs each provide API call telemetry for their respective platforms, but in isolation, they provide only a partial view. An exfiltration path that moves data from an AWS S3 bucket through an Azure Logic App to a GCP Cloud Storage bucket appears in three separate log streams and requires correlation to identify as a single data movement event. Organizations that have invested in a cloud-agnostic SIEM or data security posture management platform can perform this correlation in near-real time. Those that have not are blind to cross-cloud data movement events until manual investigation uncovers them, typically after the damage is done.

10.9 Third-Party Data Sharing

Data sharing with third parties, whether partners, contractors, regulators, auditors, or customers, is a routine operational requirement that data-centric security must accommodate without disabling. The challenge is that third-party environments are, by definition, outside the organization's direct control. The partner's cloud storage may not meet the organization's encryption requirements. The contractor's endpoint may not have the organization's DLP agent installed. The regulatory examiner's workstation may not be able to participate in the organization's key management system. Data-centric security must provide a mechanism for controlled, audited, time-limited data sharing without requiring the receiving party to adopt the organization's full security stack.

Derived keys are the primary technical mechanism for controlled third-party data sharing. Rather than sharing the original encryption key, which would give the receiving party unrestricted and permanent access to the encrypted data, the organization's KMS generates a derived key scoped to a specific data object or set of objects, valid for a specific time window, and authorized for a specific identity or certificate belonging to the receiving party. The derived key can decrypt only the objects it was scoped to and only for the permitted period. When the time window expires or the sharing relationship ends, the derived key is revoked, and the receiving party loses access to the data without taking any action. The original data object and its master key are unaffected.

Controlled Third-Party Data Sharing with Derived Keys

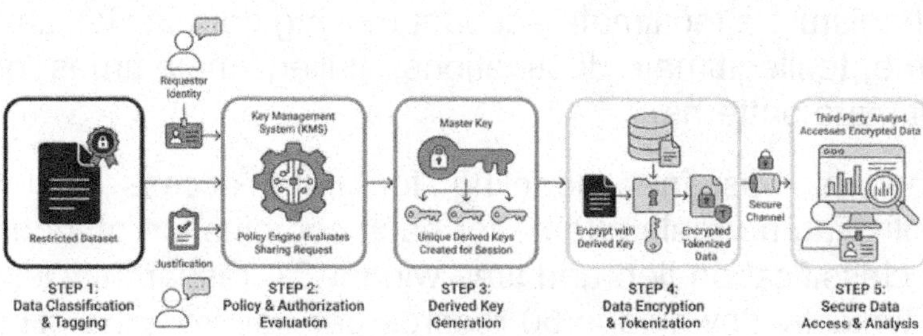

Controlled Third-Party Data Sharing with Derived Keys enables secure, policy-driven access to sensitive information without exposing master encryption keys.

Data sharing agreements must be backed by technical controls, not just contractual language. A data processing agreement with a cloud vendor that specifies encryption requirements and data residency constraints is not enforceable solely through contract terms. If the vendor's environment does not technically enforce the requirements, the contract does not prevent a violation; it only creates liability after one occurs. Technical enforcement means that data shared with a third party is encrypted under a derived key the receiving party can only use in the manner the policy permits, that access is logged at the organization's key management system rather than depending on the third party's logging, and that data residency is enforced by storing data in region-specific storage and restricting the derived key to identities that authenticate from permitted regions. Organizations that rely on contractual data sharing without corresponding technical enforcement are making a compliance representation they cannot operationally verify.

10.10 Exfiltration Detection

Classification, encryption, and access controls reduce the probability of successful exfiltration, but they do not eliminate it. Insider threats, compromised privileged accounts, and sophisticated attackers who obtain valid credentials or valid key material can still move data through authorized channels. Exfiltration detection is the discipline of identifying unauthorized data movement that uses authorized mechanisms: a legitimate account moving data at illegitimate volume, to illegitimate destinations, at illegitimate times, or in illegitimate patterns.

Behavioral baseline modeling for data access requires establishing normal patterns for each combination of identity, data classification tier, and time window. A research scientist who typically downloads 50 records per day from the clinical data repository is considered abnormal if they download 5,000 records in a single day. A finance system that typically calls the data warehouse on weekday business hours is abnormal at 2 AM on a Saturday. A cloud storage bucket that receives writes from three known pipeline sources is abnormal when writes arrive from a fourth source. Each deviation from baseline is a detection signal. No single deviation is definitive: the researcher may be performing a legitimate bulk analysis. The finance system may have a scheduled report running after hours. The bucket may have a new, legitimate pipeline. Exfiltration detection correlates multiple signals over time before escalating, because the cost of false-positive alerts, which desensitize teams and delay responses to real events, is nearly as damaging as the cost of missed detections.

Exfiltration Detection Signal Stack

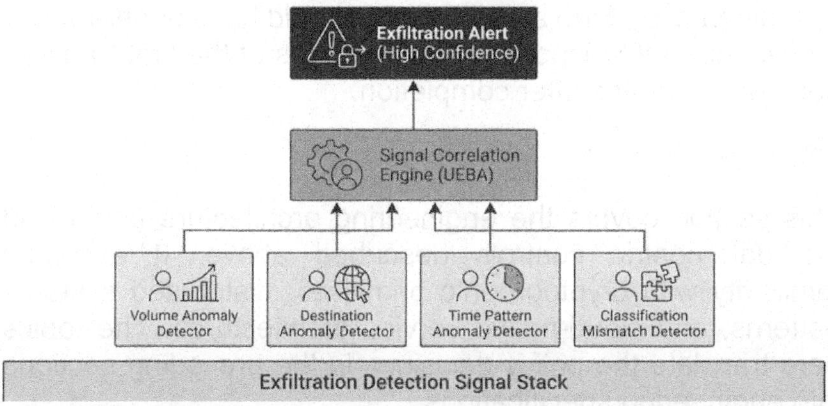

Volume-based detection identifies data transfers that exceed established thresholds for a given identity, resource, or time window. Destination-based detection identifies data transfers to endpoints that are not on the approved transfer destination list for the relevant classification tier. Classification mismatch detection identifies data labeled at one tier appearing in an environment approved only for a lower tier, which indicates either a mislabeled object or an unauthorized transfer. Time-of-day and day-of-week detection identifies transfers occurring outside the established operational pattern for the source system or identity. Each detection dimension produces signals that feed into a unified alert-scoring model, where a single anomaly generates a low-priority investigation ticket, and multiple correlated anomalies trigger an immediate SOC response.

The twenty-eight minutes in Marcus's post-mortem scenario would have triggered several of these detection signals in a mature exfiltration detection system. The volume of data moved was anomalous for the contractor identity operating. The destination, a personal cloud drive, was not in the approved transfer destination list for any classification tier. The time of day was outside the contractor's normal access window. Any one of these signals, when correlated with the contractor's recent activity and the classification of the accessed data, would have triggered an investigation. Combined, they form a pattern that should have generated an immediate SOC response within minutes of the first transfer, not ninety minutes after completion.

10.11 Geeks Only

This section covers the engineering architecture underlying the data-centric controls described above. It assumes familiarity with cryptographic primitives, distributed systems patterns, and cloud-native service architectures. The topics here translate the policy decisions in the preceding sections into engineering specifications.

KMS Architectures and Key Wrapping

A key management system is a distributed cryptographic service responsible for key generation, storage, access control enforcement, audit logging, and key lifecycle operations, including rotation and destruction. At the architectural layer, a KMS must satisfy four requirements simultaneously: high availability, because any data encrypted under a KMS-managed key is inaccessible if the KMS is unavailable; tamper resistance, because a KMS that an attacker can modify to return incorrect key material undermines all encryption based on it; auditability, because every key operation must be logged with sufficient context to reconstruct access history for forensic purposes; and access control, because the KMS must enforce policy-based access to key material at the individual key level, not just at the service level.

Key wrapping is the mechanism by which data encryption keys are stored securely within a KMS. A DEK is never stored in plaintext anywhere outside an HSM; it is always stored encrypted under a key-encrypting key. The wrapping operation is typically AES-KW or RSA-OAEP for asymmetric wrapping scenarios. In an envelope encryption pattern, the DEK is generated locally at the point of encryption, used to encrypt the data, then immediately wrapped under a key-encrypting key retrieved from the KMS. Only the wrapped DEK is stored alongside the ciphertext. To decrypt, the system retrieves the wrapped DEK, calls the KMS to unwrap it using the KEK, then uses the unwrapped DEK to decrypt the data. The DEK exists in memory only during the cryptographic operation. This pattern ensures that the KMS audit log captures every decryption attempt, that the DEK is never transmitted in the clear over any network, and that revocation of the KEK immediately prevents decryption of all objects whose DEKs were wrapped under it.

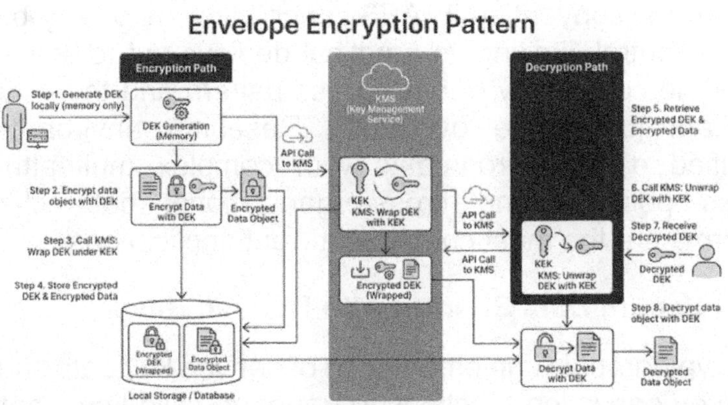

Envelope Encryption Pattern

Attribute-Based Encryption

Attribute-based encryption is a cryptographic scheme in which the ciphertext is associated with a set of attributes and the decryption key is associated with a policy that specifies which attribute combinations are authorized to decrypt. Unlike conventional encryption, where a specific key decrypts a specific ciphertext, ABE allows a ciphertext labeled with attributes such as "classification: restricted, department: research, jurisdiction: us" to be decrypted by any holder of a key whose policy matches the ciphertext attributes. The policy evaluation happens inside the cryptography: no separate policy engine call is required at decryption time. ABE is particularly relevant to data-sharing scenarios in which the same encrypted object must be accessible to multiple parties with different but overlapping attribute profiles, without requiring re-encryption for each recipient.

Ciphertext-policy ABE and key-policy ABE are the two primary variants. In CP-ABE, the policy is embedded in the ciphertext, and keys are associated with attribute sets. A ciphertext encrypted under the policy "research AND (US-clearance OR EU-research-permit)" can only be decrypted by a key that presents the required attribute combination. In KP-ABE, the policy is embedded in the key, and ciphertexts are associated with attribute sets. ABE remains computationally expensive relative to conventional AES encryption and key-based access control, limiting its practical deployment to scenarios where the complexity of the access pattern and the value of the data justify the overhead. Research environments, classified data environments with complex multi-attribute access policies, and cross-organizational data sharing platforms are the most common current applications.

Secure Data Sharing with Derived Keys

The cryptographic implementation of derived key data sharing uses key derivation functions to generate child keys that are mathematically related to a master key but cannot be used to reconstruct it. HKDF, defined in RFC 5869, is the NIST-recommended KDF for this purpose in non-password-based scenarios. The derived key is parameterized by the sharing context: the recipient identity, the permitted object scope, expressed as a set of object identifiers or a cryptographic commitment to the object set, and the validity window, expressed as a Unix timestamp range. The KMS generates the derived key, logs the derivation with all parameters, and records the derivation event against the master key's audit chain. When the derived key is revoked, the KMS records the revocation timestamp and refuses subsequent unwrap operations for the derived key. The master key and all other derived keys are unaffected.

For scenarios where the recipient is in an external environment, and the organization's KMS cannot be reached directly for each decryption operation, a sealed key pattern provides offline access within the permitted window. The KMS generates the derived key, seals it using the recipient's public key, and delivers the sealed package. The recipient can unseal the derived key using their private key and use it locally without calling the organization's KMS. The validity window is enforced by building an expiration timestamp into the sealed package and requiring the recipient's decryption library to verify the timestamp before using the key. This pattern accepts the trade-off that revocation within the validity window is not instantaneous: the recipient holds a valid key that cannot be remotely revoked until expiration. The acceptable validity window for Restricted data is typically 24 to 72 hours, limiting the revocation gap.

Telemetry from Data-Plane Services

Data-plane telemetry for exfiltration detection requires instrumentation at the storage access layer, the key management layer, and the network transfer layer. At the storage layer, every read and write operation against classified data stores should emit a structured event containing the object identifier, the classification label, the accessor's identity, the operation type, the number of bytes transferred, the source and destination network endpoints, and a monotonic timestamp. At the KMS layer, every key unwrap operation should emit an event containing the key identifier, the identity requesting the unwrap, the application or service making the request, and the result. At the network layer, flow records should capture total bytes transferred per five-minute window for all flows involving classified storage endpoints. When these three telemetry streams are correlated with the behavioral baselines for each identity and object, they produce a detection signal with lower false-positive rates than any single stream alone.

Schema normalization across these three telemetry streams is a prerequisite for effective correlation. Events from AWS CloudTrail, Azure Key Vault audit logs, GCP Cloud KMS audit logs, and on-premises HSM audit logs use different field names, timestamp formats, and identity representation schemes. A normalization pipeline that maps each source's event schema to a common data model, typically an extension of OCSF or the OpenTelemetry data model, is necessary for efficient correlation. Organizations that deploy normalization as part of their SIEM ingestion pipeline can correlate across all three layers in near-real time. Those that maintain separate log analysis processes for each platform cannot perform cross-platform correlation until logs are manually exported and compared, which is too slow for actionable exfiltration detection.

Schema Evolution in Classified Data Lakes

Data lakes that store classified data face a distinctive engineering challenge: the data's schema evolves as source systems are upgraded, new data sources are added, and analytical requirements change, but the classification policy must remain accurate even as the schema changes. A dataset classified as Restricted because it contains direct patient identifiers becomes more complex when a schema evolution adds an indirect identifier column that, combined with other available data, enables re-identification. The classification system must track not just the current schema but the semantic meaning of each field, the re-identification risk of field combinations, and the impact of schema changes on the applicable regulatory frameworks.

Schema-aware classification tools use column-level classification rather than table-level classification to handle this complexity. Each column in a classified table contains its own classification metadata, including the field type, applicable regulatory frameworks, the re-identification risk score, and permitted query types. A query that selects only non-sensitive columns from a Restricted table may run under a Confidential-level access policy. A query that selects a combination of columns whose re-identification risk score exceeds a threshold requires a Restricted-level access policy. Schema evolution events trigger an automated re-classification review: when a new column is added, the classification system inspects the column's values, applies content classification rules, and either assigns an automatic classification or flags the column for human review before the table is made available for query. This process ensures that data lakes do not accumulate unclassified columns through the ordinary evolution of their schemas.

Data Lineage for AI and ML Pipelines

AI and machine learning pipelines introduce a classification challenge with no precedent in traditional data governance: the trained model itself may encode sensitive information from the training dataset, making its classification at least as high as the highest classification of any training data it was trained on. A model trained on restricted clinical data is not a generic software artifact; it is a Restricted artifact whose outputs may reveal information about individual training subjects under adversarial membership inference attacks. The model must be classified, encrypted, access-controlled, and audited in accordance with the same policies that govern the training data.

Data lineage tracking for AI pipelines requires recording the provenance of every training dataset used to produce a model, including the version, classification, regulatory scope, and any data sharing agreements governing the training data. This provenance record must accompany the model through its entire lifecycle: training, evaluation, packaging, deployment, serving, and eventual decommissioning. Model cards, an emerging documentation standard for AI model metadata, provide a natural vehicle for embedding classification and lineage information in a format that model deployment pipelines can read and enforce. A deployment pipeline that refuses to serve a Restricted-classified model to identities without Restricted data access is the AI-pipeline equivalent of classification-based access controls that govern data at rest.

Data Lineage Tracking for ML Pipelines

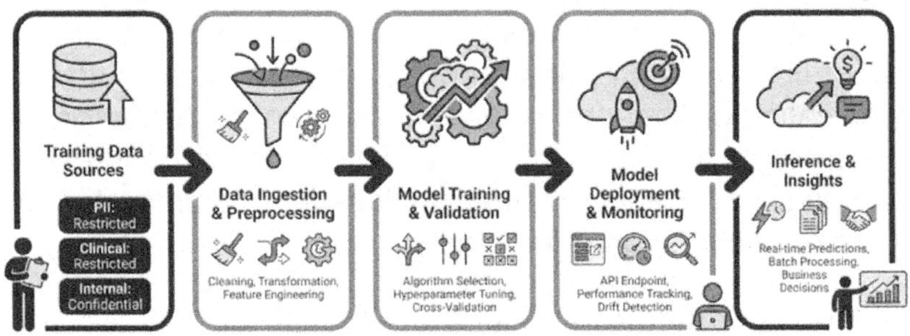

Data lineage tracking ensures accountability, compliance, and reproducibility across the entire machine learning lifecycle, from raw data sources to final business outcomes.

10.12 Manager's Decision Guide

Data-centric security requires decisions across classification, encryption, key management, and governance that span multiple teams and budget cycles. These decision criteria will help you assess where your program stands, where the critical gaps are, and in what sequence to address them.

Decision Criteria

- Classification program maturity: Determine whether your organization's data classification program is documented policy only, manual user-applied labels, or automated classification enforced at ingestion. Policy-only programs with no automated enforcement provide no technical protection. Prioritize automated classification tooling if the program relies on users manually applying labels.
- Encryption posture by classification tier: Confirm that each classification tier has a defined encryption requirement and that the requirement is technically

enforced. Ask your team to demonstrate, for a specific restricted dataset, the encryption algorithm, the key management model (cloud-managed, BYOK, or HYOK), and the mechanism by which a compromised cloud provider account would fail to access the data.

- Key management architecture review: If your organization uses cloud-managed keys for Restricted or Highly Restricted data, evaluate whether BYOK or HYOK adoption is required by applicable regulatory frameworks or by your organization's risk tolerance for cloud provider access to data. Establish a migration timeline if the current model is insufficient.
- Data flow map completeness: Ask your data security team to produce the current data flow map for your Restricted classification tier. If such a map does not exist or covers only a subset of known data stores, treat data flow mapping as a near-term priority. You cannot enforce policy on flows you cannot see.
- DLP and CASB integration: Verify that your DLP and CASB tools enforce consistent classification-based policies. Ask for a demonstration of what happens when a user attempts to copy a Restricted file to an unsanctioned cloud drive. If the answer is "we generate a log entry," the enforcement is detection-only. If the answer is "the transfer is blocked before it completes," enforcement is active.
- Exfiltration detection coverage: Confirm that exfiltration detection applies behavioral baselines per identity and per classification tier, not just volume thresholds. Ask what the detection latency would be for the Marcus scenario: a former contractor moving restricted data to a personal cloud drive. If the answer is "we would see it in the CASB logs," determine how long after the transfer that log appears and what triggers the review.

- Third-party data sharing controls: For every data sharing relationship with a partner, contractor, or vendor, verify that the sharing mechanism uses derived or scoped keys rather than sharing the original encryption key. If original keys are shared, the receiving party retains permanent access to the data regardless of whether the sharing relationship has ended.
- Regulatory jurisdiction mapping: Confirm that every Restricted data flow involving personal data has been mapped to the applicable regulatory frameworks (GDPR, CCPA, HIPAA, ITAR, CUI) and that the technical controls enforcing each framework have been verified, not just documented in a compliance matrix.

Decision Frame

Prioritize your data-centric security investment in the following sequence. First, achieve classification coverage: data you cannot label cannot be protected by label-based controls. Second, enforce encryption with organization-controlled keys for Restricted and above: if a cloud provider can decrypt your most sensitive data without your involvement, your data-centric posture has a fundamental gap. Third, build a complete data flow map: gap analysis against the map drives every subsequent priority. Fourth, unify DLP and CASB policies into a single classification-based framework: consistency between systems eliminates the gap that the Marcus scenario exploited. Fifth, build behavioral exfiltration detection on top of the flow map: you cannot detect anomalous movement without first knowing what normal looks like. These five steps, in sequence, produce a data-centric security program with the pilot-to-production pathway of a phased multi-year effort rather than a Big Bang deployment, and each step produces measurable protection outcomes independent of the steps that follow.

10.13 Case Study: Continental Assurance Group

Continental Assurance Group is a fictional multinational insurance company with operations across twenty-two countries. Its policyholder data was distributed across four cloud providers and two on-premises data centers, subject to GDPR in the European Union, CCPA in California, PDPA in Singapore, and varying state privacy laws across twelve additional U.S. jurisdictions. Managing consistent data protection across this environment was acknowledged as a challenge. Still, the organization's security program had addressed it primarily through contractual data processing agreements and cloud provider native encryption, both configured with cloud-managed keys. The technical enforcement of data residency and data classification was limited to regional bucket configurations and IAM policies that restricted bucket access to identities in specific cloud tenants.

The incident originated in the analytics environment. A data pipeline was designed to extract policyholder records from the core policy administration system, anonymize them by removing direct identifiers, aggregate them by geographic and demographic cohort, and write the aggregated results to a cloud storage bucket used by the analytics team. This pipeline had been running correctly for eight months. A configuration change to the anonymization step, intended to improve processing performance, inadvertently bypassed the de-identification logic for a subset of records: those with the policyholder's address field containing multiple entries due to a data model change in the source system. The pipeline wrote approximately 340,000 records containing full PII, including names, addresses, policy numbers, health condition codes, and beneficiary relationships, to the analytics bucket without triggering any classification validation. Because the analytics bucket was configured for Confidential classification rather than Restricted, its access controls permitted cross-region access from compute workloads in EU, Singapore, and U.S. tenants simultaneously. The misconfigured records were accessible from all three jurisdictions for eleven days before the error was detected during a routine data quality review.

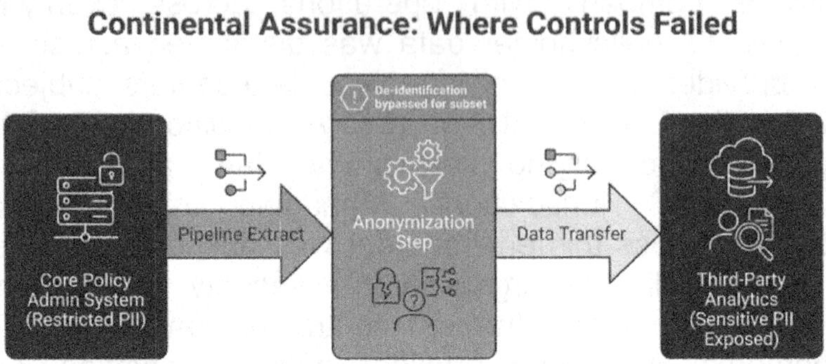

Inadequate access controls and a flaw in the anonymization script allowed sensitive data to be moved to an unsecured environment, highlighting the critical need for robust data governance.

The regulatory consequences were significant. GDPR enforcement actions in three EU member states cited both the breach of data integrity (Article 5) and the inadequacy of technical measures (Article 32). California regulators cited CCPA violations for California residents whose data appeared in the EU-accessible bucket. Singapore's PDPA enforcement authority found violations related to cross-border data transfers without appropriate safeguards. The aggregate regulatory exposure reached forty-seven million dollars before settlement discussions concluded. Beyond the direct penalties, the organization was required to implement a supervised data governance remediation program under regulatory oversight, which consumed substantial engineering resources over 18 months.

The post-incident remediation program identified three control failures that, had any one of them been operating correctly, would have prevented or significantly limited the exposure. The first was the absence of automated classification validation at the pipeline output stage: a control that inspected the classification of each record being written to the analytics bucket and blocked the write if the record contained Restricted-classified content would have caught the misconfigured records before they reached the incorrectly-configured bucket. The second was the use of cloud-managed keys for the analytics bucket. If the bucket had used BYOK with a key whose access policy restricted decryption to EU-registered identities, cross-region access by Singapore and U.S. workloads would have failed at the key-unwrap stage regardless of the IAM policy configuration. The third was the absence of data residency enforcement through microsegmentation: workloads in EU, Singapore, and U.S. cloud regions had unrestricted network access to the analytics bucket, which should have been restricted to only the pipeline processes running in the intended jurisdiction.

The remediation program implemented all three controls, along with automated data lineage tracking for every pipeline that touched Restricted classification data, behavioral anomaly detection for cross-region data access events, and a unified DLP and CASB policy framework that enforced classification-based access controls consistently across all four cloud providers. The organization also migrated all Restricted-classified data to BYOK key management, a project that took 11 months and required significant coordination with the cloud provider's key management teams. The outcome was a data security posture that addressed not just the specific failure mode that caused the incident but the systematic absence of data-centric controls that made the failure possible.

10.14 Team Conversation Guide

Run these three discussions in sequence over two to four weeks. The first surfaces the issue, the second works the trade-offs, the third commits the team to action.

Discussion 1. Classification Reality Check

Run this discussion before beginning any data-centric security initiative. Its purpose is to establish an honest baseline by surfacing where classification policy exists in documentation but not in operation. Many teams have classification frameworks that no one enforces and labels that no system reads. This conversation identifies the gap.

Discussion questions to ask the team:

- For our three most sensitive data categories, who applied the classification label on the most recent batch of records that entered those categories? Was it an automated system, a human, or does nobody know?
- If I ask our DLP system what it would do with a Restricted-labeled file being copied to a personal USB drive, what is the actual answer? Has anyone tested this recently?
- Do our cloud storage buckets and database schemas carry classification metadata that any other system reads and acts on, or are classification labels stored only in a separate spreadsheet that nobody queries programmatically?
- What would it take to classify every new data object entering the environment automatically at the point of creation? Which systems would need to be modified, and which classification would be applied to objects that automated tools cannot confidently classify?

What to listen for: Listen for whether the team distinguishes between classification that exists as policy and classification that is enforced by technical controls. Teams that conflate documentation with enforcement have a gap that will surface in the next incident. Also, listen for whether anyone can name the system that enforces classification: a specific product, a specific configuration, a specific pipeline step. If the enforcing system cannot be named, enforcement is not happening.

Follow-up actions: Assign one person to perform a live demonstration of classification enforcement within two weeks: take a known Restricted file, attempt to move it through three different channels that should be blocked, and report what actually happened. The gap between policy and reality will be immediately visible, and the team will have a concrete starting point for remediation rather than a theoretical discussion.

Discussion 2. Key Management Architecture Decisions

This discussion is for teams that have identified Restricted or Highly Restricted data stored in cloud environments with cloud-managed keys. The purpose is to work through the BYOK versus HYOK trade-off concretely, using actual data categories and regulatory requirements, rather than in the abstract. The goal is a decision, not a research project.

Discussion questions to ask the team:

- For each Restricted data category we store in cloud environments, what is the specific harm scenario if a legal process compelled the cloud provider to decrypt and disclose that data? Is that scenario in our threat model or outside it?
- Do any of our regulatory obligations, GDPR, HIPAA, ITAR, CCPA, or applicable contractual commitments require that encryption keys be held outside the cloud provider's control? Has legal reviewed this, or are we relying on a general interpretation?
- What is the operational overhead of BYOK for our highest-sensitivity data categories? Does it introduce latency to data access patterns in real-time operational systems, or is it limited to batch analytics workloads where latency is acceptable?
- If we move to BYOK for Restricted data, what happens to existing data encrypted under cloud-managed keys? Is there a migration path that does not require decrypting and re-encrypting the entire dataset in a single operation?

What to listen for: Listen for whether the team can articulate the actual regulatory requirement rather than a general preference for stronger controls. BYOK is the right answer for some regulatory environments and unnecessary overhead for others. A team that cannot connect the key management decision to a specific regulatory obligation may be gold-plating controls in one area while leaving gaps in another. Also, listen for whether the migration question produces a concrete answer or a deferral. If the team does not know how to migrate from cloud-managed to BYOK, that is a prerequisite skill that must be built before the architectural decision becomes actionable.

Follow-up actions: Establish a working group with representation from legal, data engineering, and security to produce a key management decision matrix within thirty days: one row per Restricted data category, with columns for applicable regulatory frameworks, the current key management model, the required key management model, and a migration complexity estimate. The matrix converts the architectural discussion into a project plan.

Discussion 3. Exfiltration Detection and Response Readiness

This discussion tests the team's ability to detect and respond to the Marcus scenario before it completes, not ninety minutes after. Run it with the SOC or security operations team present, because detection and response are primarily their responsibility. The question is whether the current tooling and process would have caught the transfer within the first ten minutes.

Discussion questions to ask the team:

- If a former contractor's account performed a bulk download of Restricted data to a personal cloud drive starting right now, at what point in the process would our first alert fire? What would trigger it, and who would receive it?
- Do we have behavioral baselines per identity for data access volume, destination, and time-of-day?

If a specific identity downloaded ten times their normal volume at 2 AM, would that combination of signals trigger an alert, or would it generate a log entry that nobody reviews until morning?
- What is our response playbook for a suspected data exfiltration in progress? Who has the authority to suspend an account or revoke a derived key immediately, without waiting for a committee approval? Has that authority been tested in a tabletop exercise?
- How does our detection coverage differ between on-premises data stores and cloud-hosted data stores? Are there cloud platforms where we have less visibility into data access events than we have on-premises?

What to listen for: the detection latency answer. If the team cannot state a specific detection latency for the scenario described, the detection capability is theoretical rather than validated. The target detection window for a bulk exfiltration event should be measured in minutes, not hours. Also, listen for whether the response playbook is written and tested or exists only as an understanding among team members. A playbook that requires committee approval before an account can be suspended is too slow for an active exfiltration event, and the team should know this before the event, not during it.

Follow-up actions: Run a tabletop exercise within sixty days that simulates the Marcus scenario in real time: a former contractor account begins a bulk download of Restricted data at 11 PM. How long does it take for the first alert to fire, for a human to acknowledge it, and for the account to be suspended? The gap between the target and actual response times is the priority for remediation. Make the exercise results visible to leadership so that detection and response investment is connected to a concrete risk scenario.

10.15 Takeaway

The twenty-eight minutes in Marcus's post-mortem represent a pattern that repeats across every industry and every architecture that builds security around the network rather than around the data. Network controls verified that the traffic was going to a sanctioned endpoint. They were correct. The data still left the organization's control and landed on a personal cloud drive, including classified research.

Data-centric security is the discipline that closes this gap by making the protection travel with the data rather than with the network perimeter. Classification at ingestion gives every data object a persistent identity that controls can evaluate. Encryption under organization-controlled keys means the data is unreadable to any party the policy engine has not authorized, regardless of where the file physically resides. Data flow mapping reveals the paths data actually takes rather than the paths the architecture diagram shows. Unified DLP and CASB policies enforce consistent classification-based restrictions across all transfer boundaries. Behavioral exfiltration detection catches authorized accounts behaving in unauthorized patterns before the transfer completes, rather than generating a log entry that surfaces ninety minutes later.

The pilot-to-production pathway for data-centric security follows the same principle as every other zero-trust capability: start with the highest-sensitivity data category, achieve complete coverage there, measure the outcome, and expand. A team that has never built an automated classification should not attempt to classify every data object in the environment simultaneously. A team that has never implemented BYOK should not migrate all cloud storage to organization-controlled keys in a single sprint. Each step in the sequence produces real protection for the data it covers. Each step also builds the operational muscle for the next one. The organization that completes this sequence has built something genuinely different from a perimeter-secured environment: not a stronger wall around the data, but a set of controls that travel with the data wherever it goes, enforcing the same policy at the research vault, the cloud collaboration platform, the partner portal, and the personal cloud drive that should have blocked the transfer Marcus spent that post-mortem reconstructing.

11 Automation, Orchestration, and DevSecOps in Zero Trust

Marcus Reyes had been watching the deployment counter tick up all afternoon. His team had spent six weeks preparing the Meridian Logistics Group's payment routing service for a major release: new authentication flows, updated API scopes, revised segmentation policies for the service mesh, and two dependency upgrades that had passed code review, SAST scanning, and a full staging run. At 4:47 on a Friday, the pipeline merged. By 5:02, a SOAR playbook had quarantined three production workloads in the payment tier.

The alert came from the continuous verification layer that Marcus's team had spent months building. A vulnerability scanner watching the runtime environment had matched a CVE in one of the upgraded libraries against a threat intelligence feed updated ninety minutes earlier. The CVSS score was 9.1. The SOAR playbook did exactly what it was configured to do: it isolated the affected workloads, revoked the service account tokens associated with the payment processing microservices, and opened a P1 incident ticket. No data was left in the environment. No lateral movement occurred. But 3,000 payment transactions were queued in a dead-letter queue, a customer-facing processing outage was building, and Marcus was staring at a quarantine he had not personally approved.

In the two hours that followed, Marcus learned something that no architecture diagram had fully communicated to him. The zero-trust program he had been building was not really a security program yet. It was a manual approximation of one. Policies were versioned but updated through console clicks after peer review, not through automated pipeline gates. The SOAR playbook had fired correctly, but the remediation path, patching the library, revalidating the container image, and redeploying to production, required four separate approvals routed through a ticketing system that did not know it was Friday evening. The quarantine that had protected the environment was also holding it hostage. Marcus pulled up the incident timeline on his laptop. He saw the organizational gap clearly: the architecture could detect and contain at machine speed, but it could not recover at machine speed, because recovery still required humans to navigate a process designed for a slower world.

That evening, Marcus drafted a one-page memo to himself. The memo had one thesis: the program will not scale through human approvals. At 40 deployments a week and growing, with five security-relevant policy changes per deployment, and an average container image refresh cycle of nine days, the math of manual review produced a permanent backlog that would grow faster than the team could clear it. The question was not whether to automate. The question was how to automate responsibly: how to write the policy engine as code, run it through a pipeline with the same rigor as the application code it governed, and give the SOAR layer enough context to recover, not just contain. This chapter is Marcus's answer to that memo.

11.1 Why It Matters

Zero-trust architecture is, at its core, a continuous evaluation system. Every access request is verified, every workload posture is assessed, and every policy is enforced against the current context rather than historical trust. That proposition is mathematically sustainable only if the underlying machinery, policy definitions, enforcement configurations, identity lifecycle operations, and compliance validations can run without a commensurate amount of human labor. A program that requires a person to review and approve each change becomes a bottleneck at scale, and a bottleneck in a security program is not just an operational inconvenience. It is a structural vulnerability. Changes that should take minutes take days. Patches that should be deployed in hours wait in approval queues. Drift between the intended and actual configurations accumulates during the gaps between review cycles.

The workflow-level impact of this gap is visible in incident timelines. When organizations measure the mean time to remediate vulnerabilities in production workloads, the delay is rarely due to detection. Modern scanning tools find CVEs quickly. The delay is in the handoff from detection to remediation: the human approvals, the manual deployment steps, and the validation procedures that were never automated because each one seemed too specific to template. DevSecOps practices, infrastructure-as-code, policy-as-code, and SOAR integration exist precisely to compress that handoff from hours to minutes. They shift security from a gate at the end of the delivery process to a continuous thread running through it, and they give the zero trust policy plane the automation substrate it needs to function at the speed of the environment it governs.

This is also where the pilot-to-production pathway for zero trust programs most commonly stalls. A team can build a technically sound architecture in a controlled environment, verify that it works for 100 identities and 20 workloads, and then watch it collapse under operational load as that population grows by an order of magnitude. The collapse is not in the technology. It is in the operating model. Organizations that sustain zero trust at scale are, without exception, organizations that have operationalized the policy plane as code, embedded security into their delivery pipelines, and connected their detection capabilities to automated response. That is what this chapter addresses: the engineering practices and organizational patterns that make a zero-trust program mission-aligned not just in its design but in its day-to-day operations.

11.2 Policy as Code: Treating Security Rules Like Software

A zero-trust policy is a set of rules that governs which identities may access which resources under which conditions. In most organizations, those rules live in product consoles: checkboxes in an identity provider's access policy editor, network rules in a cloud security group, segmentation configurations in a firewall management interface. These representations share a critical property: they are not reviewable with the same tools and practices used for software. A pull request for application code goes through peer review, automated testing, staged deployment, and version history. A change to an IAM policy often goes through a verbal approval and a console click, with the only audit trail being a timestamped log entry that records the outcome but not the reasoning.

Policy-as-code reverses this. Security rules are expressed in a language the toolchain can read, stored in a version control system, reviewed through the same pull request workflow as application code, tested against a suite of unit tests that verify expected allow and deny decisions, and deployed through an automated pipeline that applies the policy to the target enforcement point. The result is not just operational convenience. It is a structural change to the organization's security posture. Every policy has a history. Every change has a reviewer. Every deployment is logged with the commit hash that produced it. When an auditor asks why a specific identity has a specific access, the answer is a link to a commit, a reviewer, and an automated test that verified the policy before it went live.

Policy-as-Code Pipeline from Repository to Enforcement Point

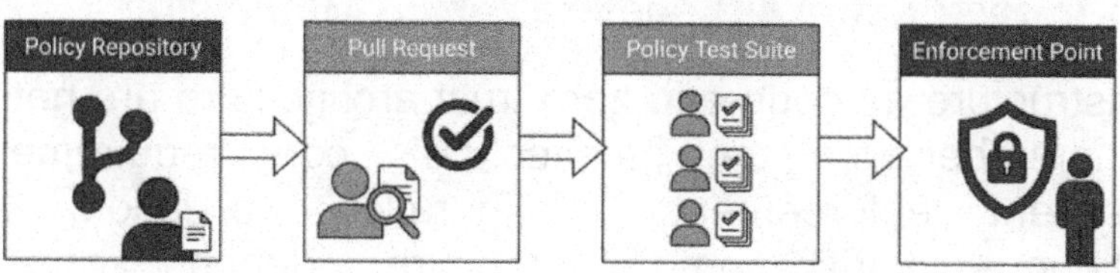

Policy-as-Code Pipeline from Repository to Enforcement Point

Open Policy Agent and its policy language Rego have become the most widely adopted implementation of this pattern in cloud-native environments. OPA acts as a decoupled policy engine that receives queries from enforcement points, evaluates them against loaded policy bundles, and returns structured decisions. The policy bundles themselves are text files checked into source control. A change to a segmentation rule is a change to a Rego file, submitted as a pull request, reviewed by a security engineer, tested by a policy test suite that asserts specific allow and deny outcomes for representative input scenarios, and then bundled and pushed to the OPA instances running at each enforcement point. The enforcement point never stores the policy locally in a form that can drift from the version in the repository. It pulls from a bundle server that always serves the latest approved version.

Policy-as-code also makes error detection tractable. When a policy change causes an unintended access denial, the blast radius investigation starts with a diff: what changed between the last known-good policy bundle and the current one? That question, which in a console-managed environment might require reconstructing intent from audit logs, is a one-command operation against the version history. The same version history that creates accountability also creates the rollback path. A policy that produces unintended behavior can be reverted in the time it takes to push a commit, and the revert carries the same audit trail as the original change.

11.3 Infrastructure as Code in a Zero Trust Pipeline

Infrastructure-as-code and zero trust architecture are natural partners because the architecture's core requirements, consistent enforcement, configuration validation, and continuous verification of posture, depend on the infrastructure being in a known, verifiable state. A firewall rule in Terraform can be diffed, tested, reviewed, and rolled back. A firewall rule configured through a cloud console during an incident response at 2 AM may or may not be documented, may or may not match the intended policy, and will not be detected as drift until the next manual audit.

In a zero trust pipeline, infrastructure-as-code governs the full stack of enforcement primitives: the network segmentation boundaries that define microsegment perimeters, the security group rules that implement those boundaries at the cloud layer, the Kubernetes NetworkPolicy objects that control pod-to-pod traffic within clusters, the identity provider configurations that define trust relationships between services, and the certificate authority configurations that govern mTLS identity for workloads. Each of these is a zero-trust enforcement primitive. Each one that lives outside of code is a primitive that can drift, be misconfigured, or be inconsistently applied across environments.

Infrastructure as Code Coverage Map for Zero Trust Primitives

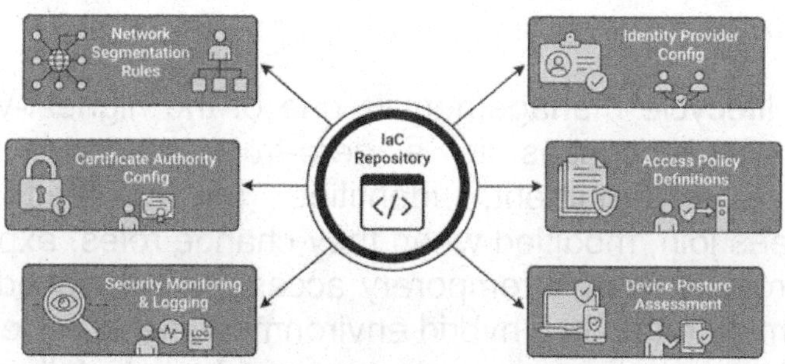

Infrastructure as Code Coverage Map for Zero Trust Primitives provides a comprehensive overview of security configurations managed through code.

The pipeline architecture for zero trust infrastructure follows the same structure as any IaC pipeline with one critical addition: zero trust policy validation is a mandatory gate, not an optional check. Before any infrastructure change is applied to a staging environment, the pipeline runs a policy-as-code evaluation to verify that the proposed change does not violate zero-trust design principles. A Terraform plan that opens a security group rule to 0.0.0.0/0 fails the gate before it reaches the environment. A change that removes mTLS requirements from a service mesh configuration fails the gate before it is applied. These gates are blocking conditions that prevent deployment, not advisory warnings that developers can dismiss.

Platform engineering teams bear primary responsibility for building and maintaining these gates. Left to individual development teams, zero-trust infrastructure validation tends to be inconsistently applied because the teams closest to the application code are also closest to the delivery deadline and farthest from the blast radius of a misconfigured enforcement primitive. Platform engineers who own the pipeline templates can embed the gates once, in the shared pipeline configuration, where they apply uniformly to every team that uses the platform. This is the organizational pattern that makes zero-trust infrastructure enforcement sustainable: it shifts the burden from individual teams making individual decisions to a platform that makes the right decision the default.

11.4 Automated Provisioning of Identity and Access

Identity lifecycle management is one of the highest-volume operational processes in a zero-trust program. In an enterprise environment, identities are created when employees join, modified when they change roles, expanded when projects require temporary access, and revoked when employment ends. In a hybrid environment, service identities, machine accounts, and application credentials follow even faster lifecycle cycles, often tied to deployment events rather than HR processes. Manual provisioning at this volume is not merely slow; it is also inefficient. It is consistently wrong. Access accumulates rather than rotating with role changes because the provisioning system only runs when someone remembers to run it.

Automated identity provisioning connects the authoritative source of truth for each identity type, HR systems for employees, service registries for workloads, device management platforms for endpoints, to the identity providers and policy engines that enforce access decisions. When an employee moves from the finance team to the engineering team, the HRMS sends a role change event that triggers an automated workflow: the old access profile is revoked, the new access profile is applied, the change is logged with the business justification from the role change record, and an access review ticket is generated for the new manager to confirm the provisioned access is correct. No human manually clicks through an IAM console. The policy is applied from the authoritative source, at the speed of the source event, with a complete audit trail.

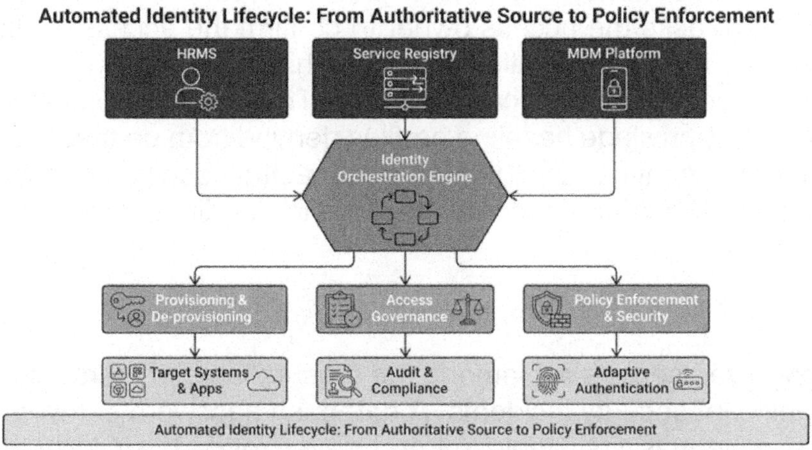

For service identities, the provisioning model differs, but the principle remains the same. Workloads should not be provisioned with static, long-lived credentials. They should be provisioned with short-lived identity tokens issued by a workload identity authority, such as SPIFFE and SPIRE, that ties the credential's validity to the workload's current execution context. A container deployed from a verified image in a specific namespace with a specific service account receives an identity token that expires within hours, is automatically rotated by the identity authority, and is cryptographically bound to the execution context that produced it. When the container is stopped, the token expires. There are no orphaned credentials to clean up. The credential lifecycle is the deployment lifecycle.

The access provisioning problem for human identities in zero-trust programs is not usually the initial grant. Most organizations have reasonable joiner processes. The failure pattern is that joiners receive access appropriate to their initial role and then receive additional access grants for projects, temporary assignments, and one-off requests, which are never revoked when the project ends. Automated provisioning addresses this by granting time-bounded access and implementing automated access review workflows. When access is granted for a project, the grant record includes a review date. When the review date arrives, an automated workflow asks the access owner to confirm the access is still needed. If no confirmation arrives within the review window, the access is automatically revoked. This pattern, combined with least-privilege baseline profiles derived from peer access patterns, continuously tightens the actual access surface without requiring security teams to audit entitlements manually.

11.5 Secret Rotation and Certificate Automation

Long-lived secrets are among the most reliable precursors to significant security incidents. A database password set when a service was first deployed and never rotated, an API key issued for an integration that persists for years because rotating it requires coordinating a deployment window, a TLS certificate renewed manually twelve months after the previous renewal: each of these is a credential that has accumulated exposure time, may have been copied across systems for operational convenience, and represents a known-valid path into the environment that no continuous verification signal can revoke until the credential is changed. In a zero-trust program, automated rotation is not a best practice; it is optional at discretion. It is a structural requirement.

HashiCorp Vault and AWS Secrets Manager represent the dominant approaches to secrets lifecycle automation in enterprise environments. Both operate on the principle of dynamic secrets: rather than storing a credential and rotating it on a schedule, the secrets manager generates a fresh credential on demand, issues it with a configured time-to-live, and either revokes it automatically when the TTL expires or allows the requesting application to renew within a lease window. A microservice that needs database credentials requests them from Vault at startup, receives credentials valid for 4 hours, and renews them or receives new ones before they expire. The database never holds a static credential bound to that service. If the service is compromised, the attacker's window is bounded by the credential TTL, not by the manual rotation schedule.

Dynamic Secret Issuance and Automatic Rotation Lifecycle

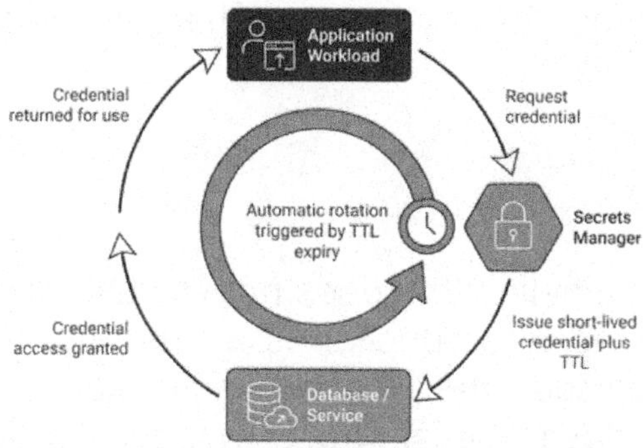

This process ensures secure access by issuing temporary credentials and automatically rotating them to minimize exposure.

Certificate automation follows a similar pattern. Manual TLS certificate management produces a predictable failure mode: certificates expire in production because the renewal process was not completed in time, causing avoidable, operationally embarrassing outages. The ACME protocol, implemented in tools like Certbot and cert-manager for Kubernetes environments, automates the entire certificate lifecycle: domain validation, certificate issuance, renewal before expiry, and rollout to the service without manual intervention. In a zero-trust environment where mutual TLS is required for service-to-service authentication, certificate automation is not optional. If mTLS certificates can expire without automatic renewal, the mTLS requirement becomes intermittently enforced, which is operationally indistinguishable from not enforced at all.

For internal PKI, where certificates authenticate workload identities rather than web server identities, the automation requirements are even more demanding. Workload identity certificates may have TTLs measured in hours rather than months, and an issuance system that requires manual intervention cannot serve that lifecycle. SPIFFE-compliant identity systems issue and rotate workload identity certificates automatically, triggered by the workload's runtime context rather than a calendar schedule. The certificate lifetime is short enough that even if a certificate is somehow extracted from a running workload, its exposure window is measured in hours, not the months or years that characterize compromised static credentials.

11.6 CI/CD Security Gates: Building Security Into the Delivery Pipeline

A deployment pipeline that does not validate security posture before deploying to production is a trust gap. Every artifact that passes through the pipeline unchecked is one that the enforcement layer must trust without evidence of its trustworthiness. DevSecOps practices close this gap by making security validation a continuous part of the delivery process rather than a periodic manual review. The goal is not to slow down delivery. It is to produce evidence of the security posture at the same cadence as delivery, so that continuous runtime verification has a trusted baseline to verify against.

Static Application Security Testing runs against source code before compilation, identifying code-level vulnerabilities such as hard-coded credentials, injection flaws, insecure API usage patterns, and dependency calls targeting known-vulnerable versions. SAST tools are fast enough to run on every commit without meaningfully extending the pipeline duration, and they catch the category of vulnerability that is cheapest to fix because it has not yet become a deployed artifact. Software Composition Analysis complements SAST by examining the application's dependency graph, matching every imported library against vulnerability databases, and producing a Software Bill of Materials that records the artifact's full dependency provenance. The SBOM is not just a security artifact. It is an operational one: when a new CVE is published against a library, the team can immediately query which deployed services contain that library, prioritizing remediation by actual exposure rather than by guesswork.

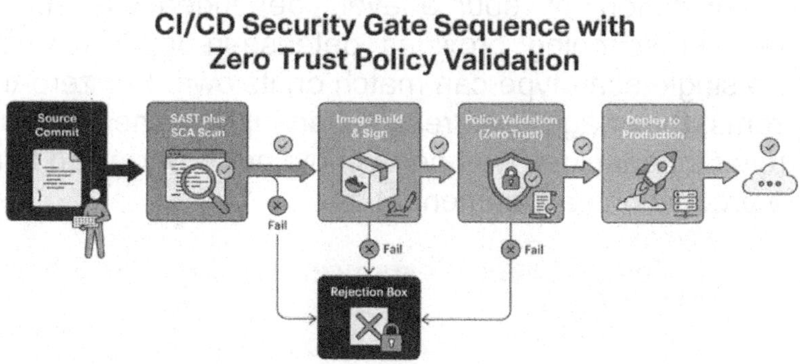

CI/CD Security Gate Sequence with Zero Trust Policy Validation

Infrastructure-as-code scanning applies the same analysis to the infrastructure definitions that will deploy the application. Tools in the Checkov, tfsec, and Terrascan families parse Terraform, CloudFormation, Kubernetes manifests, and Helm charts, evaluating them against a library of security rules that map to zero trust requirements: no public IP exposure without explicit justification, no privileged container execution, no host network mode, no secrets stored in environment variables, and mTLS enabled for all service-to-service routes. These checks run in the same pipeline as the application code, fail the build on the same severity thresholds, and produce the same audit evidence format. The result is that infrastructure configuration and application code are held to the same review and validation standard, closing the gap that exists when infrastructure is treated as a separate operational concern.

Container image scanning examines the final built artifact, the container image, against vulnerability databases at build time. This matters because SAST and SCA check the application code and its declared dependencies, but the base image may contain operating system packages with their own vulnerabilities. The build process may introduce artifacts not visible in source analysis. A base image running an outdated version of a system library is a vulnerability that source-level scanning cannot detect. Image scanning closes this gap, and the combination of source-level, dependency-level, and image-level scanning provides defense-in-depth coverage that no single scan type can match on its own. For zero-trust programs, the image scan result is an input to the admission controller that decides whether a container is permitted to run in the production environment at all.

11.7 SOAR for Incident Response: Closing the Loop at Machine Speed

Security orchestration, automation, and response platforms connect the detection layer of a zero-trust architecture to the response layer, reducing the time between alert and containment from hours to seconds. The value proposition is not simply speed, though speed matters enormously when an adversary has seconds to minutes to move laterally through a compromised environment. The deeper value is consistency. A human incident responder working at 5 PM on a Friday, under time pressure and with incomplete context, will make different decisions than the same responder would on a Tuesday morning with full documentation. A SOAR playbook makes the same decision every time, and its decision history is fully auditable.

The architecture of an effective SOAR integration for zero trust follows a three-phase pattern. The first phase is enrichment: when a continuous verification alert fires, the SOAR platform automatically gathers context from every relevant data source before any response action is taken. The identity of the affected workload or user is queried against the identity directory. The network connections of the affected workload are retrieved from the service mesh telemetry. The recent access history of the associated identity is pulled from the policy decision logs. The CVE or behavioral indicator that triggered the alert is evaluated against the threat intelligence feed to determine its current exploitation status. This enrichment takes seconds and produces a structured incident record with enough context to support an automated decision.

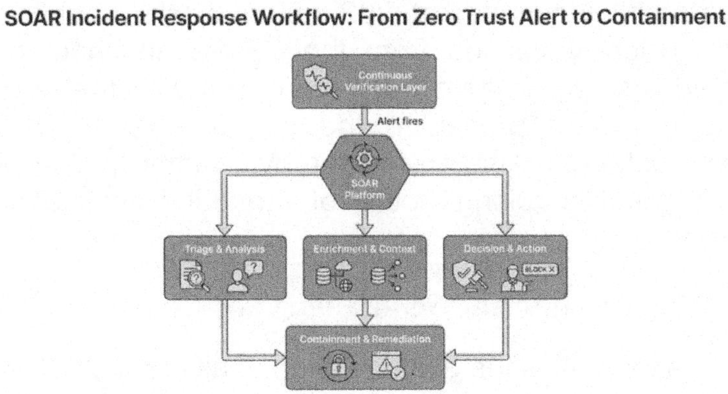

SOAR Incident Response Workflow: From Zero Trust Alert to Containment

The second phase is automated response, applied to the subset of incidents where the enriched context is sufficient to determine the appropriate action with high confidence. A workload that has contacted a known command-and-control domain is quarantined automatically: its outbound network routes are blocked, its service account tokens are revoked, and its pod is scheduled for replacement with a clean image from the verified artifact registry. A user account exhibiting impossible travel behavior is automatically stepped up to a phishing-resistant authentication challenge rather than immediately revoked, because the behavioral indicator is consistent with both compromise and a legitimate VPN connection. The playbook's decision logic reflects the organization's risk tolerance and the reliability of each signal type. That logic is reviewed and approved through the same change management process as application code.

The third phase is evidence preservation and orchestration of recovery. Automated containment stops the immediate threat, but the incident remains open. The SOAR platform continues working after containment: it captures forensic snapshots of the affected workloads, archives the relevant telemetry to an immutable storage location, generates the initial incident report from the enrichment data, and begins the recovery workflow. For the Friday afternoon scenario that opened this chapter, a mature SOAR integration would have quarantined the vulnerable workloads at 5:02, automatically identified the offending library version from the SBOM attached to the deployed artifact, triggered a pipeline run with the patched dependency, and presented the on-call engineer with a recovery action pre-staged for approval rather than a blank incident ticket requiring hours of manual investigation to diagnose.

11.8 Drift Detection and Self-Healing Infrastructure

Configuration drift is the accumulation of differences between the intended state of an infrastructure component, as expressed in code, and its actual state in the running environment. In a zero-trust architecture, drift is not merely a technical inconsistency. It is a trust gap. A security group that has been manually modified to allow a port range not specified in the Terraform configuration is a gap in the microsegmentation boundary that no policy engine knows about, because the policy engine reasons about the intended configuration, not the actual one. Drift detection closes this gap by continuously comparing the actual state of enforcement primitives against their declared state in code and alerting when they diverge.

Self-healing infrastructure takes the next step: when drift is detected, an automated system restores the intended state without human intervention. This is appropriate for the class of drift that represents either accidental modification or configuration injection by an adversary attempting to create a persistent access path. A security group rule that should not exist is deleted. A modified Kubernetes NetworkPolicy that allows broader traffic is reverted to its declared state. The self-healing action is logged with the detected drift, the declared intended state, and the timestamp of the correction, providing a complete record for both operational and compliance purposes. The organizational effect is significant: security teams spend less time hunting for drift and more time improving the intended configurations.

Drift Detection and Automated Remediation Cycle

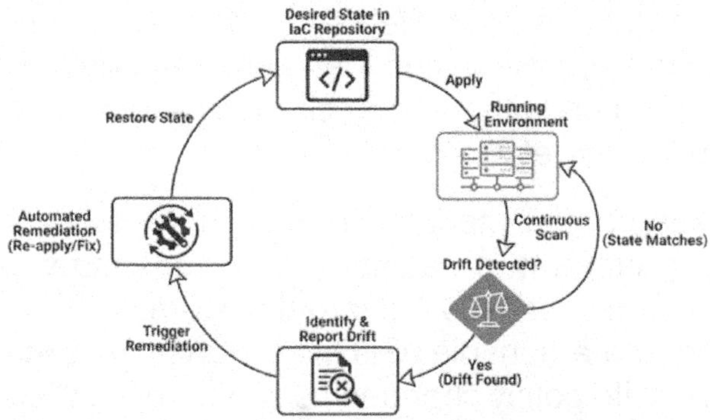

Drift Detection and Automated Remediation Cycle ensures infrastructure consistency by continuously comparing running environments to the desired state in IaC and automatically correcting any deviations.

The boundary between self-healing and autonomous modification requires careful design. Not all drift should be automatically corrected. Some drift represents legitimate emergency changes, such as a security group rule added during an active incident response that has not yet been formalized in code. Self-healing logic should distinguish between unauthorized drift, which is configuration changes not authorized through the change management process, and emergency changes that are in process. One pattern is to use change window tags: infrastructure changes made with an active incident ticket receive a temporary exception label that suspends self-healing for a defined period, during which the team is expected to formalize the change in code or revert it deliberately. When the exception window expires, self-healing resumes. This pattern maintains the integrity of the declared state while accommodating the operational reality that emergencies happen outside of normal change windows.

11.9 GitOps for Zero Trust Policy Management

GitOps is an operational model in which a Git repository serves as the single source of truth for the system's desired state, and an automated operator continuously reconciles the system's actual state with that desired state. In the context of zero-trust policy management, GitOps means that every access policy, every segmentation rule, and every enforcement configuration has exactly one authoritative representation: a file in a Git repository. Any change to the policy is made in that file, submitted as a pull request, reviewed and approved, and then automatically applied by the GitOps operator without the change author having direct access to the target system.

The access control model that GitOps introduces is itself a zero-trust pattern. In a traditional policy management model, engineers with access to the target system can change policy by logging into a console. GitOps removes that access path. Engineers write policy changes as code. The GitOps operator, a system account with tightly scoped permissions, applies the changes. No engineer needs direct write access to the policy enforcement system. The blast radius of a compromised engineer's credentials is limited to proposing policy changes, not to applying them. The additional control comes from the mandatory code-review step, which ensures that at least one other person reviews every policy change before it takes effect.

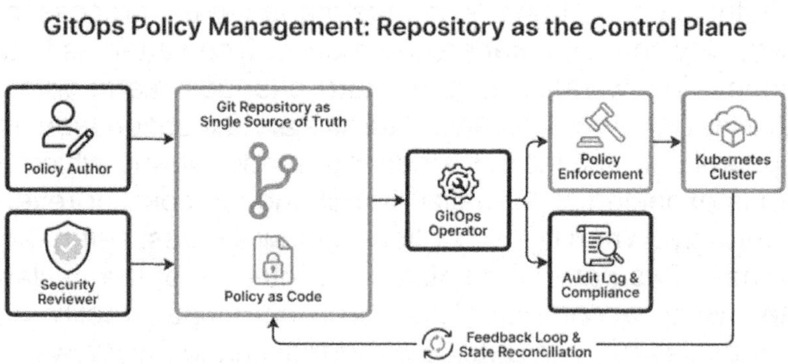

GitOps Policy Management: Repository as the Control Plane

The operational discipline required to make GitOps work with a zero-trust policy is the hardest part of the implementation. Teams accustomed to clicking through a console to make a quick policy change must learn to make that same change through a pull request and wait for the GitOps operator to apply it. The latency from commit to application is typically seconds to minutes for automated operators, but the process takes longer than a console click when a review step is required. Organizations that succeed with GitOps for policy management invest in tooling that reduces friction: pre-populated pull request templates that capture the business justification, automated policy testing that runs in seconds so reviewers have evidence before they approve, and CODEOWNERS configurations that automatically route policy changes to the appropriate security reviewer. The process is slower per change than console access, but the aggregate effect is a more reliable, more auditable, and more secure policy environment.

11.10 The Role of Platform Engineering Teams in Zero Trust Automation

Platform engineering teams, sometimes called internal developer platforms or platform enablement teams, sit at the center of the automation patterns described in this chapter. They build and maintain the pipelines, the shared policy templates, the SOAR playbook library, the GitOps operators, and the drift detection tooling that individual development teams use without needing to understand the underlying implementation. The relationship is analogous to that between a cloud provider and its customers: platform engineers build the guardrails and golden paths, and development teams walk those paths without needing to reason about every security constraint enforced along the way.

For zero-trust programs, this organizational pattern addresses a fundamental scaling problem. A security team of fifteen engineers cannot review every deployment, every policy change, and every configuration update across a development organization of two hundred engineers pushing forty deployments a day. But a platform team of fifteen engineers can build pipeline templates that uniformly embed security gates across all two hundred engineers' workflows, instrument telemetry to surface anomalies to the security team for review, and maintain SOAR playbooks that automatically contain incidents without requiring a security engineer on every call. The security team becomes a force multiplier through the platform rather than a bottleneck at every gate.

The organizational challenge is positioning. Platform engineering teams are typically funded and evaluated based on developer productivity metrics: pipeline speed, deployment frequency, and mean time to recovery from build failures. Zero-trust requirements add friction to pipelines in the form of security gates, and friction to productivity metrics appears as failure from a platform team's perspective. Organizations that successfully integrate zero trust requirements into platform engineering do so by changing the measurement framework: security gates are not friction; they are features. Pipeline completion rate, including security gates, is the target metric, not pipeline speed without them. When a security gate correctly blocks a deployment with a critical CVE, that is the platform working as designed. Responsible by design means embedding this measurement philosophy into how the platform team's performance is evaluated.

11.11 Geeks Only

This section is for the engineers who will implement the practices described above. It covers the technical specifics of Terraform and Pulumi modules for zero trust primitives, OPA policy bundle structure and Rego authoring patterns, CI pipeline gating configurations, runtime admission controllers, image signing and SBOM verification, SOAR playbook engineering, blast-radius-aware automated containment logic, and how to reason about change windows in a continuous deployment model.

Terraform and Pulumi Modules for Zero Trust Primitives

The most effective pattern for IaC-governed zero-trust infrastructure is the module abstraction: rather than writing raw resource definitions in every deployment configuration, teams define reusable modules that encapsulate the zero-trust requirements for a given infrastructure primitive. A zero-trust-network-segment module for AWS, for example, takes inputs for allowed ingress ports, source CIDR blocks, and egress destinations, and produces a VPC with the appropriate security groups, flow log configuration, and CloudWatch metrics subscription. The module's defaults enforce zero trust: no default ingress, no default egress, flow logging is always on, and security group rule descriptions are required. The consuming team cannot forget to enable flow logging because the module does not offer an option to turn it off.

Pulumi adds a dimension that Terraform does not natively offer: because Pulumi programs are written in a general-purpose language, policy validations can be expressed as code that runs as part of the stack evaluation, not as a separate tool invoked after the plan is produced. A Pulumi CrossGuard policy that prohibits public-facing resources without an explicit override tag runs at stack evaluation time, before any API call is made, and produces a structured policy violation report that identifies the non-compliant resource by name and the policy rule it violated. The report is the same format every time, produced by the same evaluation logic, and available as a structured artifact in the pipeline run log.

The module design principle most important for zero trust is immutability as the default. Zero-trust infrastructure should be designed to be replaced rather than modified. Rather than updating a running security group rule, the pipeline should replace the security group with one that has the new rule and drain traffic gracefully. This constraint, enforced through module design, eliminates the class of configuration drift that occurs when modifications accumulate on top of a resource without being reflected in the code that created it.

OPA and Rego Policy Bundles

Open Policy Agent bundles are collections of Rego policy files and data files packaged as a tarball with a manifest. The bundle server, which can range from a signed artifact in a container registry to a dedicated OPA bundle service, serves these bundles to OPA instances, which download and cache them according to a polling interval. When a bundle is updated, OPA instances detect the new version at the next poll interval and load the new policies without a restart. The practical implication is that policy changes propagate to all enforcement points within the polling interval, which can be configured as low as seconds for latency-sensitive environments.

Rego policy authoring for zero trust enforcement follows a default-deny structure. The base policy for any resource access decision returns false by default, and each allow rule expresses a specific, bounded condition under which access is permitted. The policy unit test framework, OPA's built-in test runner, verifies that specific inputs produce specific decisions. A policy that should allow a payment service identity to read from a payments database but deny reads from a user profile database should include explicit test cases for both the allowed and denied scenarios. The CI gate blocks policy changes that break existing test cases before they reach a bundle. The governance pattern for Rego policy bundles mirrors application code governance: each bundle has an OWNERS file that maps resource domains to reviewing teams, and changes to a domain's policies require review from the team with domain context to evaluate whether the change is correct.

OPA Policy Bundle Lifecycle: Author to Enforcement

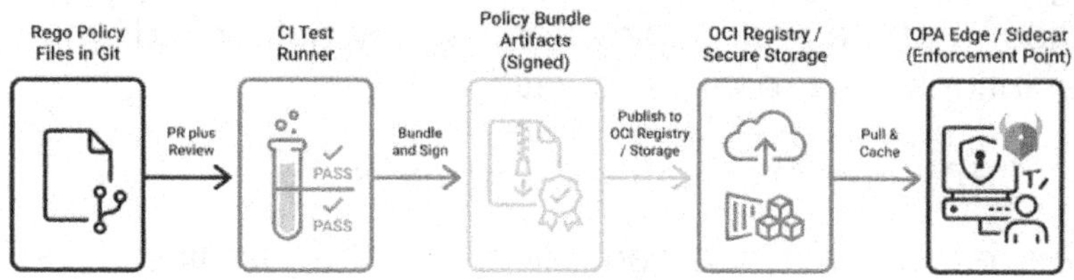

OPA Policy Bundle Lifecycle: Author to Enforcement provides a secure, automated workflow for managing and deploying policies from development to production.

CI Pipeline Gating Patterns

Effective CI security gating requires distinguishing between blocking gates, which fail the pipeline and prevent deployment, and advisory gates, which report findings but allow the pipeline to continue. The discipline is in calibrating which findings trigger which gate type. A critical CVSS vulnerability in a runtime dependency is a blocking gate: the deployment should not proceed until the vulnerability is resolved or an explicit override is granted through the change management process. A medium CVSS vulnerability in a test-only dependency is an advisory gate: the finding should be logged and tracked, but should not block a production deployment. Getting this calibration wrong in either direction creates problems: too many blocking gates lead to a culture of overrides and bypass approvals that defeats the purpose of the gate; too few blocking gates allow high-severity findings to reach production under the assumption they will be remediated later.

Semaphore files and attestation artifacts are the mechanism that carries gate results forward through the pipeline. When a SAST gate passes, the pipeline writes a signed attestation artifact that records the scanner version, the policy version, the commit hash, and the pass result. Subsequent stages verify the attestation's presence and signature before proceeding. This prevents a scenario where a gate is bypassed by re-running the pipeline from a later stage. Every stage depends on the attestation chain from all preceding stages, and the final deployment is only authorized when all attestations are present and valid.

Runtime Admission Controllers

Kubernetes admission controllers are the last line of policy enforcement before a workload is scheduled in the cluster. They intercept API requests to create or modify pods, evaluate them against policy, and either admit, reject, or mutate the request. In a zero trust context, the admission controller enforces the runtime security requirements that the CI pipeline has validated: the container image must carry a valid signature from the build pipeline, the image digest must match an entry in the approved artifact registry, the pod security context must match the security standards profile for the target namespace, and the pod must declare the service account it will use so that SPIFFE identity can be verified at startup.

OPA Gatekeeper and Kyverno are the dominant policy engines for admission control. Both work on the same principle: policy is expressed in a declarative language, loaded into the cluster as custom resources, and evaluated by the controller webhook for every admission request. The key operational property that distinguishes production-grade admission controller deployments from experimental ones is failure mode configuration. An admission controller configured in fail-open mode, which allows requests when the controller itself is unavailable, is not a security control. It is a control that disappears precisely when the cluster is under stress. Production zero trust admission controllers fail closed: when the webhook is unavailable, the request is denied. This means the admission controller must be highly available and its availability must be monitored as a security metric, not just an operational one.

Image Signing and SBOM Verification

Sigstore and its component tools, Cosign for signing and Rekor for transparency logging, provide the cryptographic infrastructure for software supply chain verification in container environments. The signing workflow runs at the end of the CI pipeline, after all security gates have passed: Cosign signs the container image digest using a key managed by the organization's KMS, and the signature is pushed to the container registry alongside the image. The Rekor transparency log records the signature event in an append-only log, enabling independent verification that the signature was recorded at a specific time. An admission controller configured to verify Cosign signatures before admitting a pod will reject any image that lacks a valid signature from the organization's signing key, whether the image was built outside the pipeline, modified after signing, or pulled from an unauthorized registry.

SBOM verification extends this to dependency provenance. During the CI pipeline, after the image is built and before it is signed, the pipeline generates an SBOM in the CycloneDX or SPDX format that enumerates every component in the image: the base layer packages, the application dependencies, and their version and provenance information. The SBOM is attached to the image as an OCI artifact and signed alongside the image. At runtime, the admission controller or a separate policy engine can query the SBOM to verify that all components meet the organization's approved component policy: no packages from untrusted registries, no components with critical vulnerabilities without an approved exception, no components with unknown provenance. When a new CVE is published, the SBOM inventory enables the operations team to query which running pods contain the affected component within seconds, rather than manually auditing deployments.

SOAR Playbook Engineering

Effective SOAR playbooks are written with the same engineering discipline as application code: they have explicit input schemas, well-defined decision branches, idempotent action implementations, and comprehensive test coverage using simulated alert data. The most common failure mode in SOAR deployments is that playbooks work correctly during tabletop exercises but fail during real incidents. The input data from live systems has fields that the playbook was not designed to handle: a null value where a string was expected, an alert format that changed when the source system was upgraded, or an API endpoint that returned a rate-limit error because the playbook was invoked three hundred times in two minutes during an automated scan.

The input normalization layer is the most important architectural element in a SOAR deployment. Before any playbook logic runs, incoming alerts should be normalized to a consistent schema that the playbooks can rely on: standardized field names, explicit null handling for optional fields, validated severity levels from a defined enum, and structured context blocks for each data source. This normalization layer serves as an interface contract between the detection and response layers. When the detection platform changes its alert format, only the normalization layer needs to be updated, not every playbook that consumes alert data.

Blast-Radius-Aware Automated Containment

Automated containment decisions should be calibrated to the blast radius of both the threat and the containment action. A playbook that quarantines a single-user workstation in response to a malware alert has a well-bounded blast radius for the containment action: only one user is affected. A playbook that quarantines a shared database server in response to an anomalous query pattern may affect hundreds of application services. The decision logic in the playbook should evaluate the downstream impact of the containment action before executing it, not after. This requires the SOAR platform to have a current dependency map of the environment: which services depend on the resource being considered for containment, and what the business impact would be of interrupting those dependencies.

One effective pattern is a containment-confidence threshold that escalates with blast radius. A single-endpoint quarantine with a high-confidence threat signal is executed automatically. A shared service quarantine with the same threat signal is staged for immediate human approval, with the enrichment data and the blast radius assessment pre-populated in the approval interface so that the approver can decide in seconds rather than minutes. The threshold matrix, mapping threat signal confidence and blast radius to automatic or escalated response, is itself a policy document that should be reviewed and approved through the change management process.

Blast-Radius Containment Decision Matrix

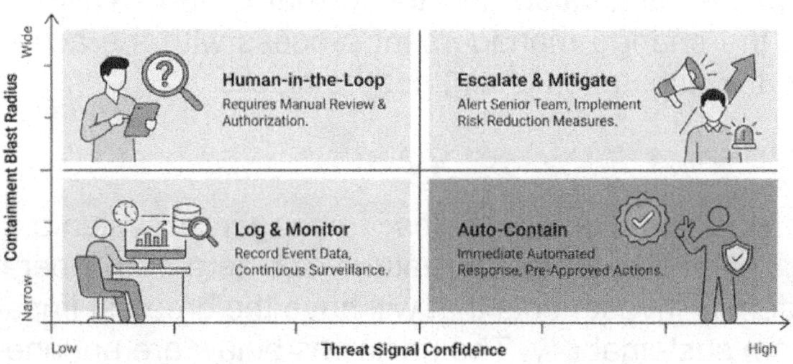

Blast-Radius Containment Decision Matrix guides automated versus human-led response based on signal reliability and potential impact.

Change Windows in a Continuous Deployment World

The traditional change window, a defined period during which infrastructure changes are permitted and outside of which they are frozen, is an artifact of a release management model that assumed changes were infrequent and each one carried significant risk. In a continuous deployment environment, the release model is inverted: changes are frequent, each one small and validated by an automated pipeline, and the risk profile of any individual change is lower than that of accumulated drift between infrequent large releases. The change window model does not disappear in this environment; it simply changes shape.

For zero-trust programs, the practical resolution is a two-tier change model. Automated pipeline deployments that include a full attestation chain, including all security gate results, SBOM, and image signatures, are permitted at any time without a change window, because the pipeline itself serves as the change authorization mechanism. Changes that require human intervention, emergency fixes applied outside the pipeline, manual configuration changes to enforcement primitives, and changes to the SOAR playbook library require explicit change approval through the normal change management process, including a change window for high-impact systems. The distinction is not between automated and manual changes per se, but between changes that carry full pipeline attestation and those that do not. This framing aligns the change management process with the actual risk driver: the attestation chain, not the clock.

11.12 Manager's Decision Guide

This decision guide helps managers evaluate their organization's automation maturity for zero-trust operations and identify investments that will have the greatest impact on program sustainability. The decisions below are not one-time choices. They represent ongoing operating model commitments that require budget, staffing, and organizational authority to sustain.

Decision Criteria

- Manual-to-automated ratio for security operations: If more than half of your zero trust policy updates, identity lifecycle changes, and compliance evidence collection steps are performed manually, your program is in a scaling risk zone. The threshold for escalating concern is not absolute volume but the trend line. If the ratio of manual steps is not decreasing quarter-over-quarter as the environment grows, the program will not scale without a proportional increase in headcount.

- CI/CD pipeline security maturity: A pipeline that deploys code without SAST, SCA, IaC scanning, and image signing is not a zero-trust pipeline. It is a fast path to production that bypasses the security controls enforced by your continuous verification layer. Evaluate pipeline maturity by asking whether a developer could deploy a container with a known critical CVE without the pipeline blocking it. If the answer is yes, the pipeline is not a security gate.
- Policy-as-code adoption: Evaluate what percentage of your zero trust policies, including identity access rules, segmentation configurations, and admission control policies, are managed through version-controlled code with automated testing and peer review. Any policy managed exclusively through a console cannot be reviewed, tested, rolled back, or audited at the level required for a mature zero-trust program.
- SOAR coverage for high-priority response scenarios: Identify your top five incident response scenarios for the zero trust environment: credential compromise, workload behavioral anomaly, lateral movement detected, supply chain indicator found in production, and unauthorized configuration change detected. Evaluate the percentage of each scenario that is automated via SOAR versus requiring manual intervention. The goal is not 100% automation across all scenarios, but a clear, documented boundary between what the machine handles and what requires human judgment.
- Automated compliance evidence generation: Before your next audit, trace the path from a compliance requirement to the evidence artifact that satisfies it. If that path runs through a manual evidence-collection process requiring log pulls, screenshots, and configuration exports from

multiple systems, you have a compliance-automation gap that will grow more expensive to fill as your environment grows. Pipeline-generated attestations, GitOps audit trails, and automated policy evaluation reports are the raw material for continuous compliance evidence.

- Platform engineering investment: Evaluate whether your organization has a dedicated platform engineering function responsible for the zero-trust automation substrate. If security automation responsibilities are distributed across development teams without a central owner, expect inconsistent implementation and accumulating technical debt within the automation layer.

Decision Frame

The central decision this chapter asks managers to make is not which tool to buy. Tool selection is engineering work. The decision is organizational: will zero trust policy management be treated as operational code, subject to the same engineering discipline as the application code it governs, or will it remain a manual practice managed through console access and periodic review cycles? The answer determines whether the program will scale. Organizations that commit to the code-first model will invest in platform engineering, policy testing, GitOps operators, and SOAR playbook development. They will accept the initial velocity cost of building pipeline infrastructure and training teams to work through it. Organizations that maintain the manual model will find themselves hiring more security operations staff to keep pace with growth, accumulating compliance debt as audit evidence collection lags behind deployment velocity, and managing a growing inventory of configuration drift that no automated system can correct. The operational clarity that comes from treating the policy plane as code is not only about security. It is about organizational sustainability.

11.13 Case Study: The Supply Chain Compromise That Did Not Spread

CloudVault Analytics ran a SaaS platform serving 2,400 enterprise customers with data analytics and business intelligence products. The engineering organization operated at a pace that reflected its commercial success: 400 deployments per day across three cloud providers, supported by an automated CI/CD pipeline built incrementally over three years. The security team was competent, the pipeline was fast, and the deployment metrics looked excellent. What the metrics did not measure was the provenance of the dependencies the pipeline pulled and built into production.

The compromise entered through a component that had been in CloudVault's dependency tree for fourteen months: a widely used open-source charting library with a strong community reputation and a maintainer account that was no longer protected by a strong authentication factor. An attacker executed an account takeover of the library maintainer's package registry credentials and published a new library version that contained a dormant backdoor. The version number was incremented within the minor version range that CloudVault's dependency pinning policy allowed for automatic update. The build pipeline pulled the new version, built it into the application container, passed all existing tests, and deployed it to production across all three cloud environments within four hours of the malicious version's publication. The backdoor persisted for six weeks because the monthly manual security review, designed for a deployment cadence measured in weeks, had no mechanism to evaluate the provenance of individual dependency updates in a pipeline producing 400 deployments daily.

The remediation that CloudVault implemented after the incident addressed three structural gaps. The first gap was dependency signature verification. After the incident, every build was required to verify the cryptographic signature of each dependency against a known-good baseline. The known-good baseline was established from a signed SBOM snapshot taken at the last successful security review and updated only through an explicit review-and-approve process. A dependency version that was not present in the baseline or whose signature did not match the registry record could not be incorporated into a build. This single control would have caught the compromised library version before it entered the build pipeline, because the new version's signature would not match the baseline entry for the same package.

The second gap was runtime behavioral monitoring with automated containment. CloudVault deployed a runtime security agent across all production workloads that monitored system call patterns and network connections against a behavioral baseline established during a clean deployment period. When the backdoor in the charting library activated a network connection to an external address not in the service's established connection profile, the runtime agent detected the anomaly within ninety seconds, and a SOAR playbook quarantined the affected pods, revoked their service account tokens, and isolated their network routes. The blast radius of the compromise was contained to the three pods active when the anomalous connection attempt occurred. No customer data was exfiltrated because the service account tokens for the affected pods had read-only access to pre-aggregated analytics data, not to the raw customer data store. Least privilege, enforced through automated identity provisioning, bounded the impact of the quarantine.

The third gap was the generation of continuous evidence of compliance. The six-week dwell time of the original compromise was enabled by the monthly manual review cadence, which could not keep pace with the deployment velocity. After the incident, CloudVault instrumented its pipeline to generate a signed attestation record for every build, capturing the SBOM, security gate results, image signature, and policy evaluation results. These attestation records were automatically shipped to a compliance evidence store, which generated a daily compliance posture report that mapped each deployment to its attestation chain. When the next audit arrived, the evidence for every deployment in the audit window was immediately available in a structured, queryable format. The audit preparation process, which had previously taken three weeks of manual evidence collection, was completed in two days. The incident that revealed the automation gap ultimately led to a compliance evidence system that made the program more auditable, more defensible, and more mission-aligned than before the compromise.

CloudVault Post-Incident Supply Chain Verification Architecture

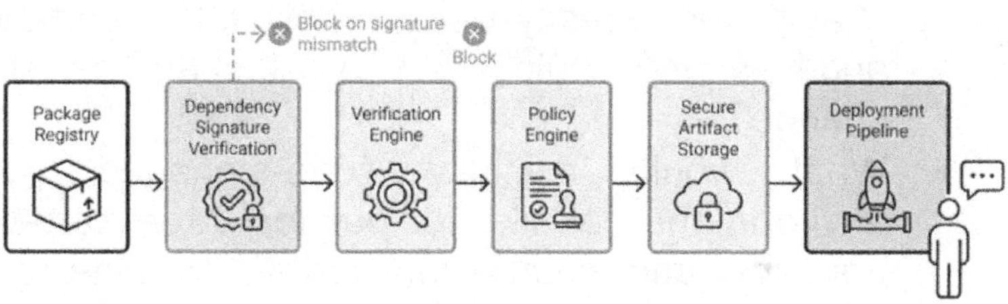

CloudVault Post-Incident Supply Chain Verification Architecture

11.14 Team Conversation Guide

Run these three discussions in sequence over two to four weeks. The first surfaces the issue, the second works the trade-offs, the third commits the team to action.

Discussion 1. Policy Coverage: What Lives in Code and What Does Not

Before this conversation, ask your engineering leads to inventory zero trust policies by management method: which policies are managed as code in a version-controlled repository, which are configured through a product console, and which exist in a hybrid state where the console configuration is not reflected in any code repository. The inventory does not need to be complete to be useful. A sample across three or four policy domains, identity access rules, network segmentation, admission control, and secret configurations, will reveal the pattern.

Discussion questions to ask the team:

- For the policies we sampled, what percentage are currently managed as code versus console configuration?
- For the console-managed policies, what would it take to express them in code? Is the blocker a tooling gap, a skills gap, or a process question about who owns the code repository?
- If a policy managed through a console was changed incorrectly tonight, how would we know, and how long would it take to detect and revert the change?
- What would a fully code-managed policy environment look like for our top three zero-trust domains, and what is the realistic timeline to get there?

What to listen for: Whether the team frames console management as a temporary state or a permanent operating model. Teams that frame it as temporary have a path forward. Teams that frame it as permanent have accepted a scaling risk they may not have fully evaluated. Also listen for "we tried that, and it was too slow" responses, which often indicate that the policy-testing tooling was not built before the policy-as-code process was adopted, making the process feel slower without delivering the safety benefits that motivate the discipline.

Follow-up actions: Schedule a working session to define the minimum viable policy-as-code implementation for one high-value domain. Deliver a working pull request workflow for that domain within thirty days and use it as the template for the next domain. Scaling policy-as-code through iterative domain migration is faster than migrating all domains simultaneously.

Discussion 2. Pipeline Security: Gates That Block Versus Gates That Report

Before this conversation, have someone on the team attempt to deploy a container image with a known low-severity CVE and a separate image with a known critical CVE through your existing pipeline. Note what happens at each severity level: does the pipeline block, report, and continue, or does it do none of these? This exercise surfaces the actual gate behavior, which is often different from the documented intended behavior.

Discussion questions to ask the team:

- What did we learn from the test deployment? Did the pipeline behave as expected?
- For each type of security finding, SAST, SCA, IaC, and image scan, do we have a documented policy that maps severity levels to blocking versus advisory gate behavior?
- Who owns the gate policy, meaning who decides what severity level triggers a block, and how is that policy reviewed and updated?
- If our pipeline gates have false positives that developers learn to override, what is the actual security value of those gates at that point?

What to listen for: Evidence that gate overrides have become routine. When developers override security gates so frequently that the gates are treated as procedural friction rather than as meaningful controls, the gate's security value is effectively eliminated. This is a calibration problem, not a culture problem. Overly sensitive gates that flag findings developers know are not actionable produce override cultures. The solution is better calibration, not removing the gates.

Follow-up actions: Establish, at a minimum, a quarterly gate calibration review cycle in which the team reviews the override rate for each gate type and adjusts thresholds based on observed false-positive rates and the actual severity of findings overridden. Document the calibration decisions and the reasoning behind them so that the gate policy has the same governance as the policies it enforces.

Discussion 3. Incident Response Automation: What the Machine Handles and What We Handle

Before this conversation, map your three most likely zero-trust incident scenarios: a compromised user credential, a workload behavioral anomaly, and a detected policy drift. For each scenario, trace the current response process from alert to containment, identifying which steps are automated and which require human action. Time each step in the manual process to establish a baseline.

Discussion questions to ask the team:

- Looking at our current response timeline for each scenario, where is most of the time spent, and what is driving that time?
- For the manual steps in our current process, which ones are genuinely judgment calls that require human assessment, and which ones are mechanical steps that could be automated without loss of safety?
- If we automated the mechanical steps, what would our mean time to containment look like for each scenario?
- What would have to be true about our environment and our SOAR playbooks for us to trust automatic containment for a workload behavioral anomaly without human approval?

What to listen for: the team's comfort level with automated containment. Teams that are uncomfortable with full automation for high-impact actions are often right: the discomfort reflects a real gap in playbook maturity, signal reliability, or blast radius assessment. The goal is not to maximize automation at the expense of safety. It is to automate steps where automation is more reliable than human response, and to preserve human judgment when the context is too complex or the blast radius too large for a machine decision to be trusted.

Follow-up actions: Define one incident scenario where automated containment is both technically feasible and organizationally safe, build the playbook, test it in a staging environment against simulated alerts, and deploy it with a 30-day monitoring period in which a human reviews every automated action after the fact. Use the review data to calibrate confidence in the playbook before extending the pattern to the next scenario.

11.15 Takeaway

Marcus's Friday evening taught him the difference between a zero-trust architecture and a zero-trust operating model. The architecture, the policy engines, the microsegmentation boundaries, and the continuous verification signals were working exactly as designed. The operating model, the process for responding to what the architecture detected, was still built for a world where deployments were weekly and security reviews were monthly. The gap between architecture and operating model is where mature zero-trust programs fail to scale, and no additional control technology can close it. It closes through automation.

Policy-as-code, infrastructure-as-code, CI/CD security gates, automated identity provisioning, secret rotation, SOAR playbooks, drift detection, and GitOps operators are not separate projects. They are components of a single operating model in which the policy plane is governed with the same engineering discipline as the application plane. When policy changes are code, they are reviewable, testable, reversible, and auditable. When pipelines enforce security gates, deployment velocity and security posture reinforce each other rather than trading off. When SOAR connects detection to response, the program operates at the speed of the threat, not the approval queue.

The platform engineering investment that makes this possible is not a cost center. It is the infrastructure that supports a sustainable, scalable zero-trust program. Organizations that build this infrastructure early find that security operations become less labor-intensive as the environment grows, not more. Compliance evidence is generated as a byproduct of normal operations. Incident response is faster, more consistent, and better documented. The pilot-to-production pathway for new zero-trust controls is shorter because the automation substrate is already in place. Responsible by design, in the context of DevSecOps, means that security is not added to the delivery process after the fact. It is woven through every stage from commit to runtime, producing evidence of its correctness as it goes.

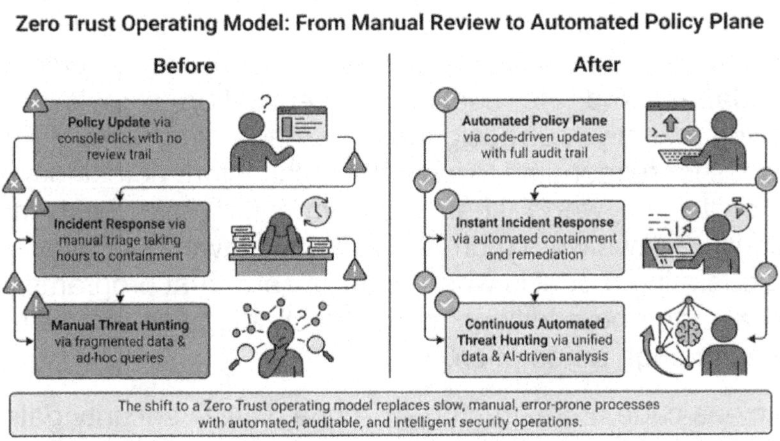

Chapter 11 completes the program view by addressing governance and compliance: how the automated evidence produced by the operating model described in this chapter gets translated into maturity assessments, regulatory compliance demonstrations, and board-level reporting that makes the zero trust program defensible, fundable, and sustainable over the multi-year arc of the transformation.

Marcus Reyes had two browser tabs open, and neither one was making his morning easier. In the left tab was the draft quarterly executive risk review that his CISO expected on her desk by Friday. In the right tab was the evidence request list from the external auditors running the organization's SOC 2 Type II examination, which had arrived that same morning with a fourteen-day response deadline. The two documents had nothing in common on the surface. The risk review needed plain-language narratives about threat exposure, maturity progress, and residual risk. The auditor's list wanted log exports, policy screenshots, access certification records, and signed attestations about specific controls. Marcus had spent the first thirty minutes of the morning staring at both lists in parallel, and somewhere in that comparison a pattern had started to emerge.

The auditors wanted evidence that privileged access was reviewed quarterly. The executive risk review had a slide asking whether the organization's access governance program was functioning. The auditors wanted network segmentation diagrams showing isolation of cardholder data environments. The risk review had a section on blast radius reduction and microsegmentation progress. The auditors wanted logs demonstrating that authentication was enforced with multi-factor verification for remote access. The risk review tracked MFA adoption rates as a leading indicator of progress in zero-trust implementation. Marcus grabbed a notepad and started drawing a rough matrix. The same control facts, system telemetry, and architectural decisions appeared in both documents. The question was whether his team was capturing those facts once and using them in both places, or capturing them twice, under different names, with different owners, at different times, in different formats, with different evidence trails.

He already knew the answer. His team had a compliance team that prepared for the SOC 2 audit and a security team that prepared for the executive review. The compliance team ran manual evidence collection sprints twice a year. The security team pulled dashboard screenshots quarterly. Neither team knew exactly what the other was collecting. Some of the same log exports were being pulled three separate times from the same SIEM platform: once for the auditors, once for the quarterly security metrics, and once for a regulatory self-assessment that the legal team ran annually. The duplication was not just wasteful. It was introducing inconsistency. A policy screenshot taken in February and a log export taken in March might reflect the same control, but capture it in different states if something had changed in between. An auditor or a board member comparing those two artifacts could reasonably ask whether the control was stable. The answer would require someone to investigate why the artifacts looked different. That investigation had happened twice in the past year.

By the time Marcus finished the notepad matrix, he had sketched the outline of a governance architecture that did not yet exist but should. If the zero trust controls the organization had spent two years implementing were documented in a control catalog that mapped each control to every applicable framework simultaneously. If the telemetry those controls generated flowed into a continuous monitoring platform that produced evidence on demand rather than during periodic sprints, then the same system state would satisfy both the auditor and the board without any reconciliation overhead. A dashboard refresh would replace the compliance sprint. The quarterly risk review would pull from the same source as the annual audit. Marcus wrote two words at the top of the notepad: unified governance. That was the chapter he needed to write for his organization, and this was where it started.

11.16 Why It Matters

Every organization that has made meaningful progress on zero trust eventually confronts the same governance problem. The technical work of deploying identity-aware access controls, microsegmentation boundaries, and continuous verification pipelines produces a rich body of evidence about actual security posture. That evidence exists inside dashboards, policy engines, SIEM platforms, and cloud security tooling. The organization is collecting it constantly, whether or not anyone has designed a process for its use. The question is whether that evidence is organized and surfaced in ways that serve every audience that needs it: the security operations team monitoring daily, the engineering team tracking implementation progress, the compliance function preparing for annual audits, the executive team assessing enterprise risk, and the board receiving periodic briefings on material risks to the business.

Governance, in the context of zero trust, is the mechanism that converts engineering implementation into institutional accountability. A policy engine that enforces least-privilege access is a technical achievement. A control catalog that maps the policy engine to NIST SP 800-53, HIPAA Security Rule, and SOC 2 CC6 criteria simultaneously is a governance achievement. The difference is not in the technology; it is in whether the organization has designed the surrounding processes, documentation structures, and reporting workflows to make the technical fact legible and auditable. Organizations that skip governance design invest in technology that protects the environment but cannot prove it, cannot measure it, and cannot sustain the investment when leadership changes and budget pressure arrives. Governance is what transforms a pilot into a program and a program into a durable institutional capability.

The workflow-level impact of poor governance is not abstract. When compliance evidence is collected manually and periodically, the security team diverts engineering bandwidth into artifact-gathering exercises that could be automated. When control documentation is maintained separately from the technical implementation, drift between the documented control and the runtime control becomes inevitable and invisible until an auditor or an incident exposes it. When executive reporting is prepared manually from raw telemetry, the metrics it contains reflect snapshot states rather than continuous posture trends, making it harder for executives to distinguish genuine security improvements from favorable sampling. Mission-aligned governance architecture solves all three problems simultaneously by making the evidence continuous, the documentation synchronized with the runtime, and the reporting derived from the same authoritative source as the compliance evidence.

11.17 Zero Trust Maturity Models

A maturity model provides a structured vocabulary for describing where an organization currently is relative to where it wants to be, and a path for getting from one state to the next. In the zero trust domain, two maturity frameworks have become reference points for most serious programs: the CISA Zero Trust Maturity Model and the DoD Zero Trust Reference Architecture maturity levels. Both frameworks share a common insight: zero trust is not a binary achievement but a continuous progression. An organization does not implement zero trust one day and consider itself done. It advances through recognizable stages of capability, and each stage provides measurably better security outcomes than the one before.

The CISA Zero Trust Maturity Model structures that progression across five pillars: Identity, Devices, Networks, Applications and Workloads, and Data. Each pillar represents a domain of security capability that must mature independently while remaining integrated with the others. Within each pillar, CISA defines three capability stages: Traditional, Advanced, and Optimal. The Traditional stage describes most organizations at the beginning of a zero trust journey, where perimeter-based controls dominate, identity management is manual and fragmented, and visibility is limited. The Advanced stage represents the meaningful deployment of zero-trust controls, with automated provisioning, policy-driven access, and integrated telemetry, but with remaining gaps in cross-pillar integration and edge cases. The Optimal stage represents a fully automated, continuously verified, data-centric architecture in which policy decisions are driven by real-time risk signals across all five pillars simultaneously.

The value of the CISA model for a governance program is not that it provides an exact checklist of what to implement in each stage. The value is that it provides a shared language for communicating current state, target state, and progress to audiences with different levels of technical depth. An executive asking "how far along are we on zero trust" gets a precise, reproducible answer when the organization has assessed itself against each pillar using the CISA stages. A board asking about investment priorities gets a specific answer about which pillars are furthest behind and what closing those gaps requires. The model converts a sprawling multi-year technical program into a governable set of advancement objectives.

CISA Zero Trust Maturity Model: Five Pillars and Three Stages

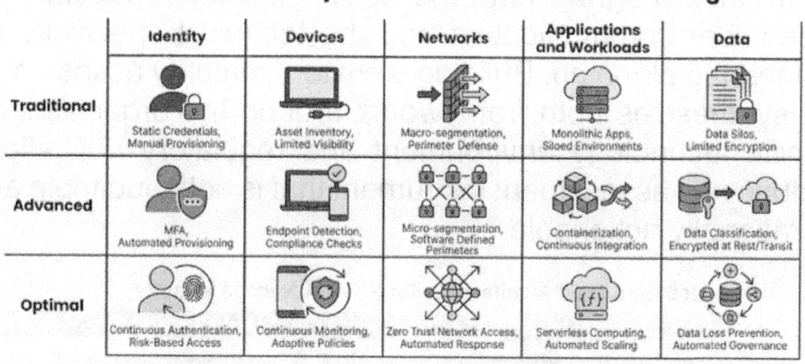

The CISA Zero Trust Maturity Model outlines a progression from traditional perimeter-based security to an adaptive, continuous risk-based approach across five key pillars.

The DoD Zero Trust Reference Architecture maturity framework takes a complementary approach. Where CISA organizes maturity by pillar, the DoD framework defines maturity levels by the degree of integration and automation across the full architecture. DoD Level 1, called Target, represents the baseline capability set that all DoD components were required to achieve by fiscal year 2027 under the department's zero trust strategy: identity management with MFA, device health validation, network micro-segmentation, data tagging, and integrated visibility. DoD Level 2, called Advanced, represents full automation of access decisions, cross-domain integration of policy enforcement, and behavioral analytics driving real-time risk decisions. The DoD framework is explicit that Level 2 is aspirational and that most organizations will spend years advancing through Level 1 before Level 2 targets become operational.

Commercial organizations that are not DoD contractors can still use the DoD framework productively. The framework's discipline about defining specific capability targets and technology integration patterns provides engineering teams with concrete implementation goals that the more abstract CISA model leaves open. Using both frameworks in parallel, the CISA model for pillar-based governance reporting and the DoD model for capability-level engineering targets, gives a governance program both the business-level vocabulary for stakeholder communication and the technical specificity for engineering planning. Building a custom maturity assessment that synthesizes both frameworks around the organization's specific technology environment and regulatory obligations produces an assessment instrument that is both auditable and operationally actionable.

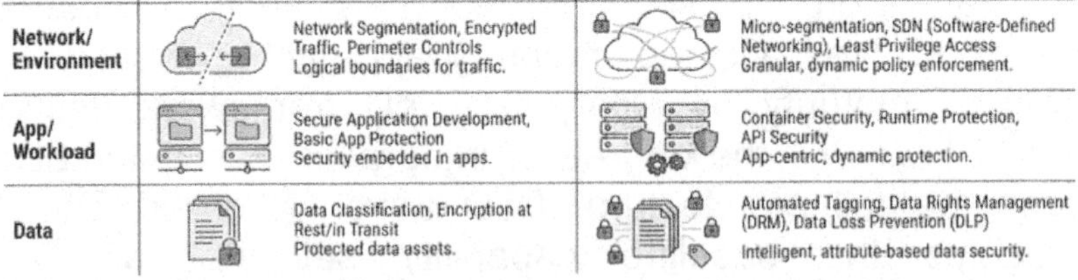

The Department of Defense implements Zero Trust progressively, from foundational security to dynamic, risk-adaptive defenses.

11.18 Mapping Zero Trust Controls to Compliance Frameworks

Every major compliance framework governing enterprise security either explicitly references zero-trust architecture or directly describes the security requirements that zero-trust controls satisfy. HIPAA, PCI-DSS 4.0, SOX, GDPR, SOC 2, and FedRAMP were all written before zero trust was formalized as an architectural discipline. Still, their underlying requirements, access control, monitoring, data protection, audit logging, and risk management, map cleanly onto zero-trust control categories. The opportunity for organizations subject to multiple frameworks is to design the zero trust architecture once, in a way that satisfies the intersection of all applicable requirements, and then generate compliance evidence from that single architecture rather than operating separate compliance programs in parallel.

HIPAA Security Rule requirements under the Access Control and Audit Controls standards map directly to zero-trust identity and access management capabilities. The requirement that covered entities implement policies and procedures for granting access to electronic protected health information maps to a zero-trust, least-privilege access policy. The requirement for unique user identification maps to the elimination of shared credentials. The requirement for audit controls that record and examine activity maps to the telemetry layer produced by zero-trust architectures as a byproduct of continuous verification. An organization that has implemented a zero-trust identity platform with continuous monitoring is not just satisfying HIPAA controls; it is producing HIPAA audit evidence continuously rather than assembling it before each examination.

PCI-DSS 4.0 introduced requirements that align even more explicitly with zero trust principles than its predecessors. Requirement 7, which governs access to system components and cardholder data, requires that access be granted on a need-to-know basis, with least-privilege principles enforced, directly matching zero-trust policy-based access control.

Requirement 8, governing identification and authentication, requires MFA for all access to the cardholder data environment, which is a baseline zero-trust control. Requirement 10, governing audit log requirements, requires logging all access to system components and cardholder data, thereby satisfying the requirements of a zero-trust telemetry architecture. The key governance insight is that a zero-trust architecture deployed correctly around cardholder data environments satisfies the technical requirements of Requirements 7, 8, and 10 simultaneously, through the same policy enforcement and logging infrastructure.

SOX IT general controls, specifically those addressing logical access controls, change management, and computer operations, map to zero-trust capabilities at the process level. The logical access control objective under SOX requires that access to financial systems is appropriately restricted and that access rights are reviewed periodically. Zero-trust continuous access certification, in which access rights are reviewed and reconfirmed through automated workflows rather than periodic manual reviews, more rigorously satisfies this objective because certification is continuous rather than point-in-time. SOC 2 Trust Service Criteria under the CC6 logical and physical access category align with zero trust access management controls in similar ways. CC6.1, which requires identifying and authenticating users before granting access, and CC6.3, which requires removal of access when no longer needed, are satisfied by identity lifecycle management controls that are foundational to any mature zero-trust program.

GDPR Article 32 requires that organizations implement appropriate technical measures to ensure security appropriate to the risk, including the ability to restore availability and access promptly and a process for regularly testing and evaluating the effectiveness of security measures. Zero-trust architectures satisfy multiple Article 32 dimensions: least-privilege access limits the scope of data exposure in the event of a breach, continuous verification provides ongoing testing of access controls, and the telemetry layer provides the monitoring infrastructure needed to detect and respond to unauthorized access quickly. FedRAMP, which draws its security control requirements from NIST SP 800-53, aligns with zero trust controls across the Access Control, Identification and Authentication, System and Communications Protection, and Audit and Accountability control families.

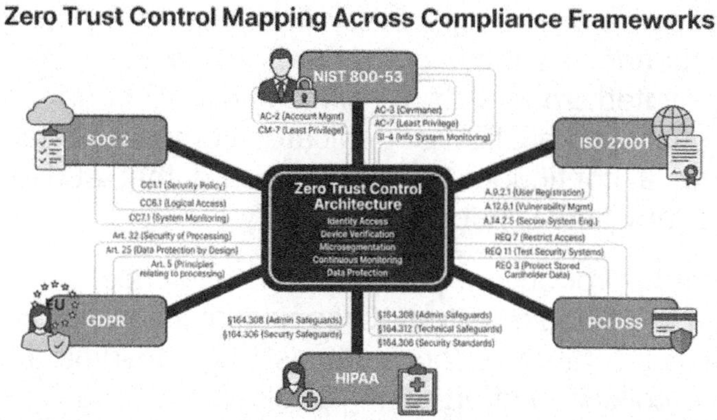

Zero Trust Control Mapping Across Compliance Frameworks

11.19 Evidence Automation and Continuous Controls Monitoring

The traditional approach to compliance evidence collection is an episodic sprint. An audit date is announced, a compliance analyst is assigned, a list of evidence requests is circulated, engineers are asked to pull screenshots and log exports, and the assembled artifacts are packaged and submitted. This approach has three structural problems. First, it is expensive in engineering time, diverting technical staff from productive work to artifact retrieval at the worst possible moment. Second, it is inherently point-in-time, representing the state of controls at the time of collection rather than demonstrating sustained compliance over the audit period. Third, it is fragile because any change to the environment between when evidence is collected and when the auditor reviews it creates a reconciliation problem.

Evidence automation replaces the episodic sprint with a continuous collection pipeline. The core insight is that every zero-trust control that operates correctly produces logs, policy decisions, and configuration states that constitute compliance evidence in machine-readable form. A policy engine that enforces least-privilege access produces a decision log for every access attempt that is evaluated. An identity platform that enforces MFA produces authentication event records for every login session. A microsegmentation platform enforces network policies and logs every flow decision. A privileged access management system records every session initiated under elevated credentials. All of these event streams, flowing continuously into a SIEM or dedicated compliance data lake, constitute a continuous evidence record that auditors can query for any time window they specify.

The pilot-to-production pathway for evidence automation follows a predictable pattern. Organizations typically begin by identifying the three or four controls that consume the most evidence-collection effort during audit sprints, usually access certification, MFA enforcement verification, and privileged-access logging. They instrument those controls with automated export pipelines that forward event data to a centralized evidence repository with immutable retention. They build query templates that generate the specific artifact formats auditors request, such as user access reports, authentication failure summaries, and privileged session inventories, on demand from the stored event data. Once those initial controls are automated, the pattern is extended to the remaining evidence categories. Within two to three audit cycles, the compliance sprint is replaced by an evidence dashboard refresh that takes hours rather than weeks.

Continuous Evidence Collection Pipeline

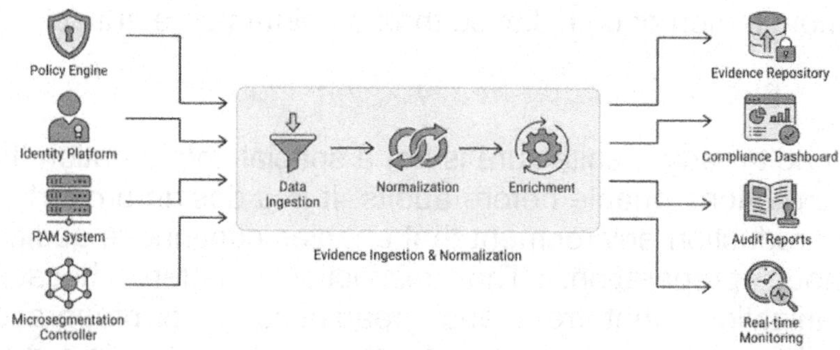

This pipeline facilitates ongoing data gathering, standardization, and reporting to maintain continuous compliance and security.

Continuous controls monitoring extends evidence automation by adding real-time evaluation of control effectiveness rather than just collecting evidence that controls were invoked. A monitoring platform that ingests policy engine decision logs can alert when the deny rate for a specific access category changes materially, which may indicate either a policy misconfiguration or a change in user behavior that warrants investigation. A monitoring platform that ingests MFA enforcement logs can track MFA bypass rates over time and alert when bypass exemptions exceed a defined threshold. These real-time signals give the security operations team visibility into control health between audit cycles, allowing them to detect and remediate control drift before it accumulates into a compliance gap.

The organizational impact of transitioning from manual evidence collection to continuous monitoring is substantial. Compliance teams that previously spent six to eight weeks per audit cycle on evidence gathering report reducing that effort to one to two weeks once the automation pipeline is mature, with the remaining effort focused on evidence review and auditor communication rather than collection. Engineering teams that previously allocated significant capacity to audit preparation can reallocate it to implementation work. And the quality of the compliance evidence improves: continuous telemetry demonstrates sustained control operation over the entire audit period rather than a snapshot at collection time, which is a more persuasive demonstration of compliance than a point-in-time artifact.

11.20 Building an Audit-Ready Architecture

An audit-ready architecture is not a special configuration that organizations enable before audits. It is a design property of the production environment that enables continuous auditing without preparation. The distinction matters because organizations that treat audit readiness as a preparation activity rather than an architectural property end up in a cycle of technical debt: they run in a configuration that is not quite audit-ready until an audit approaches, scramble to clean it up, pass the audit, and then let configuration drift accumulate again because the audit-ready configuration is more restrictive than what teams find operationally convenient.

The foundational design principle of an audit-ready architecture is that every security control is implemented as a version-controlled, automatically enforced, and continuously monitored policy artifact. Access policies live in a policy engine and are deployed through a GitOps pipeline that records every change with a timestamp, author, and change justification. Network segmentation rules live in infrastructure-as-code repositories, reviewed through a pull request process, and deployed automatically via CI/CD pipelines. Identity access rights are managed through role-based access control definitions that are codified, reviewed at defined intervals, and provisioned through automated workflows that reject any manual deviation from the defined role structure. When an auditor asks for evidence that a specific access control was in place during the audit period, the response is a pull from the policy engine's change log showing that the policy has been continuously in place with no deviations during the period requested.

Responsible by design means that audit evidence is not an afterthought but a designed output of every technical implementation decision. When an engineering team deploys a new application workload, the deployment process should require that the workload's access policy be defined in the policy catalog, its network segment be declared in the segmentation template, its authentication requirements be specified in the identity platform, and its logging configuration be validated before deployment completes. A pre-deployment compliance checklist enforced by the CI/CD pipeline, rather than performed manually, ensures that every new workload enters the environment in a known, documented, and auditable state. Workloads that have not been provisioned through this process are prioritized for remediation rather than treated as normalized exceptions.

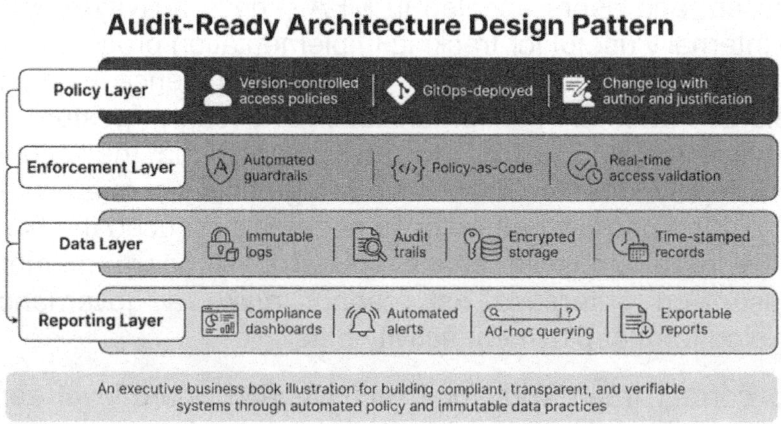

Supply chain and third-party assurance require specific attention in an audit-ready architecture because third-party access is often a gap in both the technical controls and the evidence record. Vendors, contractors, and managed service providers who access the production environment create access events that must be captured in the same evidence pipeline as internal user events. Zero-trust architecture simplifies this requirement by governing third-party access through the same policy engine as all other access. A contractor who authenticates through the identity platform, passes device health verification, and receives a time-limited access token that expires at the end of the session generates an access event record that is structurally identical to the record generated by an internal user. The evidence trail for third-party access is produced automatically by the same infrastructure, satisfying auditor requirements for third-party access logging without requiring separate tools or processes.

11.21 Executive Risk Reporting and Board-Level Zero Trust KPIs

The challenge of reporting zero trust progress to executive and board audiences is not primarily a technical communication challenge. It is a metrics design challenge. The instinct of engineering teams is to report deployment activity: policies deployed, segments created, agents installed, and users enrolled in MFA. These activity metrics are internally useful for tracking implementation progress, but they are uninformative to an executive audience trying to evaluate whether the organization's security posture is actually improving. A CISO who tells a board that the security team deployed twelve new network segments last quarter has not told the board whether those segments reduced the blast radius of a plausible attack scenario. A board member focused on enterprise risk cannot make a governance decision from deployment activity data.

The correct framework for executive and board-level zero trust reporting distinguishes between leading indicators, which are deployment and capability metrics that predict future security improvement, and lagging indicators, which are outcome metrics that measure actual security posture. Leading indicators include MFA enforcement coverage (percentage of authentication events requiring a second factor), privileged access scope (number of standing privileged accounts compared to just-in-time access grants), and policy coverage (percentage of network flows governed by explicit allow or deny policies rather than implicit permissiveness). Lagging indicators include mean time to detect (how quickly anomalous access patterns are flagged), mean time to contain (how quickly a flagged session can be terminated and the affected credential revoked), blast radius metrics (the number of resources reachable from a typical access tier if a credential in that tier is compromised), and lateral movement attempt rate (how frequently east-west traffic matching lateral movement patterns is blocked by segmentation policy).

Privilege sprawl ratio, defined as the ratio of unique access rights granted across the user population to the minimum set of rights required to support all documented business functions, is a particularly useful board-level metric because it is directly interpretable. A privilege sprawl ratio of 3.0 means the environment has granted three times as many access rights as are necessary, which translates directly into a threefold increase in the blast radius of a minimal-privilege architecture. Reducing that ratio is a measurable security improvement that correlates with engineering work the board can observe. It also correlates with compliance requirements under HIPAA, PCI-DSS, and SOC 2, making it a cross-framework governance metric rather than a framework-specific compliance metric.

Zero Trust KPI Framework: Leading and Lagging Indicators

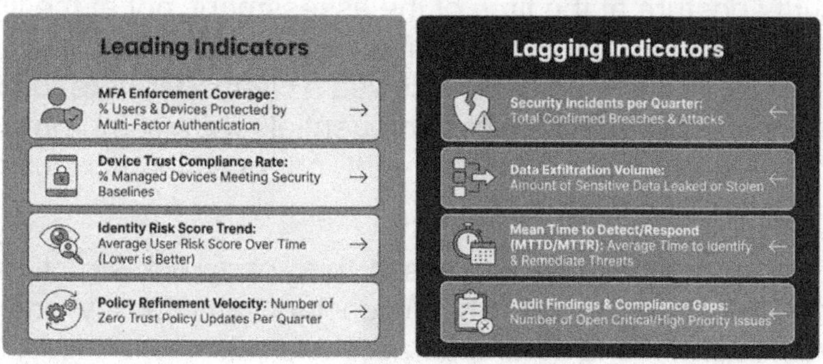

Zero Trust KPI Framework: Leading and Lagging Indicators guide strategic adjustments for continuous security improvement.

Board-level reporting should present two or three lagging indicators with trend data spanning at least four quarters, alongside a maturity-stage assessment for each CISA pillar indicating which have advanced since the prior report. The narrative accompanying the metrics should translate each metric into a business risk statement. A mean time to detect that improved from 36 hours to 8 hours is not just a technical improvement: it is a statement that the window during which an attacker can operate undetected in the environment has been reduced from a business day and a half to a single work shift, which materially reduces the organization's exposure to the most costly category of breach, the long-dwell-time breach where the attacker has time to establish persistence, escalate privileges, and exfiltrate large data volumes before detection. That translation from technical metric to business risk is what makes executive reporting mission-aligned rather than technically accurate but strategically uninformative.

11.22 Third-Party Assurance and Supply Chain Security

Third-party risk is one of the most consistent failure patterns in enterprise security, and it is one of the dimensions that governance frameworks have been slowest to address effectively. Traditional vendor risk management programs assess third-party security posture at contract initiation through questionnaires, periodic reviews, and occasionally SOC 2 reports. These assessments reflect the vendor's security posture at the time of the assessment, not at the time the vendor's personnel are actively accessing the production environment. A vendor who received a clean SOC 2 report six months ago but whose access credentials were compromised last week presents a threat that no questionnaire-based assessment would detect.

Zero-trust architecture addresses third-party risk structurally rather than administratively. When the same policy engine governs vendor and contractor access as all other access, the access rights granted to third parties are explicitly scoped to the resources and time windows required for the specific engagement rather than provisioned broadly and expected to be managed by the vendor. A contractor performing a one-week assessment of a specific application receives a time-limited access token for that application in the relevant network segment for the specified period, with MFA required for every session and all access events logged in the same evidence pipeline as internal user events. The vendor's access is verifiable, auditable, and automatically terminated at the end of the engagement, without requiring manual deprovisioning.

Third-Party Access: Traditional vs. Zero Trust Governance

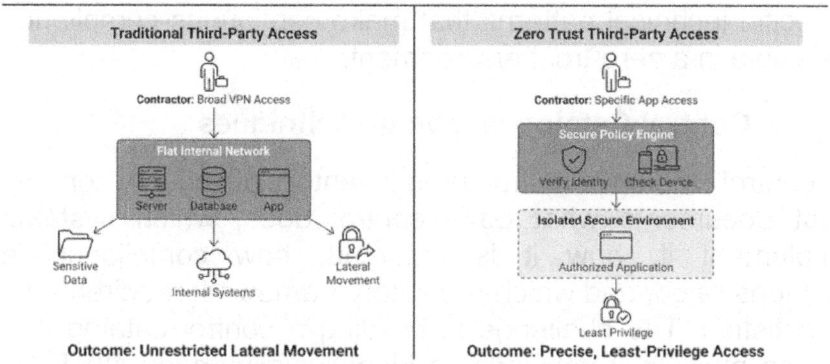

Third-Party Access: Traditional vs. Zero Trust Governance

Supply chain security extends the third-party assurance model to the software and component supply chain. Organizations that build applications using open-source libraries, commercial components, and vendor APIs inherit the security posture of those components at the code level, not just the network access level. A zero-trust governance program that addresses only human access control while leaving software supply chain risk unaddressed has a material blind spot. The governance framework should include controls for software composition analysis, which identifies known vulnerabilities in imported dependencies, and for verifying the provenance and integrity of build artifacts through signed build pipelines. When a software component is compromised upstream, the governance framework should enable rapid identification of all applications that consume that component and prioritize remediation based on the sensitivity of the data those applications handle.

11.23 Geeks Only

This section is written for engineers, architects, and technically oriented team leaders who want to understand the implementation mechanics of governance automation at the control-catalog and evidence-pipeline level. If you are a manager who has absorbed the preceding sections, you can move to the Manager's Decision Guide. If you are an engineer responsible for instrumentation, policy management, or compliance automation, the following sections describe the specific technical patterns that make continuous compliance possible in a zero-trust environment.

Control Catalog Mapping Techniques

A control catalog is a structured inventory of security controls that documents what each control does, which systems implement it, how it is enforced, how compliance is demonstrated, and which regulatory framework requirements it satisfies. The challenge in building a control catalog that spans multiple frameworks is identifier alignment: NIST SP 800-53 uses the AC family for access controls, HIPAA uses 164.312(a) for access control requirements, PCI-DSS uses Requirements 7 and 8, and SOC 2 uses CC6. The same zero-trust access-control capability must be mapped to all four frameworks, using each framework's own control identifier.

The standard approach is to build the catalog in a normalized schema that separates the control implementation fact (what the system actually does) from the framework mapping (which requirement identifier that fact satisfies). A control record in the catalog might read: "Access Policy Enforcement: The Policy Decision Point evaluates every access request against the current user role, device posture score, and requested resource classification before issuing an access token." The framework mapping block then lists: NIST AC-2 (Account Management), NIST AC-3 (Access Enforcement), HIPAA 164.312(a)(1), PCI-DSS Req 7.2.1, SOC 2 CC6.1, FedRAMP AC-2 enhancement 1. Auditors from any of the four frameworks who request evidence for their specific control identifier receive the same underlying evidence package, extracted from the same policy engine log, with the appropriate framework identifier cited in the cover sheet. The compliance team maintains the mapping block; the implementation team maintains the implementation fact; neither team needs to understand the other's vocabulary to serve the auditor correctly.

Control mapping at scale requires a catalog tool rather than a spreadsheet. Organizations that manage fewer than fifty distinct controls can function with a well-structured spreadsheet, but the mapping complexity grows non-linearly as the number of controls and frameworks increases. With 100 controls across four frameworks, each with two to five requirement mappings per control, the catalog contains 800 to 2,000 mapping relationships that must be kept synchronized as frameworks are updated. Purpose-built GRC (Governance, Risk, and Compliance) platforms provide the data model, workflow automation, and evidence attachment capabilities needed to manage this complexity reliably. The choice of platform matters less than the discipline of maintaining the catalog as a living artifact that is updated whenever a control is added, modified, or deprecated, and whenever a framework requirement is revised.

Evidence Collection from Policy Engines and SIEM

Policy engine decision logs are the richest source of compliance evidence in a zero-trust architecture because they record every access decision with full context: who requested access, from which device, with what posture score, to which resource, under which policy, and whether the request was allowed or denied. Extracting compliance evidence from these logs requires understanding the log schema of the specific policy engine in use. Most modern policy engines expose logs in structured JSON format through either a syslog forwarding interface or a native API. A SIEM integration that ingests policy engine logs at full fidelity creates a queryable record of every access decision over the retention period.

SIEM query patterns for common compliance evidence requests follow predictable forms. For MFA enforcement evidence, the query selects all authentication events with a successful event type and filters for the presence of a second-factor confirmation field. The result set, exported as a CSV with timestamp, user identifier, application name, and MFA method fields, constitutes the evidence package for MFA enforcement requirements. For privileged access logging evidence, the query selects all sessions initiated with elevated credential contexts, such as those issued by a PAM system, including fields for session start time, session end time, commands executed (when command logging is enabled), and the justification ticket associated with the elevation request. For access review evidence, the query selects all access right modification events within a specified period, showing which access rights were reviewed, confirmed, or revoked, and which identity performed the review action.

The SIEM evidence pipeline requires attention to log completeness and fidelity. Common gaps include policy engines that do not log denied access attempts (logging only allowed decisions), identity platforms that log successful authentications but not authentication failures with error codes, and cloud access brokers that log access events with anonymized or tokenized identifiers that cannot be correlated back to individual users without a separate lookup table. Each of these gaps produces evidence that is incomplete in ways that may satisfy a basic auditor request but will not withstand detailed investigation. Engineering the telemetry layer for compliance completeness means explicitly documenting what is logged, what is not logged, and why, and then making a governance decision about whether the gaps represent acceptable limitations or must be remediated.

SIEM Evidence Query Architecture for Compliance

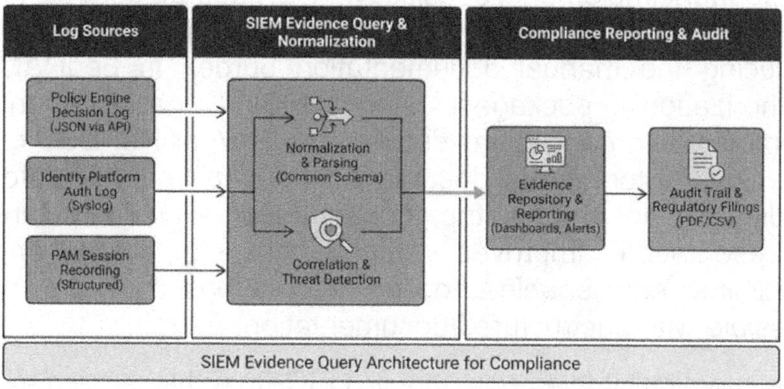

Attestation Pipelines and OSCAL

An attestation pipeline is an automated workflow that produces a signed, timestamped assertion that a specific control was in a specific state at a specific time, and chains that assertion to the underlying evidence in a way that allows the assertion to be independently verified. The concept is analogous to a software code signing process: just as a code signing certificate asserts that a specific build system produced a binary at a specific time and has not been modified since, an attestation record asserts that a control was operating correctly at a specific time and that the evidence supporting that assertion has not been altered. Attestation pipelines are the technical mechanism that converts continuous monitoring telemetry into statements that carry the same evidentiary weight as manually signed audit artifacts.

OSCAL, the Open Security Controls Assessment Language developed by NIST, provides a machine-readable format for expressing security control catalogs, system security plans, assessment plans, and assessment results. An organization that maintains its control catalog in OSCAL format can exchange control documentation with federal agencies, auditors, and GRC platforms that support the format without manual translation. OSCAL component definitions allow an organization to document how a specific technology component implements specific NIST SP 800-53 controls in a format that assessment tools can process automatically, reducing the manual documentation burden in FedRAMP authorization packages and FISMA assessments. Commercial organizations subject to HIPAA or PCI-DSS can use OSCAL for internal documentation even when regulatory bodies do not require the format, because the structured representation improves control catalog maintenance discipline and enables tooling integrations that are not possible with unstructured documentation.

The practical implementation of OSCAL within a zero-trust governance program begins with the system security plan. An OSCAL system security plan (SSP) is a JSON or XML document that describes the system boundary, the applicable control catalog, and the implementation statement for each control, in a format that automated tools can validate and cross-reference. Organizations that build their SSP in OSCAL can automate validation that every required control has an implementation statement, that every implementation statement references at least one component or service that implements it, and that every component referenced in the SSP has a corresponding entry in the component definition library. This structural validation replaces a significant portion of the manual document review that currently occurs in authorization and audit preparation processes.

OSCAL Control Catalog: From Documentation to Automated Assessment

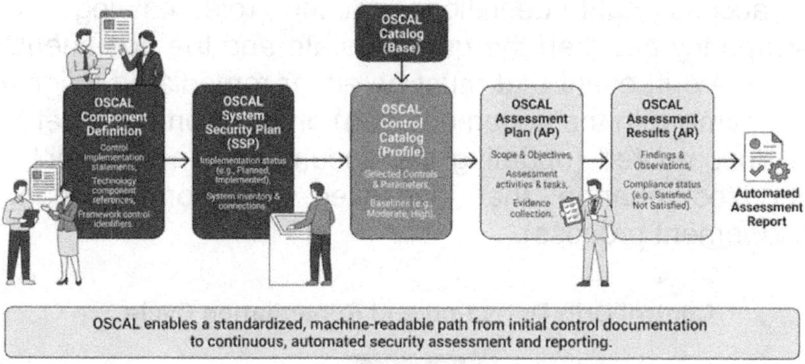

Drift Between Documented and Runtime Controls

Control drift occurs when the actual state of a security control in the production environment diverges from the documented state in the control catalog and system security plan. Drift is not a rare exception: it is the default trajectory in any environment where the control documentation is updated manually, separate from the control deployment process. Every time an engineer modifies a policy rule through a direct console action rather than through the GitOps pipeline, every time a network segment rule is adjusted to fix an operational issue without a change ticket, and every time an access right is granted manually rather than through the automated provisioning workflow, the runtime environment moves away from the documented state without a corresponding update to the documentation.

Detecting drift requires a reconciliation process that compares the documented control state against the runtime control state at a defined frequency. For policy engine controls, this means querying the policy engine's current ruleset through its API and comparing the retrieved ruleset against the version in the GitOps repository. For network segmentation controls, this means querying the microsegmentation controller for the current segment membership and policy rules, and comparing them against the infrastructure-as-code definitions. For identity access controls, it means querying the identity platform for current role assignments and comparing against the access right definitions in the role catalog. Any discrepancy between the queried state and the documented state is a drift event that must be either remediated (aligning the runtime with the documentation) or intentionally accepted and documented (updating the documentation to reflect an authorized change that bypassed the normal change management process).

Control Drift Detection and Remediation Cycle

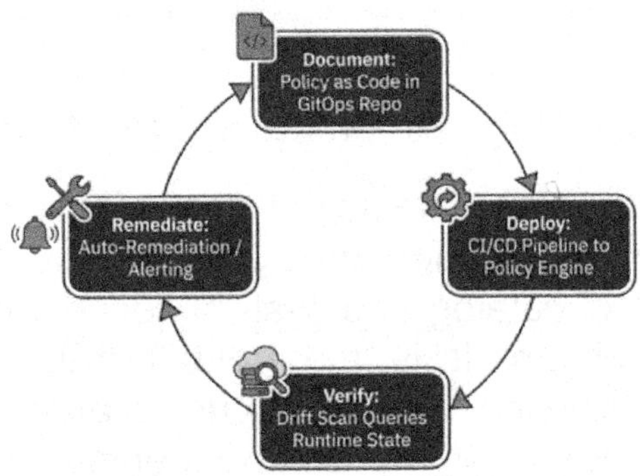

Automated processes ensure runtime configurations align with defined security policies, detecting and correcting deviations.

Audit Query Patterns and Evidence Retention

Auditors working with a continuous evidence repository rather than a manually assembled package will typically request evidence using one of four query patterns. The first is a point-in-time state query: show me the access rights for user group X as of date Y. This requires that the identity platform's access right records are retained with timestamps that allow historical state reconstruction, not just current state. The second is a change audit query: show me all changes to the network segment policy between dates X and Y, with the author and justification for each change. This requires a policy-change log with mandatory author and justification fields, not just the technical-change record. The third is an event coverage query: show me the count of authentication events that required MFA versus those that did not, for the period January through December. This requires that authentication event logs be retained for the full audit period and that the log schema include a field distinguishing MFA-enforced from MFA-exempt sessions. The fourth is an exception query: show me all access requests that were granted despite a device posture score below the policy-defined threshold. This requires that policy engine decision logs include the posture score at decision time and that exception decisions are tagged with a distinct decision code.

Evidence retention requirements vary by framework and jurisdiction. PCI-DSS requires audit log retention for at least twelve months, with three months immediately available for analysis. HIPAA requires the ability to provide documentation for six years. FedRAMP-authorized systems must retain audit records in accordance with the agency's NARA-approved records schedule, which is three years for most security-related records. An evidence retention policy that satisfies the most demanding applicable requirement, typically six years for HIPAA-covered organizations, will satisfy the shorter requirements of other applicable frameworks simultaneously. The retention architecture should distinguish between hot storage (immediately queryable, covering the most recent twelve months for operational use), warm storage (queryable within hours, covering the prior two years for audit use), and cold storage (restorable within days, covering the remainder of the retention period for regulatory compliance).

Chain of custody for compliance evidence is the documentation trail that demonstrates that evidence has not been altered between collection and audit review. For automated evidence pipelines, a chain of custody is established by cryptographic hashing of log records at ingestion, with the hash values stored separately from the evidence records. A log record whose hash does not match the value stored at ingestion time has been altered and cannot be relied upon as evidence. Evidence export processes should include a hash validation step that verifies the integrity of every record included in the export package before delivery, with a signed attestation that validation was performed. Auditors and regulators increasingly understand that automated evidence with cryptographic integrity verification is more reliable than manually assembled paper documentation, and governance programs that can demonstrate this property of their evidence pipeline are at an advantage in audit proceedings.

11.24 Manager's Decision Guide

Governance for zero trust requires decisions across three dimensions: what to measure, how to measure it, and how to report it to the audiences that require accountability. This guide walks through the decision criteria that determine whether your governance architecture is ready to support a mature zero-trust program. Work through each criterion with your compliance lead, security architect, and CISO before committing to a governance model, and use the decision frame to evaluate your current readiness.

Decision Criteria

- Maturity baseline: Has the organization completed a structured self-assessment against the CISA Zero Trust Maturity Model across all five pillars? A governance program that cannot answer "where are we now" for each pillar cannot set credible targets, track progress, or justify the next phase of investment. Commission a baseline assessment before any governance reporting begins.
- Framework scope: Have you identified every compliance framework to which your organization is subject, including both federal mandates and commercial regulatory requirements, and confirmed that your zero trust control catalog maps to each framework's requirements? Organizations that discover a missing framework mapping during an audit face emergency remediation under time pressure, which is far more expensive than designing the mapping in advance.

- Evidence automation maturity: What percentage of your compliance evidence is generated automatically by continuous telemetry versus collected manually in periodic sprints? If more than half of your evidence is still collected manually, your compliance costs will remain high, and your evidence quality will remain inconsistent. The pilot-to-production pathway for evidence automation should be an immediate priority.
- Control drift posture: Does your team have a documented process for detecting drift between the control catalog and the runtime environment, and what is the current frequency of drift detection scans? An audit that reveals a discrepancy between your documented controls and your actual runtime configuration is a worse outcome than discovering that discrepancy internally and remediating it first. Implement drift detection before the next audit cycle.
- Executive reporting quality: Do your current executive and board-level zero trust reports include lagging indicators that measure actual security posture improvement, or do they report only deployment activity? If your reports describe deployed tools and enrolled users, revise them to include at least two lagging indicators with trend data before the next board presentation.
- Third-party access governance: Does your organization have visibility into every active third-party credential and access session in the production environment, with automatic expiration for time-bounded engagements? Third-party access that is not governed by the policy engine constitutes both an evidence gap and a security gap. Identify any third-party access paths that bypass the policy engine and remediate them as a priority.

- Cross-framework efficiency: Is your organization maintaining separate compliance programs for different frameworks, with different evidence collection processes, different documentation formats, and different teams responsible for each? If yes, the consolidation opportunity through a unified control catalog is likely worth six to twelve months of governance restructuring effort, with substantial ongoing savings in audit preparation cost and engineering bandwidth.

Decision Frame

A governance architecture is ready to sustain a mature zero trust program when the organization can demonstrate continuous evidence collection for at least seventy percent of applicable control requirements, has completed a maturity baseline across all CISA pillars within the past twelve months, has implemented drift detection with a documented remediation workflow, and has at least one lagging indicator in executive reporting with at least two quarters of trend data. Organizations that meet all four criteria have operationalized governance. Organizations that meet two or three criteria are in the transition phase and should prioritize whichever missing criterion most directly enables their next audit cycle. Organizations that meet fewer than two criteria should treat governance architecture as a foundational investment on par with the zero trust technical controls themselves, because a technically excellent zero trust implementation that cannot be demonstrated to an auditor or a board provides less institutional value than it should.

11.25 Case Study: Sentinel Aerospace Consolidates Two Compliance Programs Into One

Sentinel Aerospace is a defense industrial base company with 6,500 employees operating across three domestic facilities and a network of supplier relationships that includes both domestic and international partners. The company held active Department of Defense contracts subject to CMMC Level 2 certification requirements, and its commercial aerospace customer base, representing about forty percent of revenue, had begun demanding SOC 2 Type II reports as a condition of contract renewal. For two years, Sentinel ran parallel compliance programs: a federal compliance team focused on CMMC practices and NIST SP 800-171 controls, and a commercial compliance team focused on SOC 2 Trust Service Criteria. Both teams conducted separate evidence-collection exercises, maintained separate documentation sets, and contracted with separate audit firms for their respective examinations.

The duplication was expensive and operationally disruptive. Engineering teams received overlapping evidence requests from both compliance functions, sometimes within the same quarter, covering the same underlying controls from two different documentation formats. A log export that satisfied the federal compliance team's evidence request for AC.1.001 (Limit information system access to authorized users) was not in the correct format for the SOC 2 compliance team's CC6.1 evidence package, even though the same policy engine configuration met both requirements. The engineering team was generating the same underlying data twice in different formats, which cost approximately 600 engineering hours per year and introduced reconciliation complexity when the two exports reflected slightly different system states.

The governance consolidation began with a control catalog project that took four months. The compliance architect, working with the security engineering team, built a unified control catalog that mapped each of the 110 CMMC Level 2 practices and the applicable SOC 2 Trust Service Criteria to the same set of zero trust control implementations. The catalog identified 34 controls that appeared in both frameworks under different identifiers but were satisfied by identical technical implementations. It also identified 12 controls where the frameworks diverged, requiring separate implementation evidence. The 34 shared controls represented the consolidation opportunity: a single evidence package, generated from a single telemetry query, could satisfy both frameworks simultaneously.

Sentinel Aerospace: Unified Control Catalog Architecture

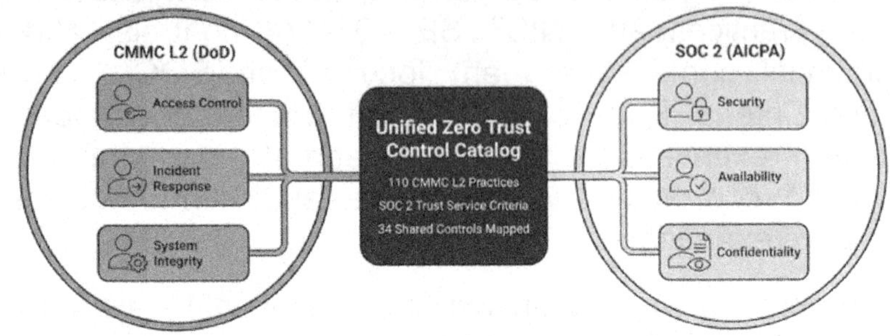

Sentinel Aerospace: Unified Control Catalog Architecture

The evidence automation pipeline was the second phase of the project. Using the control catalog as the source of record, the engineering team built SIEM query templates for each of the 34 shared controls and for each of the 12 framework-specific controls. The templates were parameterized by audit period, so that any date range could be submitted to generate the corresponding evidence package. Evidence exports were cryptographically hashed at generation and stored with the hash in an immutable evidence repository. When the CMMC examination opened, the compliance team submitted a hash-verified evidence package generated by template queries. When the SOC 2 audit opened six months later, the same templates were applied to the same evidence repository for the overlapping period, with the SOC 2 control identifiers substituted for the CMMC practice identifiers on the cover sheet.

The consolidation results were measurable across multiple dimensions. Annual audit preparation time fell by approximately sixty percent, from an estimated 1,400 combined engineering and compliance hours across both audit cycles to approximately 560 hours. The reduction came entirely from eliminating duplicate evidence collection, duplicate documentation maintenance, and duplicate auditor communication. Engineering teams reported that the reduction in compliance interruptions during the second half of each year materially improved sprint velocity in quarters previously dominated by audit preparation. On the security posture dimension, the unified control catalog produced an unexpected benefit: when the catalog was reviewed for the first time as a unified document, the compliance team identified six controls that appeared in both frameworks but had been implemented with slightly different configurations for each, creating undocumented inconsistencies in how the same policy was applied to federal-contract systems versus commercial-customer systems. Normalizing those six controls to a single implementation improved both security posture and compliance consistency.

The lesson Sentinel Aerospace's governance team documented for peer defense contractors was precise: the decision to run separate compliance programs for CMMC and SOC 2 was made at the beginning of the compliance journey before the organization understood how much the two frameworks overlapped at the control implementation level. That decision made sense as a risk management choice when the frameworks were unfamiliar; it became an expensive structural choice once the organization had enough operational experience to see the overlap clearly. The recommendation was to build the unified catalog at the beginning of any multi-framework compliance program, even if it requires an initial investment in catalog development, because the ongoing savings in engineering and compliance effort compound across every subsequent audit cycle.

11.26 Team Conversation Guide

Run these three discussions in sequence over two to four weeks. The first surfaces the issue, the second works the trade-offs, the third commits the team to action.

Discussion 1. Assessing Our Maturity Position

This discussion asks the team to evaluate the organization's current zero-trust maturity against a structured framework rather than relying on intuitive assessments. Before the session, review the CISA Zero Trust Maturity Model's five pillars and three stages, and have the team briefly research what the Advanced stage looks like for each pillar. The goal is to produce an honest, evidence-based maturity assessment that can serve as the baseline for governance reporting.

Discussion questions to ask the team:

- For each of the five CISA pillars, Identity, Devices, Networks, Applications, Workloads, and Data, what stage would we honestly assess ourselves at: Traditional, Advanced, or Optimal? What specific evidence supports each assessment, and where are we relying on assumptions rather than documented facts?
- Which pillar represents the largest gap between where we are and where our compliance requirements implicitly or explicitly require us to be? What would it take to advance that pillar by one stage in the next twelve months?
- What metrics would we need to instrument to demonstrate maturity stage advancement to an auditor or executive who has not observed the implementation process directly? Which of those metrics are we currently capturing, and which are not yet instrumented?
- If we assessed our zero-trust maturity using both the CISA model and the DoD Reference Architecture framework simultaneously, would we get consistent answers? Where do the two frameworks differ in priorities or capability definitions?

What to listen for: Listen for assessments that are more optimistic than the evidence supports. Teams that assess themselves as Advanced when they cannot name specific evidence for each Advanced capability are describing an aspirational state rather than an actual state. Both are useful to know, but they require different governance responses. Also listen for uncertainty about which evidence would be needed to substantiate a maturity claim: that uncertainty identifies where instrumentation investment is needed before the next maturity assessment.

Follow-up actions: Document the maturity assessment formally, citing evidence for each pillar stage assessment and the identified gaps. Use the assessment as the starting point for governance reporting and as the baseline against which future assessments are measured. Schedule a maturity reassessment every 6 months to track progress across the pillars.

Discussion 2. Designing Metrics That Measure Security Improvement

This discussion addresses the trade-off between vanity metrics and outcome metrics in zero-trust reporting. Most teams have been measuring and reporting deployment activity because it is easy to measure. The goal of this conversation is to identify two or three outcome-oriented metrics that the team can instrument and report as leading or lagging indicators of actual improvements in security posture.

Discussion questions to ask the team:

- If a board member asked us to demonstrate that our zero trust implementation has reduced the blast radius of a plausible credential compromise, what specific data would we provide, and how would we calculate blast radius in quantitative terms? Is that data currently available from our systems?
- What is our current mean time to detect an anomalous access pattern, and how would we know if that metric improved by fifty percent? What instrumentation and process changes would be

required to move from a rough estimate to a precisely measured, continuously tracked figure?

- How do we currently measure privilege sprawl, and does our measurement include not just human users but also service accounts, API keys, and machine identities? What would a privilege sprawl ratio look like for our environment, and what target ratio would represent meaningful improvement?
- Which of our current security metrics would a knowledgeable board member reject as deployment activity rather than security outcome data, and what outcome metric would replace each one? How long would it take to instrument those replacement metrics?

What to listen for: Listen for metrics that can be instrumented from existing telemetry sources versus those that require new data collection infrastructure. The most accessible improvements are typically metrics derived from SIEM queries against data already being collected but not yet analyzed for this purpose. Also listen for disagreement about what constitutes improvement. If the team cannot agree on a target value for a metric, the metric is not yet well-defined enough to serve as a governance instrument.

Follow-up actions: Commit to instrumenting at least one lagging indicator before the next executive reporting cycle. Build the dashboard query, validate it against known data, and include the metric with at least one prior period for trend context in the next board presentation. Identify the two remaining lagging indicators that would complete a minimal viable outcome reporting set and schedule their instrumentation.

Discussion 3. Eliminating Compliance Duplication

This discussion is for organizations subject to multiple compliance frameworks. It is designed to surface the specific duplication overhead that a unified control catalog and evidence pipeline would eliminate. Before the session, ask each team member who interacts with a compliance framework to estimate how many hours per year they spend on compliance-specific evidence collection, documentation maintenance, and auditor communication that is not already captured in normal operational processes.

Discussion questions to ask the team:

- Across all compliance frameworks we are subject to, how many distinct evidence collection requests have we received in the past twelve months, and how many of those requests covered the same underlying controls under different framework names? If we mapped those requests to our actual technical controls, how many unique control facts do we think they represent versus how many framework-specific presentations of the same facts?
- What automated compliance evidence can our current infrastructure generate today without any engineering changes, and what is the gap between what we can generate automatically and what auditors currently request from us? Which of those gaps would require new instrumentation versus simply building a new query against existing data?
- How do we structure architecture reviews and engineering documentation to support governance requirements without creating documentation overhead that slows engineering velocity? Where is the boundary between documentation that the governance program genuinely needs and documentation that has accumulated as audit theater, adding no actual compliance or security value?
- If we built a unified control catalog that mapped all applicable framework requirements to our specific technical implementations, how long would it take to build that catalog, and what would be the payback period in audit preparation time saved?

What to listen for: Listen for engineers who describe compliance work as separate from their normal responsibilities, something that happens to them rather than something integrated into their development and operations process. That perception reflects a governance architecture that has not yet integrated compliance into the engineering workflow. Also listen for estimates of audit preparation time that seem unusually high: those estimates often include significant effort that could be eliminated by automation and would not have surfaced without a direct discussion.

Follow-up actions: Run a one-page audit of compliance evidence requests from the past twelve months and map each request to the underlying technical control it represents. Use that mapping to estimate the consolidation opportunity in a unified catalog. Present the analysis to compliance and engineering leadership as the business case for catalog investment, with a specific hourly estimate of ongoing savings.

11.27 Takeaway

Marcus Reyes's notepad moment, the recognition that the auditor's evidence list and the executive risk review were asking for the same control facts under different headings, is where every mature governance program begins. The technical work of zero trust implementation generates a continuous stream of facts about access decisions, policy enforcement, device posture, and data protection. Governance is the discipline of capturing those facts once, organizing them in a structure that maps to every applicable framework simultaneously, and surfacing them for each audience in the format that audience needs. When that architecture is working correctly, compliance preparation does not require an engineering sprint. Executive reporting does not require data-gathering. Auditors receive evidence packages that reflect the same system state that the security operations team monitors daily.

The maturity models, CISA's five-pillar framework, and the DoD's capability levels provide the vocabulary for making zero-trust progress legible to stakeholders who cannot read infrastructure-as-code repositories or policy-engine logs. They convert engineering implementation into a shared measurement framework that engineers, compliance teams, executives, and boards can all reference without translation. A quarterly executive review that reports maturity-stage advancement across five pillars, along with two lagging indicators showing blast-radius reduction and mean time to detect improvement, tells a complete story about the organization's security posture in less than a page. That brevity is made possible by the governance architecture underneath it, which ensures that every number in that page is derived from the same authoritative, continuously collected, cryptographically verified evidence that the external auditor reviewed six months earlier.

Operational clarity at the governance level requires making the documentation and the runtime the same thing. When policy-as-code, infrastructure-as-code, and identity-as-code are not aspirational goals but operational realities, the control catalog is always synchronized with the production environment, drift is detected automatically, and the compliance evidence pipeline operates continuously without human intervention. Organizations that reach this state discover that governance transforms from an overhead function into a competitive advantage: the ability to onboard new compliance frameworks quickly, demonstrate security posture to prospective customers and partners on demand, and sustain executive trust through consistent, transparent, and auditable reporting becomes a differentiator in markets where security posture is increasingly a commercial consideration. Chapter 12 synthesizes everything the book has covered into the program-building framework that makes this level of operational clarity achievable at scale.

- Building and Scaling Your Zero Trust Program

Marcus Reyes had been in this conference room before, but not in this chair. Two years ago, when Meridian Logistics Group's new CISO handed him the security architect role and tasked him with building a zero-trust program from scratch, he sat across the table from leadership and answered questions. Today, he was presenting to the new CIO, a woman named Sandra Park, who had come from the healthcare sector and, within forty-eight hours of arriving, had asked for a current-state assessment of every infrastructure security program. Marcus had three weeks to prepare. He spent the first week organizing documentation. He spent the second week building the roadmap deck. He spent the third week figuring out how to explain two years of architectural transformation to someone who would rightly ask whether it was working and what it would cost to finish.

The presentation he built was not a technology inventory. Sandra Park did not need a list of tools. She needed to understand what the organization was protecting, how it was protecting it, what had changed in the past two years, what remained to be done, and what it would cost to get there. Marcus structured the brief around four phases: the assessment phase that established baseline maturity in year one, the foundation phase that deployed identity consolidation, initial microsegmentation, and continuous verification pilots across the first eighteen months, the acceleration phase that was currently underway extending zero trust controls to cloud workloads and the contractor ecosystem, and the optimization phase that lay three years out, where automated policy enforcement, AI-assisted anomaly detection, and self-healing network segmentation would close the remaining gaps. He had numbers for each phase. He had risk metrics that showed what each phase of completion had actually changed in the organization's threat profile. He had an honest accounting of what had not gone according to plan.

Sandra Park stopped him fourteen minutes into the presentation, not because she was dissatisfied but because she wanted to go deeper on one particular slide. It showed a reduction in mean time to detect over the course of the program, from 26 hours at baseline to 9 hours at the end of the foundation phase, with a trend toward a 4-hour target in the acceleration phase. She asked how those numbers were calculated. Marcus walked her through the telemetry architecture, the behavioral baseline system, the policy engine logging infrastructure, and the quarterly incident simulation cadence they used to validate detection capability rather than assume it. She listened, asked three follow-up questions about the methodology, and nodded. "This is what I need," she said. "This is how we talk to the board about security investment."

That moment in the conference room is the destination this chapter is designed to reach. Not the technology, not the architecture, not the engineering decisions, but the point at which a security program can present itself to organizational leadership in the language of outcomes, evidence, and forward investment rather than jargon, tools, and aspirational architecture diagrams. Building a zero-trust program is a technical problem. Scaling it, sustaining it, and defending it across budget cycles and leadership transitions is an organizational problem. This chapter addresses both. Everything that preceded it in this book describes what to build. This chapter describes how to build an organization capable of delivering on that vision, and how to keep that organization moving over the three-to-five-year horizon that a genuine zero-trust transformation requires.

11.28 Why It Matters

A technically correct zero-trust architecture that cannot survive a budget review, a leadership change, or a year of organizational competing priorities is not a security asset. It is a pilot that gradually decays. The gap between "we have zero trust technology deployed" and "we have a functioning zero trust program" is where most enterprise security transformations fail. The technology components are purchased, deployed, and piloted. Then a key champion leaves. Then a budget cycle introduces cuts that seem targeted at new programs rather than established operations. Then a merger adds fifty thousand endpoints and an entirely different identity infrastructure, and there is no governance mechanism to absorb the integration work into the existing program. The architecture exists. The program does not. And within eighteen months, the technology components are operating independently, without shared policy, without coordinated telemetry, and without the organizational accountability structure that was supposed to make them coherent.

The mission-aligned security program does not treat these organizational risks as secondary concerns. It treats them as first-order architectural inputs. The governance structure that maintains program accountability through leadership transitions is as much a design artifact as the policy engine. The budget justification framework that translates security outcomes into board-level risk language is as important as the microsegmentation deployment plan. The staffing model that identifies which capabilities require deep internal expertise versus which can be sourced through managed services is as consequential as the tool selection decisions it makes. None of this is soft. All of it is engineering in the organizational domain, and it requires the same discipline, the same measurement rigor, and the same willingness to iterate on evidence that the technical domains demand.

The organizations that complete genuine zero trust transformations share a pattern: they built a program, not a project. Projects have end dates, fixed budgets, and defined deliverables. Programs have governance, evolving objectives, continuous measurement, and organizational structures that allow them to absorb setbacks without collapsing. The transition from project mindset to program mindset is the single most important shift a security leader can make when facing a multi-year transformation. This chapter provides the frameworks, language, and decision tools to make that transition successfully and sustain the program once the initial energy of the launch phase has settled into the harder work of multi-year execution.

11.29 Phased Program Design: From Assessment to Optimization

A zero-trust program that attempts to deploy comprehensively in the first year will almost certainly fail. The number of simultaneous dependencies, the organizational change management burden, the skills development required, and the sheer volume of legacy systems that must be inventoried, assessed, and integrated mean that a big-bang approach produces either a scope reduction that hollows out the architecture or an execution failure that damages the credibility of the program for years. The alternative is a phased design that delivers genuine, measurable security improvement at each stage while building the technical and organizational foundation that makes the subsequent phase possible.

Four phases characterize the programs that succeed. The assessment phase, typically six to nine months, establishes the baseline. The foundation phase, typically twelve to eighteen months, deploys the core architectural components. The acceleration phase, typically twelve to eighteen months, extends coverage and integrates advanced capabilities. The ongoing optimization phase automates, refines, and continuously validates the program. Each phase should have defined entry criteria, measurable outcomes, and explicit exit gates that the organization must pass before committing resources to the next phase. The phase structure is not bureaucratic formality. It is how a complex multi-year transformation maintains operational clarity across the entire lifecycle.

Phase 1: Assessment (Months 1 to 9)

The assessment phase answers a specific question: what does the current environment actually look like, not the documented architecture, but the live operational reality? Organizations that skip or abbreviate this phase spend enormous resources in the foundation phase, discovering gaps that a thorough assessment would have caught. The assessment must cover all five zero trust pillars: identity, devices, networks, applications, and data. For each pillar, the assessment establishes a baseline maturity score against a recognized framework such as the CISA Zero Trust Maturity Model, documents the specific gaps between the current state and the target state, and identifies the technical dependencies that sequence the remediation work.

The assessment phase also produces the inventory artifacts that subsequent phases depend on. Identity inventory: every human and machine identity, its authentication mechanism, its current privilege level, and its ownership. Device inventory: every managed and unmanaged endpoint, its compliance posture, and its network connectivity pattern. Application inventory: every application, its authentication integration, its data classification, and its current access control model. Network inventory: every segment, every trust relationship, every firewall rule, and every cloud security group. Data inventory: the highest-sensitivity data assets, their locations, access patterns, and the current controls protecting them. Without these inventories, the foundation phase is guesswork.

Zero Trust Program Phase Structure

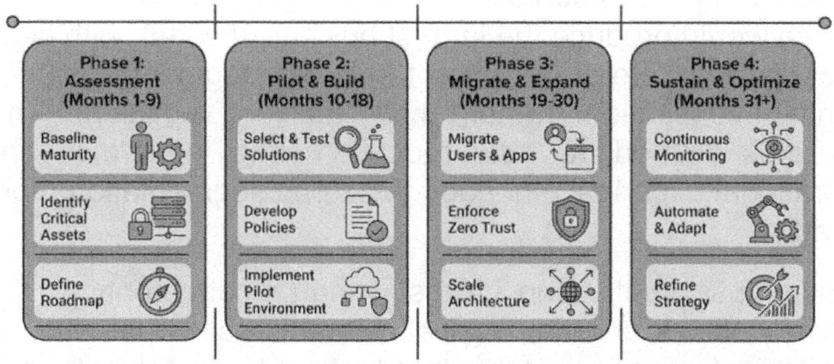

A structured, multi-phase approach to implementing and maturing a Zero Trust security framework.

Phase 2: Foundation (Months 10 to 27)

The foundation phase deploys the core architectural components that enable the subsequent phases. It does not attempt to achieve full coverage. It achieves depth in the highest-priority areas while establishing the operational patterns that the engineering team will replicate across the broader environment in the acceleration phase. The three foundation pillars are identity consolidation, initial microsegmentation, and continuous verification.

Identity consolidation means establishing a single authoritative identity plane with phishing-resistant authentication for all interactive users, machine identity governance for all service accounts and certificates, and lifecycle automation that connects HR and IT provisioning events to real-time access changes. This pillar is the prerequisite for everything else: without a clean, authoritative identity inventory and a functioning policy plane, device trust signals cannot be correlated, network segments cannot enforce identity-aware access, and continuous verification has no baseline to verify against.

Initial microsegmentation means deploying segmentation boundaries around the highest-risk and highest-value asset clusters first, typically financial systems, intellectual property repositories, and the identity infrastructure itself. The goal is not pervasive microsegmentation in year two. It is demonstrable segmentation around the assets where a breach would produce the largest business impact, with clear evidence that lateral movement between segments requires identity verification rather than network position. This produces immediate risk reduction and generates the operational experience needed to extend segmentation more broadly.

Continuous verification pilots mean instrumenting the telemetry stack and running behavioral baseline detection in monitor-only mode across the highest-traffic identity and device populations before enforcing policy in block mode. The pilot phase discovers the false-positive profile that determines enforcement readiness. Organizations that skip the monitor-only phase and enforce immediately typically generate enough access disruption to trigger a rollback, which sets the program back by months and erodes the credibility needed for subsequent phases.

Phase 3: Acceleration (Months 28 to 45)

The acceleration phase extends the patterns established in the foundation phase across the broader environment while integrating more advanced capabilities enabled by the foundation phase. Extension targets in a typical enterprise include the contractor and partner ecosystem, cloud workloads across all tenants, IoT and OT device populations, and the application layer where legacy authentication mechanisms still operate. Advanced capabilities typically introduced in this phase include AI-assisted anomaly detection tuned to the behavioral baselines established in the foundation phase, automated generation of compliance evidence, and the first elements of self-service access request and approval workflows that reduce the operational overhead of the zero-trust policy enforcement layer.

The acceleration phase is also when the program typically encounters the most significant organizational resistance, because it touches systems and workflows that the foundation phase deliberately avoided. Legacy application owners who accepted that "identity consolidation" was someone else's problem discover in the acceleration phase that their applications must integrate with the policy engine. Network teams whose segments were not part of the initial microsegmentation scope find that the acceleration phase extends enforcement into their operational territory. The program management disciplines established in earlier phases, particularly the executive communication cadence and the stakeholder governance structure, enable navigation of these encounters without derailing the program.

Phase 4: Optimization (Ongoing)

The optimization phase has no end date. It is the operating mode of a mature zero-trust program: continuous measurement, continuous refinement, periodic reassessment of maturity, and ongoing investment in capabilities to keep the architecture current as the threat landscape, the regulatory environment, and the business technology portfolio evolve. Organizations that treat phase four as a lower-cost steady state are correct that the major capital expenditure is behind them, but incorrect if they assume that the program can sustain itself without ongoing investment. The threat the zero-trust program is designed to counter is actively evolving. A program that stops evolving becomes a compliance artifact rather than a security mechanism.

11.30 Vendor-Neutral Technology Evaluation

The zero-trust technology market is large, growing, and populated by vendors whose marketing language has converged on a common set of terminology. Every identity vendor claims to support zero trust. Every network security vendor claims to enable microsegmentation. Every endpoint detection vendor claims to provide continuous verification. The convergence of language means that evaluating technology against vendor claims is nearly useless. Evaluating technology against architectural requirements, integration standards, and operational evidence is the only framework that produces defensible procurement decisions.

Zero Trust Technology Evaluation Framework

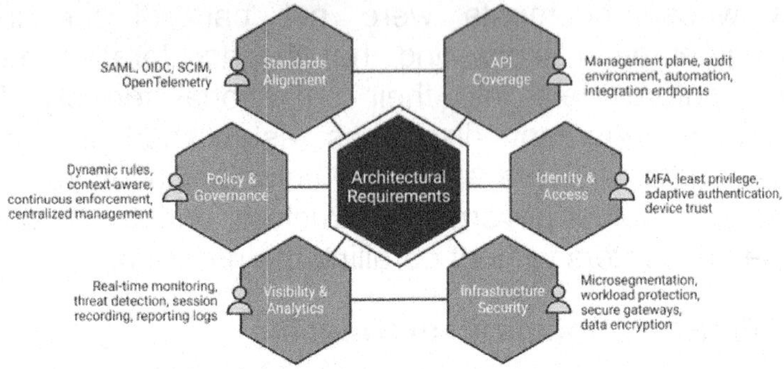

Zero Trust Technology Evaluation Framework provides a structured approach for assessing and selecting security solutions aligned with modern security principles.

The architecture requirements dimension asks whether a candidate solution implements the zero-trust capability you actually need, not the capability the vendor chooses to demonstrate in a proof of concept. A policy engine that works beautifully with the vendor's own identity product but requires a proprietary connector to your existing directory is not implementing zero-trust architecture. It is creating a new perimeter around the vendor's ecosystem. The evaluation must test against your specific environment, your existing identity sources, your device inventory, and your network topology, not against a sanitized lab environment.

The standards alignment dimension asks whether the product speaks the open protocols that allow it to integrate with the rest of the architecture without vendor-specific glue code. For identity, this means SAML 2.0, OIDC, and SCIM. For telemetry, it means OpenTelemetry or a compatible schema that allows ingestion by your SIEM without proprietary agents. For network policy, it means support for open configuration APIs that your automation tooling can call without SDK lock-in. Standards alignment is not pedantry. It is the practical foundation of vendor-neutral architecture, allowing you to replace any component without rebuilding the stack around it.

The operational fit dimension is the one that vendor evaluations most consistently underweight. A solution that requires three full-time engineers to operate, generates fifteen thousand alerts per day at an untuned baseline, and requires four-week professional services engagements to change a policy template is not operationally fit for a team of eight managing it alongside a dozen other security functions. The operational fit assessment requires hands-on trial in your environment, not a vendor demo. It means running the solution in your lab for 30 to 60 days with your actual data, alert volume, and team, then honestly assessing whether the operational model is sustainable.

The migration pathway dimension asks whether the vendor supports a phased deployment model that allows you to run the new solution alongside your existing tools during transition, preserve existing policy logic during migration, and establish feature parity with current capabilities before decommissioning legacy components. Vendors that require a forklift replacement of existing infrastructure as a condition of deployment impose a risk and disruption cost that does not belong in a responsible zero-trust program. Pilot-to-production pathway discipline means demanding a migration plan from every vendor before committing to a contract.

11.31 Staffing and Skills Development

The skills required to build and operate a zero-trust program are not uniformly distributed across the security profession. Identity engineering, policy plane architecture, microsegmentation design, behavioral analytics tuning, and automated runbook development are specializations that require significant investment to develop or acquire. Most organizations entering a zero trust transformation have teams built for the perimeter era: firewall engineers, VPN administrators, antivirus platform managers, and perimeter network specialists. These are capable professionals whose skills are partially transferable but will need substantial investment in development to operate effectively in the zero-trust operating model.

Zero Trust Program Staffing Model

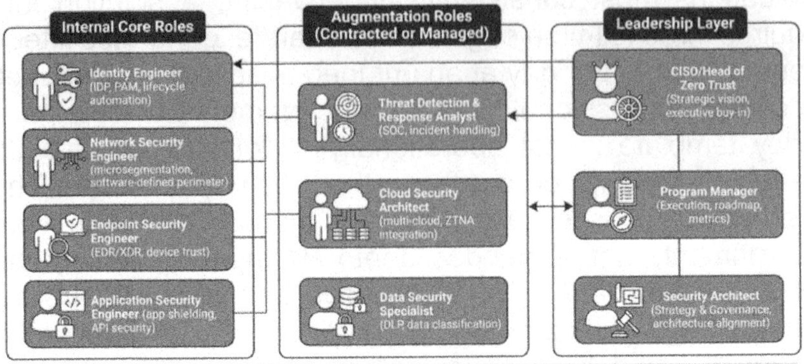

Zero Trust Program Staffing Model requires a blend of specialized internal expertise, scalable external support, and strategic oversight for effective implementation.

The staffing model for a zero-trust program distinguishes between roles that must be owned internally and roles that can be sourced externally without creating capability dependency risk. Internal ownership is required for four core functions: identity engineering, network security engineering, security operations analysis, and cloud security engineering. These functions require deep familiarity with the organization's specific environment, history, and operational constraints. External professionals operating without that context can design the architecture and validate the implementation. Still, they cannot own the daily operational decisions that determine whether the architecture functions as intended.

Skills development for existing team members is more cost-effective than external hiring for most organizations, but it requires structure and accountability. A skills assessment, conducted at the beginning of the assessment phase, maps current team capabilities against the competency requirements for each zero-trust domain. The gaps identified by the skills assessment drive a training investment plan that combines vendor certifications where relevant, vendor-neutral zero-trust curriculum, and hands-on lab development time with actual program tooling. The most effective skills development programs pair formal training with structured mentorship from more experienced practitioners, either from within the organization or sourced through a professional development partnership with a consulting firm working on the program.

Hiring for genuine zero-trust expertise is difficult and competitive. The talent market for identity engineers, policy plane architects, and behavioral analytics specialists is tight, and candidates who can demonstrate genuine depth, rather than surface familiarity with vendor certifications, command significant compensation premiums. The organizations that compete most effectively for this talent are those that can articulate a specific, multi-year program with visible technical challenges, genuine engineering autonomy, and a career development path. Candidates who are serious about zero-trust architecture as a professional focus are not motivated solely by compensation. They are motivated by the quality of the technical problem.

11.32 Executive Communication and Board Reporting

The bridge between the zero-trust program and the organizational leadership that funds and governs it is built on a specific communication discipline: translating technical security metrics into business-risk language that executives can act on. Security professionals who present dashboards of vulnerability counts, patch compliance percentages, and policy rule totals to executive audiences are communicating in the wrong currency. Executives make decisions about risk, investment, and organizational priority. The communication framework that sustains a multi-year zero trust program must speak directly to those decision categories.

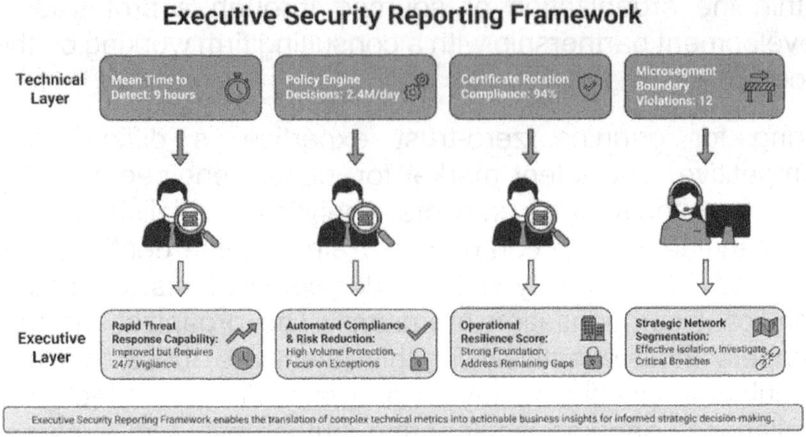

The executive communication cadence should be quarterly for the standing update and annual for the formal program review. The quarterly update is fifteen to twenty minutes and covers four topics: progress against current-phase milestones, the three or four security metrics that best express program effectiveness in business risk terms, any significant incidents or near-misses, and how the zero trust architecture performed relative to the pre-program baseline, and any resource or scope changes needed in the current quarter. The format should be fixed so that executives can immediately orient to each update without relearning the structure.

Board-level reporting is distinct from executive communication in both frequency and framing. Boards receive security updates quarterly or semi-annually in most organizations, and the framing appropriate for a board audience is explicitly one of governance and fiduciary risk rather than operational detail. The board communication should address three questions: what is the current threat environment and how does it relate to the organization's specific risk profile; what is the current state of the zero trust program relative to the board-approved risk tolerance; and what investment decision, if any, does the board need to make to maintain or accelerate program progress. Boards are not equipped to evaluate firewall rule counts or certificate expiration statistics. They are equipped to evaluate whether the organization's security investment is proportionate to its risk exposure and whether the current program trajectory is likely to achieve the risk reduction it was funded to produce.

Maintaining executive sponsorship through program setbacks is a specific challenge that requires deliberate management. Zero-trust transformations encounter setbacks. A major identity consolidation migration breaks a business-critical application. A microsegmentation deployment blocks access to a partner data feed that nobody knew existed. An acceleration phase deployment doubles the alert volume in the security operations center for three weeks while the team tunes the new detection logic. These are not program failures. They are expected characteristics of complex technical transformation. The executive communication strategy must establish this expectation early, provide a clear framework for distinguishing between a significant setback and a normal implementation friction event, and consistently demonstrate that the program's response to setbacks is structured and evidence-based rather than reactive.

11.33 ROI Measurement: Quantifying the Zero Trust Investment

The return-on-investment case for zero trust is real and quantifiable, but it requires a measurement methodology that goes beyond "we deployed these tools, and nothing bad happened." The three primary ROI categories for zero-trust investment are reducing breach probability, avoiding compliance costs, and improving cyber insurance coverage. A secondary category, operational efficiency, becomes measurable in the later phases of the program, when automation reduces the workload of manual security operations. Each category requires a distinct measurement approach and a distinct evidence standard.

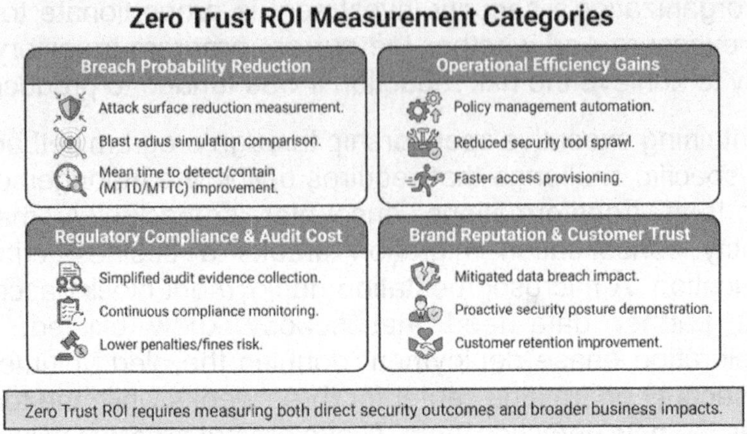

Breach-probability reduction is the most direct ROI category but also the most methodologically complex to measure. The approach that produces defensible numbers uses attack simulation and blast-radius measurements before and after program phases. Before the foundation phase, conduct a tabletop or technical breach simulation that traces the path from a common initial access vector, typically a compromised credential, to the organization's highest-value data assets. Measure the number of distinct access steps required, the number of trust boundaries crossed, and the time a skilled attacker would require to complete the path. After the foundation phase, repeat the simulation. The measurable reductions in path viability, lateral movement opportunity, and time-to-detection constitute empirical evidence of a reduction in breach probability. This approach produces numbers that a security-literate board member or CFO can evaluate.

Compliance cost avoidance is often the most financially legible ROI category for organizations operating in regulated industries. The measurement methodology compares the direct cost of compliance audit preparation before and after the deployment of the zero-trust program's evidence-automation capabilities. Before automation, compliance evidence collection for a major audit in a typical enterprise required hundreds of person-hours of manual work: pulling access logs, generating user access reviews, producing network segmentation evidence, and documenting control implementation. After deploying automated compliance evidence generation, this same evidence set is produced continuously and delivered automatically when an audit request is received. The person-hours saved, multiplied by the loaded labor cost, is a direct compliance-cost-avoidance figure that requires no actuarial assumptions.

The impact of cyber insurance is increasingly concrete as insurance underwriters have developed sophisticated zero-trust assessment frameworks. Major cyber insurance carriers now offer premium reductions of 15 to 30 percent for organizations that can demonstrate specific zero-trust control maturity: phishing-resistant MFA for all privileged access, micro-segmented network environments, privileged access management with session recording, and endpoint detection and response with behavioral analytics. The measurement methodology documents current controls against insurer questionnaire criteria, maps each deployed zero-trust capability to specific insurance policy terms, and calculates the premium reduction negotiated based on evidence of program maturity. Organizations that have completed this negotiation report that the premium reduction over a three-year policy term frequently offsets a significant portion of the investment in the zero-trust program.

11.34 Federal Program Management Considerations

Federal agencies operating zero trust programs face a set of external constraints that commercial enterprises do not: statutory deadlines, congressional reporting requirements, appropriations cycles that do not align with multi-year program timelines, acquisition regulations that shape how technology can be procured and deployed, and Authority to Operate processes that must accommodate the continuously adaptive nature of zero trust architecture. Understanding these constraints is not optional for federal security leaders. It is the operating environment within which every technical decision must be made.

The Office of Management and Budget's M-22-09 memorandum, issued in January 2022, established specific federal zero-trust architecture requirements, with completion targets for fiscal year 2024 across all five pillars of the CISA Zero Trust Maturity Model. These requirements are not aspirational guidance. They are compliance obligations with oversight mechanisms, including agency attestation requirements, Inspector General assessment processes, and congressional reporting tied to annual defense authorization. Federal agency program managers who treat the OMB requirements as a compliance checkbox rather than an architectural roadmap will find the compliance attestation process increasingly difficult as the maturity model assessments become more rigorous.

Federal Zero Trust Compliance Landscape

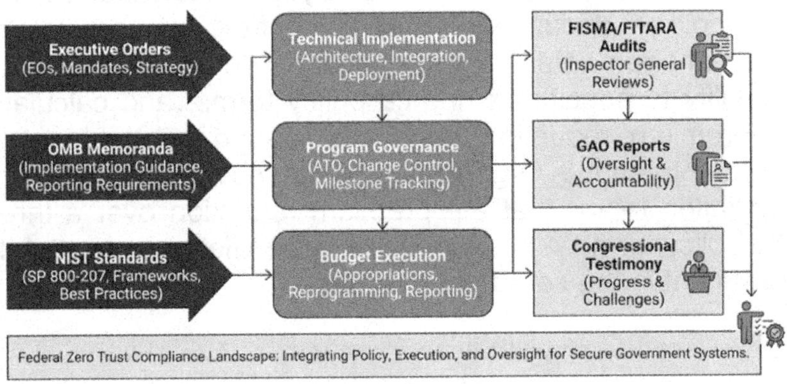

The Authority to Operate process poses a specific challenge for zero trust programs because ATO is typically granted to a bounded system with a fixed configuration baseline. In contrast, zero-trust architecture is deliberately adaptive, continuously updating policies based on new telemetry signals, evolving behavioral baselines, and threat-intelligence feeds. The emerging approach in the federal space is the continuous Authorization to Operate, which replaces the point-in-time ATO with a continuous monitoring framework that maintains authorization based on real-time control validation rather than periodic snapshots. Federal agencies pursuing zero trust should work with their Authorizing Official early in the program to establish a continuous ATO framework rather than attempting to retrofit traditional ATO processes onto an architecture whose policy configuration changes in real time.

Congressional reporting requirements create a specific communication discipline for federal program managers. Congressional oversight, whether through appropriations committee hearings, Inspector General reports, or Government Accountability Office program reviews, demands that program status be expressed in terms of milestones achieved, funding expended, and outcomes delivered rather than technical architecture descriptions. Federal zero-trust program managers who build their milestone tracking and outcome measurement systems with congressional reporting in mind from the beginning will find the reporting process substantially less burdensome than those who build their internal program management systems in technical terms and then attempt to translate them for reporting.

11.35 Commercial Program Considerations

Commercial enterprises face a different set of external constraints than federal agencies, but they are no less real. Board risk reporting requirements, audit committee oversight, cyber insurance underwriting processes, and M&A due diligence all create external accountability mechanisms that shape how a zero-trust program must be designed, documented, and communicated. The commercial enterprise that builds its zero-trust program documentation system with these external audiences in mind will be better prepared for moments when external scrutiny intensifies.

Board risk reporting for security has evolved substantially in recent years. The Securities and Exchange Commission's cybersecurity disclosure rules, effective for large accelerated filers since December 2023, require public companies to disclose material cybersecurity incidents within four business days and to provide annual disclosures of their cybersecurity risk management processes, strategy, and governance. For zero trust program managers at publicly traded companies, these disclosure requirements create both an obligation and an opportunity: the obligation to maintain defensible, board-level documentation of the security program's design and effectiveness, and the opportunity to position the zero trust program as direct evidence of the governance practices that the disclosure rules require companies to describe.

Commercial Zero Trust External Accountability Map

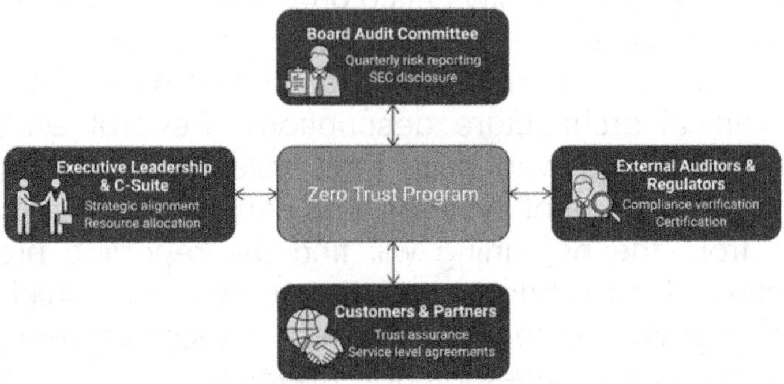

Commercial Zero Trust External Accountability Map

Mergers and acquisitions pose some of the most acute challenges for zero-trust programs that commercial enterprise security leaders face. When an organization acquires a target company, the combined entity immediately has two identity systems, two network architectures, two device management philosophies, two data classification frameworks, and potentially two entirely different security program maturity levels. The zero-trust program that has invested in standards-based, vendor-neutral architecture is substantially better positioned to absorb an acquisition than the program built on proprietary vendor ecosystems, because standards-based integration points can accommodate the target's systems without requiring an immediate forklift migration. The acquisition integration roadmap should be a specific artifact in the zero trust program documentation, even when no acquisition is imminent, because having thought through the integration model in advance dramatically reduces the time required to produce a credible integration plan when an acquisition actually occurs.

The acquisition integration model for zero trust has a specific sequence. The priority is identity: establishing visibility into the acquisition target's identity within thirty days of close, even if full consolidation takes eighteen months. The second priority is network segmentation: establishing a controlled connectivity model between the acquired entity and the parent organization's environment rather than allowing open connectivity that would void the segmentation investments in the existing program. The third priority is device management: understanding the endpoint posture of the acquired entity and establishing a compliance baseline before extending trust signals that assume device health. Data classification and application integration follow as the environments are stabilized. The specific timelines depend on the acquisition size and complexity, but the sequence is consistent.

11.36 Sustaining Momentum Across Leadership and Budget Transitions

The leadership transition is the most common cause of interruptions to zero-trust programs. A CISO who championed the program departs. A new CIO arrives with different technology preferences and a fresh mandate to reduce vendor sprawl. A change in administration or corporate ownership reshuffles the entire leadership layer above the program team. In each of these scenarios, the program's survival depends on how thoroughly it has embedded itself in the organization's governance structures, documented its rationale, and built relationships beyond its immediate champion.

The governance mechanisms that sustain a program through leadership transitions are the Architecture Review Board, the documented program charter, the established metrics cadence, and the cross-functional stakeholder network. The Architecture Review Board is a standing governance body with explicit authority to approve zero trust architecture decisions, technology selections, and scope changes. Its membership should span business units, legal, compliance, and the IT organization, not just the security team. When the security leadership changes, the ARB continues to serve as the program's institutional memory of architectural decisions and the governance body that maintains continuity during the transition period.

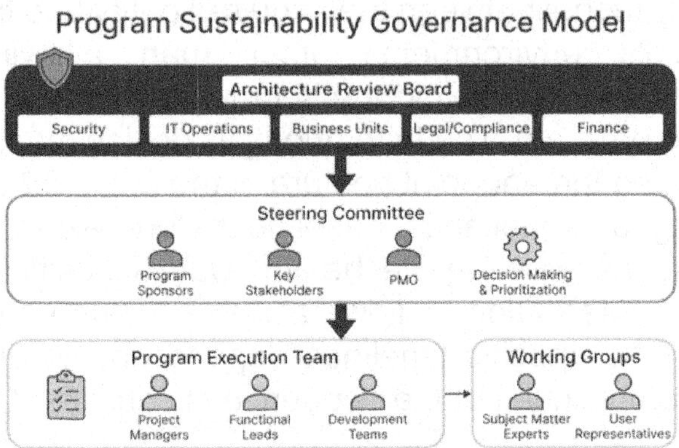

A structured framework for ensuring long-term program viability through strategic alignment, resource optimization, and continuous adaptation.

Budget cycle management requires a specific discipline that differs from the technical program management disciplines that security engineers are typically comfortable with. Annual budget cycles create a recurring moment of vulnerability for multi-year programs because the incremental investment required in year two or three of a zero-trust program is easy to cut when a budget reduction is needed, and the program's benefits are not yet fully realized. The defense against this pattern is a benefit realization calendar that explicitly projects which measurable outcomes will be achieved in the next budget period, calibrated to what the approved investment will buy, so that a corresponding explicit reduction in projected outcome accompanies any reduction in investment.

The benefit realization calendar translates the technical program roadmap into a schedule of business outcomes that leadership can track and validate. If the current-year investment funds the deployment of microsegmentation around the financial systems environment, the benefit realization calendar shows when that deployment will be complete, what it will measurably change about the blast radius of a credential compromise in the financial systems zone, and how that change will be validated. When the next budget cycle arrives, the program manager can point to completed deliverables with validated outcomes rather than projected future benefits. Programs that consistently deliver what they promise, and measure that delivery transparently, are substantially more defensible in budget reductions than programs whose value remains theoretical.

Team continuity is the third dimension of program sustainability, alongside leadership continuity and budget continuity. Zero trust programs develop significant institutional knowledge in the engineering team: the specific quirks of the organization's identity infrastructure, the history of policy decisions and why they were made, the integration patterns that required creative solutions, and the operational muscle memory that allows the team to respond effectively when something breaks at two in the morning. This knowledge is organizational capital, and it depreciates rapidly when engineers leave. Retention strategies for zero-trust program engineers should be explicit, not assumed. Engineers doing interesting, challenging work on a well-resourced program are generally retainable, but retention requires active management rather than assuming that interesting work is automatically sufficient.

11.37 Geeks Only

This section addresses the engineering infrastructure of the zero trust program itself: the systems, processes, and technical disciplines that enable it to operate at scale, rigorously measure its effectiveness, and keep the engineering organization aligned across a multi-year, multi-team transformation. Readers who are primarily interested in the program management and executive communication dimensions may proceed to the Manager's Decision Guide. Engineers who own pieces of the implementation, the security operations center, the identity platform, the network segmentation infrastructure, the cloud security posture, or the automation stack will find in this section the operational frameworks that translate the program architecture into daily engineering practice.

Building the Engineering Scorecard

The engineering scorecard is the operational measurement system that sits below the executive-facing metrics and provides the engineering organization with a shared, objective view of program progress at the component level. Where executive metrics express program outcomes in business risk terms, the engineering scorecard expresses progress in technical performance terms that engineers can directly influence: identity coverage percentage, policy engine decision latency, microsegmentation rule count versus target count, certificate rotation SLA compliance, behavioral baseline coverage percentage, and detection capability scores from quarterly purple team exercises.

Scorecard design must balance comprehensiveness with operational usability. A scorecard with forty metrics is not a management tool; it is a report. The engineering scorecard should include 10 to 15 key indicators, organized by zero-trust pillar, for the team to review weekly. Each indicator should have a current value, a target value, a trend direction, and a red-yellow-green status based on defined threshold bands. The threshold bands require careful calibration: too tight and the team spends all its time managing the dashboard rather than building the architecture; too loose and the scorecard fails to surface real problems before they become program-level concerns.

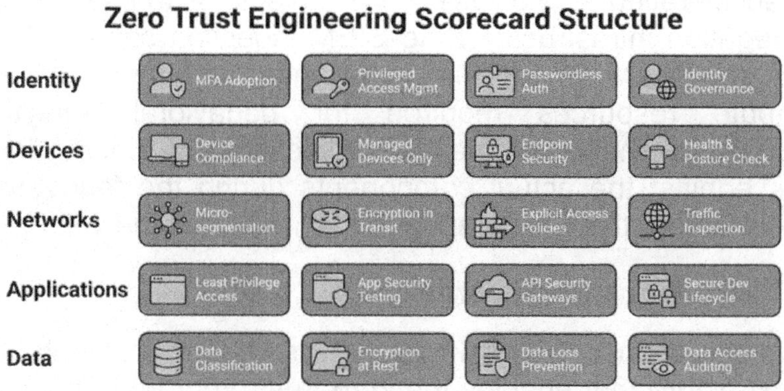

Zero Trust Engineering Scorecard Structure

Capacity Planning for Zero Trust Operations

The zero trust operations model generates significantly more security events, policy decisions, and audit artifacts than the perimeter security model it replaces. A policy engine making 2 million access decisions per day generates 2 million log entries that require storage, indexing, and periodic analysis. A behavioral analytics system maintaining baselines across 50,000 identities is consuming memory and compute resources that must be planned for and provisioned. A microsegmentation environment with ten thousand policy rules requires a change management process that validates every proposed rule change against the full policy set before deployment. None of this operational load is automatic. It requires capacity planning, infrastructure provisioning, and operational runbook development before the architecture is deployed at scale.

The capacity planning process for zero-trust operations begins with event-volume projection. For each architectural component, estimate the event rate at the scale the program will eventually reach: the policy engine decision rate, the identity provider authentication event rate, the endpoint detection event rate, the network flow log volume, and the compliance artifact generation rate. Apply a growth factor of thirty to fifty percent to account for program expansion during the acceleration phase. Use these projections to size the log aggregation infrastructure, the SIEM ingestion capacity, the long-term storage tier for compliance artifacts, and the compute resources required for behavioral analytics processing. Then, validate the projections by running load tests against the actual components during the foundation phase, before the acceleration phase multiplies the volume.

Runbook Engineering

The operational runbook library is the documentation infrastructure that enables the zero-trust program to operate reliably even when the engineers who built it are unavailable. Every routine operation in the zero trust environment should have a runbook: the process for provisioning a new application into the policy engine, the steps for onboarding a new identity provider into the federation layer, the procedure for responding to a policy enforcement incident, the approach for validating microsegment boundary integrity after a network change, and the method for rotating compromised certificates at scale. Runbooks are not just documentation. They are the mechanism by which the program's operational knowledge is preserved, transferred, and improved over time.

Runbook engineering as a discipline means applying the same rigor to operational procedure design that engineers apply to code: version control, peer review, testing against real scenarios, and regular currency review to ensure that the runbook reflects the current state of the environment. Runbooks that are written once and never updated are worse than no runbooks, because they create false confidence that a documented procedure exists while providing guidance that may be actively wrong. The runbook engineering process should include a quarterly review cycle in which every runbook is validated against the current environment by having an engineer who did not write it execute the procedure and document any discrepancies.

Cross-Team RACI

The zero trust program spans multiple engineering teams that have historically operated in separate organizational silos: the identity team, the network security team, the cloud security team, the endpoint security team, and the security operations center. The RACI matrix, specifying who is Responsible, Accountable, Consulted, and Informed for each program function, is the governance artifact that prevents these organizational boundaries from becoming gaps in program execution. Without explicit RACI documentation, ambiguity about who owns the policy between the identity provider and the policy enforcement point leads each team to assume the other is handling it, and neither is.

The most important RACI decisions in a zero-trust program involve the interfaces between teams: who owns the policy that governs access to a resource when the access request originates from an application managed by one team but authenticates against an identity infrastructure managed by another. Resolving these interface questions in the RACI matrix before the architecture is deployed, rather than during an incident response when the policy is failing, and two team leads are having an unproductive argument about whose problem it is, is the organizational discipline that separates effective zero-trust programs from technically competent but organizationally fragmented ones.

Tooling Consolidation and Deprecation Pathways

The zero-trust program will eventually operate alongside a legacy security tooling portfolio whose components are being progressively replaced. The tooling consolidation strategy manages this transition by defining the deprecation timeline for each legacy component, the migration pathway from legacy functionality to zero-trust equivalent, and the criteria that must be met before a legacy tool is decommissioned. Running legacy tools in parallel with zero-trust components for extended periods is expensive and operationally complex. Still, premature decommissioning of legacy tools before their replacement is fully validated creates an availability and security risk. The consolidation strategy defines the criteria that govern when parallel operation ends and deprecation begins.

Tooling Consolidation and Deprecation Decision Flow

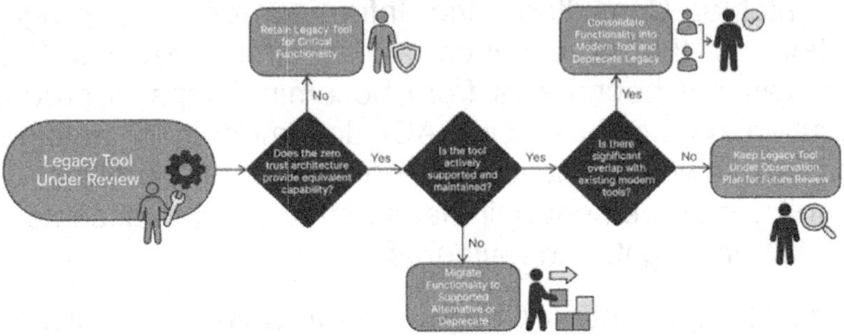

Tooling Consolidation and Deprecation Decision Flow

The Architecture Review Board

The Architecture Review Board is the governance mechanism that prevents the zero-trust architecture from drifting through a series of locally rational but globally incoherent decisions made by individual teams under time pressure. Every significant architectural change in the zero-trust environment should pass through the ARB: new technology acquisitions, changes to the policy-engine rule structure, additions to the microsegmentation topology, modifications to the identity provider federation configuration, and changes to the behavioral analytics baseline methodology. The ARB is not a bureaucratic obstacle. It is the mechanism that catches the interaction effects between components that individual teams cannot see from within their own operational scope.

ARB effectiveness requires a specific operational design. Meetings should be scheduled regularly, at least monthly, with a clear agenda, defined decision-making authority, and documented outcomes. The ARB should have explicit criteria for which changes require full ARB review and which can be approved by a delegated authority. It should have a defined escalation path for urgent changes that cannot wait for the next scheduled meeting. And it should maintain a decision log that records the rationale for significant architectural choices, so that when a new team member joins the program or a leadership transition occurs, the history of why the architecture looks the way it does is accessible rather than lost with the individual engineers who made the decisions.

11.38 Manager's Decision Guide

The decisions that determine whether a zero-trust program succeeds are not primarily technical. They are organizational: how the program is structured, funded, governed, communicated, and sustains itself through the transitions and pressures that every multi-year transformation encounters. The decision criteria below help program leaders make the organizational decisions that determine program sustainability, drawing on the frameworks developed throughout this chapter.

Decision Criteria

- Phase commitment: Fund phases sequentially rather than committing the entire multi-year budget upfront. Each phase exit gate requires measurable outcomes before the next phase investment is authorized. This approach maintains accountability and allows scope adjustment based on what each phase reveals about the environment.
- Staffing model: Determine which zero trust functions require internal ownership and which can be sourced through managed services or consulting. Identity engineering, security operations, and cloud security require internal depth. Architecture design review, red team validation, and 24x7 coverage can be sourced externally without creating capability dependency risk.
- Technology selection timing: Do not select technology before completing the assessment

phase. The assessment phase reveals the specific technical requirements, integration constraints, and operational fit criteria that must drive vendor evaluation. Selecting technology before the assessment is complete guarantees misalignment between what was purchased and what the environment needs.

- Executive communication cadence: Establish the quarterly update and annual program review as fixed commitments before the program launches, not after the first setback occurs. Programs that establish their communication rhythm proactively have dramatically better executive sponsorship retention than programs that report only when there is something to report.
- Governance structure: Stand up the Architecture Review Board before the foundation phase begins. The ARB needs time to develop its operating rhythm and decision authority before it faces the complex trade-off decisions that the foundation phase generates. An ARB established after the foundation phase, which is already underway, is chronically behind.
- Benefit realization tracking: Build the benefit realization calendar at the start of each phase, not at the end. The calendar forces explicit commitment to the measurable outcomes the current phase of investment will produce, making both the program team and leadership accountable for specific outcomes rather than activities.
- Sustainability mechanisms: Before starting the acceleration phase, explicitly audit the program's resilience to three specific scenarios: the departure of the primary technical champion, a twenty-percent budget reduction, and a complete leadership transition above the program team. If the program does not survive any of these scenarios in its current form, the governance and documentation

infrastructure needs further investment before the scope expands.

Decision Frame

When evaluating whether the zero trust program is ready to advance to the next phase, the decision should require affirmative answers to four questions: Have the current-phase milestones been achieved and validated by evidence rather than activity? Has the benefit realization for the current phase been measured and reported to executive leadership? Have the technical dependencies for the next phase been fully assessed, and have the resource requirements been confirmed? Has the Architecture Review Board reviewed and approved the next-phase architecture design? If any of these questions cannot be answered affirmatively, the next phase of investment should wait. Advancing on a weak foundation does not accelerate the program. It imports the foundation's deficiencies into the next phase, multiplying their impact.

11.39 Case Study: NISA's Three-Year Zero Trust Transformation

The National Infrastructure Security Agency entered its zero-trust transformation facing a challenge that many federal agencies recognize: a 15,000-endpoint environment spanning four cloud tenants and twelve field offices nationwide, operating on a mix of modern managed workstations and legacy systems that had accumulated over decades of incremental infrastructure investment. The agency had 223 applications in its portfolio, of which fewer than 40% had documented authentication integration points. Its identity infrastructure was a federation of four separate directory systems that had grown together organically rather than by design. The impetus for the transformation was not a major breach. It was the OMB M-22-09 mandate, which set specific architectural requirements with fiscal year 2024 deadlines and created the congressional reporting obligation that made the program's progress visible to oversight committees for the first time.

The assessment phase lasted six months and produced findings that the program leadership described as "clarifying and uncomfortable." The identity inventory revealed 14,000 orphaned accounts across the four directory systems, a number the operational security team had not known was possible, given the deprovisioning processes nominally in place. The device inventory found that 31% of endpoints connecting to agency resources were either unmanaged or running management agents that had not checked in for more than 90 days. The application inventory discovered that 68 applications used local authentication stores that were entirely outside the scope of the existing single sign-on deployment, meaning that credential compromise on any of those applications would not trigger a signal in the agency's identity monitoring infrastructure. The network inventory found that east-west traffic between field office segments and the headquarters environment flowed through a single legacy firewall configuration with 3,000 rules, of which the current team could explain the business justification for fewer than half.

NISA Three-Year Transformation Timeline

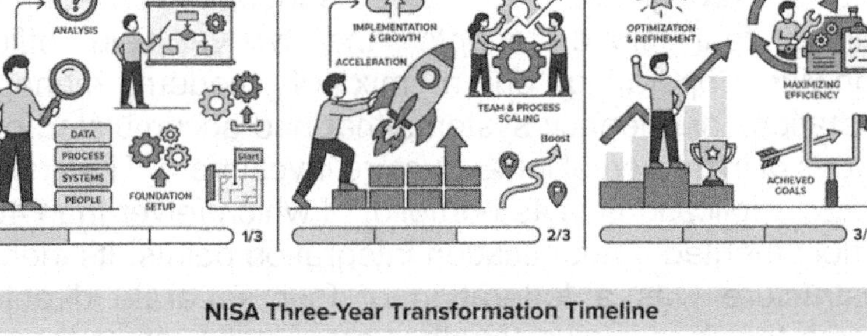

NISA Three-Year Transformation Timeline

The foundation phase ran for eighteen months and covered three primary workstreams in parallel. Identity consolidation reduced the four directory systems to a single authoritative source with federation connectors for the three legacy systems that could not be fully migrated within the phase timeline. Phishing-resistant authentication was deployed to all privileged users in month twelve and to all general staff by month fifteen, with the final field office populations completed in month seventeen. The orphaned account remediation was completed in month eight, generating the single largest measurable risk reduction event of the foundation phase: fourteen thousand non-human, unmonitored access vectors removed from the environment in ninety days. Microsegmentation was deployed around the agency's financial management systems, classified information processing environment, and identity infrastructure. Continuous verification pilots ran in monitor-only mode for 90 days before enforcement was enabled, a decision the program team credits with avoiding access disruption events that would have occurred if enforcement had been enabled immediately.

The foundation phase experienced one significant setback in month twenty-two. A certificate authority migration as part of the identity consolidation workstream caused a three-week service disruption to the legacy authentication integration for 31 field office applications. The disruption did not result in a security incident, but it generated enough operational impact that three senior agency officials formally requested a program review. The program leadership responded with a structured incident analysis, a documented root cause, a remediation plan with specific completion dates, and a revised testing protocol for all future certificate infrastructure changes. The review board accepted the analysis, and the program continued. The incident became the agency's primary internal training case for the importance of staged migration planning and pre-deployment testing in environments with legacy authentication dependencies.

The acceleration phase extended zero-trust controls to the agency's IoT sensor network, which monitored physical infrastructure at the twelve field offices, and to the contractor ecosystem, which included forty-seven firms with personnel whose access patterns ranged from daily to quarterly. IoT integration required deploying network access control at the field office segments rather than treating IoT devices as network-adjacent endpoints. This decision required approval from the Architecture Review Board and a six-month parallel operation period before the legacy DMZ approach was deprecated. The contractor ecosystem integration required building a federated identity onboarding process that allowed contractor identity providers to assert into the agency's policy engine without requiring the agency to own contractor credentials, a design that reduced the credential management overhead per contractor firm from sixteen hours to approximately two hours.

At the end of year three, the agency submitted its OMB M-22-09 attestation, with documentation covering all five pillars of the zero-trust maturity model. The program had operated on a total budget of twenty-eight million dollars across twelve cross-functional teams, and the mean time to detect had reduced from seventy-two hours at baseline to four hours. The average blast radius of simulated breach scenarios, measured through quarterly purple team exercises, had decreased by eighty-nine percent. The compliance evidence automation deployed during the acceleration phase reduced annual audit preparation effort by an estimated 300 person-hours. Three peer agencies formally requested access to the NISA program playbook, which the agency published through an inter-agency sharing agreement. The program director presented the three-year results to the agency's oversight committee as evidence of what a federally funded zero-trust transformation looked like when managed as a program rather than a project: phased, measured, honest about setbacks, and accountable to outcomes rather than activities.

11.40 Team Conversation Guide

Run these three discussions in sequence over two to four weeks. The first surfaces the issue, the second works the trade-offs, the third commits the team to action.

Discussion 1. What Does Our Phase One Actually Need to Prove?

This first discussion pushes the team to be specific about what the assessment and foundation phases are accountable for delivering. Most program teams can describe what they plan to do. Fewer can articulate what they need to prove, in measurable terms, to justify investment in the subsequent phase. Run this conversation at the start of the program or before the next phase authorization meeting.

Discussion questions to ask the team:

- If we complete the assessment phase and find that the most urgent gap is not what we expected, how does that change our foundation phase priorities? Are we willing to reprioritize based on evidence, or have we already committed to a fixed plan?
- What is the one measurable outcome from the foundation phase that would make the case for acceleration phase investment undeniable to our executive sponsor? How do we design the foundation phase to guarantee we can demonstrate that outcome?
- What technical dependencies between zero trust pillars in our specific environment must be satisfied before we can deploy the next pillar? Have we explicitly mapped those dependencies, or are we assuming a generic deployment sequence will work in our environment?

What to listen for: Teams that are deeply attached to a specific tool selection or a particular vendor's deployment model before the assessment is complete have substituted vendor preference for architectural thinking. Teams that can articulate specific, measurable foundation-phase outcomes in business-risk terms are operating at the right level of rigor.

Follow-up actions: After the discussion, have the team document the top three measurable outcomes that the foundation phase must deliver, in specific terms: not "improve identity posture" but "reduce orphaned account count to zero and demonstrate phishing-resistant authentication coverage for all privileged users." Those commitments form the foundation for the exit-gate phase.

Discussion 2. How Do We Evaluate Technology Without Getting Locked In?

This second discussion surfaces the team's implicit assumptions about the technology evaluation process. Zero-trust programs that allow vendor relationships to drive architectural decisions rather than the other way around consistently end up with fragmented architectures that serve vendors' integration stories rather than the organization's security model. Run this conversation before any major technology evaluation begins.

Discussion questions to ask the team:

- If we removed all the vendor briefings and marketing materials from our evaluation process and evaluated candidate solutions exclusively against our documented architectural requirements and integration standards, would our current frontrunner still be in first place?
- What would it cost in engineering effort to replace the tool we are considering if we discovered in three years that it did not perform as expected, or the vendor changed its pricing model significantly? Is that migration cost acceptable given what we would gain in the interim?
- Which of our existing tools are the new zero trust components expected to integrate with, and have we actually tested that integration in our specific environment, or only in a vendor-provided demonstration environment?

What to listen for: Teams that struggle to articulate their architectural requirements independently of vendor capabilities have not rigorously completed the assessment phase. Teams that have tested integrations in their own environment rather than in vendor demonstrations are operating with evidence rather than assumptions.

Follow-up actions: Establish a formal evaluation process that requires hands-on testing in the organization's actual environment, not a vendor proof-of-concept lab, before any significant technology commitment. Assign one engineer the explicit role of "integration skeptic" in every evaluation, responsible for documenting all integration assumptions that have not been tested against production-equivalent systems.

Discussion 3. How Do We Keep the Team Motivated Through Year Three?

This third discussion addresses the program's organizational sustainability at the team level. Multi-year transformations create sustained pressure on the engineers executing them: the repetitive work of extending the same patterns to new systems, the frustration of organizational resistance, and the challenge of maintaining quality standards under delivery pressure. Run this conversation at the start of each program year and whenever team energy appears to be flagging.

Discussion questions to ask the team:

- What is the most technically interesting problem in the next phase of the program, and which team member is best positioned to own that problem? Are we structured to put the right people on the problems that will keep them engaged, or do we default to assignment patterns that optimize for familiarity over growth?
- What visible milestone in the next six months, something the entire organization will notice rather than just the security team, can the program celebrate in a way that makes the team's work legible to the broader organization?
- If one of our key zero trust engineers were recruited away by a competitor offering a twenty-percent

salary increase, what would it take to retain them? Are those retention factors currently in place, or do they require action we have not taken?

What to listen for: Teams whose members cannot identify an exciting technical problem in the current or next phase are telling you the program has become execution-only work without sufficient technical challenge to sustain engagement. Teams that cannot identify a visible organizational milestone within the next six months are operating in a way that makes their work invisible, posing a risk to retention and recognition.

Follow-up actions: Build a six-month celebration calendar at the start of each program year that maps specific program milestones to specific recognition events, visible to the engineering team and to the broader organization. Identify the top two technical challenges in the next phase and explicitly assign the engineers who will benefit most from owning them, treating technical challenge assignment as a retention mechanism rather than an administrative task.

11.41 Takeaway

Two years into the Meridian Logistics Group's zero-trust program, Marcus Reyes sat across from a new CIO and presented evidence of a transformation in progress. He had numbers that told a coherent story about risk reduction. He had a roadmap that connected every architectural decision to a measurable outcome. He had an organization capable of building what the architecture required, sustaining it through setbacks, and explaining its value in language that leadership could act on. He had not finished. Three more years of work lay ahead. But the program was structured to deliver those years regardless of which individuals held which roles, because the governance, documentation, measurement cadence, and stakeholder network were embedded in the organization rather than dependent on any individual.

That is what this book has been building toward over twelve chapters: not a technology checklist, not a vendor evaluation guide, not an aspirational architecture diagram, but the complete picture of what it means to build a zero-trust program that actually works in the operational reality of a modern hybrid enterprise. The identity architecture addressed in the early chapters is the foundation. The microsegmentation and continuous verification patterns in the middle chapters are the structure. The AI-assisted detection, IoT and OT integration, data-centric security, and automation frameworks in the later chapters are the advanced capabilities that a mature program progressively incorporates. This chapter is the organizational and programmatic scaffolding that holds all of it together across the multi-year timeline that genuine transformation requires.

The organizations that complete zero trust transformations share a commitment to treating security architecture as a program discipline rather than a project discipline. They measure outcomes rather than activities. They communicate in the language of business risk rather than technical specifications. They invest in the governance structures and team capabilities that sustain the program through the inevitable disruptions of organizational life. And they approach the work with the intellectual honesty to distinguish between what the architecture currently provides and what it will eventually provide, committing to the journey rather than claiming the destination before it has been reached.

The Closing section of this book addresses the question that every completed chapter has been building toward: what does a zero-trust future actually look like for the organizations that build it, and what does the security profession need to become to design, operate, and defend that future effectively? The architectural patterns, program frameworks, and organizational disciplines developed across these twelve chapters form the foundation for that future. The work of building it starts with the assessment phase. And the assessment phase starts with the honest question that Marcus Reyes asked on a Monday morning two years ago:

Where are the real trust boundaries, and what happens when they fail?

12 Closing

Marcus Reyes walked into the Monday morning incident review four months into his role at Meridian Logistics Group with a cup of coffee and a reasonable expectation of a quiet week. He walked out of it with a problem he could not hand off, a root cause that pointed directly at architectural assumptions, and the beginning of a multi-year program to rebuild trust as something that is verified rather than inherited. That arc from reckoning to a working program is what this book has traced, chapter by chapter, through the decisions that define a real zero-trust implementation.

The three pillars that organized this journey are not abstract principles. They are the operational levers that determine whether a zero trust program actually reduces risk or merely satisfies a checklist. Identity verification made trust explicit at Meridian, where it had always been implicit. Microsegmentation converted a flat, traversable interior into a set of bounded zones where lateral movement runs into enforcement, not open corridors. Continuous verification transformed authentication from a one-time event into a persistent process that adjusts access dynamically as context changes. These three pillars do not function sequentially; they reinforce each other simultaneously, and the compounding effect of all three working together converts a security architecture investment into workflow-level impact across the organization.

The federal compliance dimension of this work is not a bureaucratic overlay on top of the architecture. EO 14028, OMB M-22-09, NIST 800-207, FISMA, and CMMC describe zero trust requirements in terms that align closely with the architectural logic the book has developed throughout. Organizations that build the architecture for sound security reasons find that the compliance posture follows. Organizations that build toward compliance without the architectural foundation find that they satisfy requirements on paper while leaving the underlying risk largely unchanged.

The distinction between those two outcomes is between operational clarity and compliance theater, and it is one that auditors, federal reviewers, and breach post-mortems will eventually expose.

The momentum behind zero trust adoption is not slowing. Federal mandates are tightening their timelines and their specificity. Commercial frameworks are raising their access control requirements with each revision cycle. The organizations that begin this work now, even imperfectly, are building institutional knowledge and architectural muscle that will compound in value as the requirements continue to evolve. The organizations that wait are not just delaying a compliance deadline; they are deferring the learning that makes adaptation possible. A program that follows the pilot-to-production pathway described throughout these chapters builds that learning incrementally, running controlled pilots that surface real friction before committing to enterprise-wide enforcement, so that production evidence rather than vendor projections informs each phase of expansion.

The threat landscape is also changing in ways that make zero trust architecture more necessary, not less. AI-era attack capabilities are lowering the cost and increasing the sophistication of credential compromise, phishing, and social engineering. The attacker who reached Meridian's cloud tenant through a contractor account with valid credentials represents a pattern that is becoming more common, not less. Identity-first architecture is the response to that pattern, because it removes network location as a trust signal and replaces it with verified identity, device health, and session context. When the attacker's stolen credentials reach an identity-aware policy enforcement point rather than a flat trusted interior, the blast radius shrinks from a network to a session.

Mission-aligned security programs share a characteristic that distinguishes them from compliance-driven ones: they are built to protect what the organization actually does, not just to satisfy a framework's requirements. Meridian's supply chain data, its contractor ecosystem, its hybrid infrastructure spanning on-premises data centers and cloud tenants: these are not generic enterprise assets. They are the operational substrate of a specific business, and the zero-trust controls that protect them need to be calibrated to that specificity. The generic zero-trust reference architecture provides the framework; the mission-aligned implementation fills it with the organization's actual risk surface, user workflows, and operational constraints.

Responsible by design means that security controls are architectural decisions, not afterthoughts. It means that identity verification is built into the access model from the first pilot, not added during remediation. It means that microsegmentation boundaries are drawn with knowledge of application dependencies, not applied broadly and tuned after production impact. It means that continuous verification policies are calibrated against real user behavior and real session risk, not set to maximum friction and gradually relaxed in response to complaints. The responsible design posture requires more deliberate up-front work. Still, it produces implementations that are both more secure and more sustainable than those assembled under emergency pressure or a compliance deadline.

By the time Marcus's team completed the first production phase of Meridian's zero-trust program, the Monday-morning incident review had become a different kind of meeting. The anomaly that had once been invisible behind a trusted perimeter was now generating alerts at the policy enforcement layer, where access attempts that did not match verified identity and session context were denied and logged before they reached a protected resource. The contractor ecosystem, once the weakest link in the perimeter model, was now operating under the same identity and access controls as the internal workforce. Segmented zones had replaced the flat interior that had enabled 12 hours of undetected lateral movement, imposing enforced boundaries. The program was not finished; zero-trust programs do not finish. But it had reached a state where the architecture was working as designed, and the security posture was measurably better than the one Marcus had inherited.

The architectural responsibility that zero trust places on practitioners and decision-makers is real. It requires sustained investment, organizational alignment, and the willingness to redesign workflows that people have been using for years. It also produces organizations that are more resilient, more adaptable, and better positioned for whatever the next decade of threat evolution brings. The journey Marcus began at Meridian is not unique to Meridian. It is the journey that every hybrid enterprise running on implicit trust will need to take, and the organizations that begin it deliberately, with operational clarity about where they are starting and where they are going, will be better positioned at every stage. Build the architecture. Verify the identity. Enforce the boundaries. Re-evaluate the trust. Do it continuously and by design.

13 References

Accenture. *Zero Trust: Rethinking Security for the Cloud-First Enterprise*. Accenture, 2024. (CH1, CH8)

Amazon Web Services. *Building Zero Trust Architectures on AWS*. Amazon Web Services, 2023. (CH3, CH4, CH6)

Baker, Thomas, and Priya Nair. "Operationalizing Zero Trust in Large Enterprises." *IEEE Security & Privacy*, vol. 21, no. 5, 2023, pp. 45–55. (CH4, CH8)

Cisco Systems. *Zero Trust Security: A Comprehensive Approach to Secure Access in Hybrid Environments*. Cisco Press, 2023. (CH2, CH3, CH4)

Cybersecurity and Infrastructure Security Agency. *Zero Trust Maturity Model, Version 2.0*. U.S. Dept. of Homeland Security, 2023. (CH2, CH3, CH8)

Cybersecurity and Infrastructure Security Agency. *Applying Zero Trust Principles to Enterprise Identity and Access Management.* U.S. Dept. of Homeland Security, 2024. (CH3, CH6)

CrowdStrike. *Zero Trust in Practice: Identity, Endpoints, and Workloads*. CrowdStrike, 2024. (CH3, CH5, CH6)

Deloitte. *Zero Trust Reality Check: Lessons from Enterprise Implementations*. Deloitte Insights, 2024. (CH1, CH4, CH8)

ENISA. *Zero Trust Architecture: Security Considerations for the EU*. European Union Agency for Cybersecurity, 2023. (CH2, CH4, CH9)

European Union Agency for Cybersecurity. *Micro-Segmentation and Zero Trust in Cloud and 5G Environments*. ENISA, 2024. (CH4, CH5, CH7)

Forrester Research. *The State of Zero Trust, 2024*. Forrester Research, 2024. (CH1, CH2, CH8)

Gartner. *Zero Trust Network Access: Strategic Roadmap for 2025*. Gartner, 2024. (CH3, CH4, CH8)

Google Cloud. *BeyondCorp Enterprise: Zero Trust for Modern Workforces*. Google Cloud, 2023. (CH3, CH4, CH6)

Google Cloud. *Implementing Zero Trust in Multi-Cloud and Hybrid Environments*. Google Cloud, 2024. (CH4, CH7, CH8)

IBM Security. *Zero Trust Blueprint: A Practical Guide for Enterprise Security Architects*. IBM, 2023. (CH2, CH4, CH8)

International Organization for Standardization. *ISO/IEC 27002:2022 and Zero Trust Controls Mapping Guide*. ISO, 2023. (CH4, CH5, CH9)

International Organization for Standardization. *ISO/IEC 27557:2024 Zero Trust Architecture—Concepts and Requirements*. ISO, 2024. (CH2, CH4, CH9)

Johnson, Erica, and Luis Romero. "Zero Trust for OT and Critical Infrastructure." *Journal of Cybersecurity*, vol. 10, no. 2, 2024. (CH5, CH7, CH9)

KPMG. *From Perimeter to Principle: Zero Trust as an Operating Model*. KPMG, 2023. (CH1, CH2, CH8)

McAfee Enterprise. *Securing the Everywhere Enterprise with Zero Trust*. McAfee, 2023. (CH3, CH4, CH6)

Microsoft. *Zero Trust Adoption Report 2024*. Microsoft, 2024. (CH1, CH3, CH8)

Microsoft. *Zero Trust Guidance Center: Identity, Devices, and Data*. Microsoft, 2023. (CH3, CH5, CH6)

National Institute of Standards and Technology. *Zero Trust Architecture (SP 800-207)*. NIST, updated ed., 2023. (CH1, CH2, CH4)

National Institute of Standards and Technology. *A Zero Trust Architecture Model for Access Control in Cloud-Native Applications in Multi-Cloud Environments (SP 800-207A)*. NIST, 2023. (CH3, CH4, CH7)

National Institute of Standards and Technology. *Implementing a Zero Trust Architecture (SP 1800-35)*. NIST, 2025. (CH4, CH6, CH8)

National Security Agency. *Advancing Zero Trust Maturity throughout the Enterprise*. National Security Agency, 2023. (CH2, CH4, CH8)

Okta. *Identity at the Center of Zero Trust: 2024 Trends Report*. Okta, 2024. (CH3, CH6)

Palo Alto Networks. *Zero Trust Network Security for Cloud-Native Applications*. Palo Alto Networks, 2023. (CH4, CH5, CH7)

Palo Alto Networks. *Practical Microsegmentation for Zero Trust Architectures*. Palo Alto Networks, 2024. (CH4, CH5)

PwC. *Zero Trust as a Business Transformation Lever*. PwC, 2024. (CH1, CH2, CH8)

SANS Institute. *Zero Trust Architecture: Design Patterns and Anti-Patterns*. SANS Institute, 2023. (CH2, CH4, CH8)

SANS Institute. *Implementing Zero Trust for Remote Workforces*. SANS Institute, 2024. (CH3, CH6, CH7)

Symantec Enterprise Division. *Data-Centric Zero Trust: Protecting Information in Motion and at Rest*. Broadcom Software, 2023. (CH5, CH6)

Thompson, Rachel, and Omar El-Sayed. "Zero Trust and Continuous Authorization in Cloud-Native Systems." *ACM Queue*, vol. 21, no. 1, 2023. (CH4, CH7)

U.S. Department of Defense. *Zero Trust Strategy and Implementation Roadmap: Progress Update 2024*. U.S. Dept. of Defense, 2024. (CH2, CH4, CH9)

World Economic Forum. *Zero Trust Cybersecurity: A Framework for Resilient Digital Economies*. World Economic Forum, 2023. (CH1, CH2, CH9)

Zscaler. *Zero Trust for Modern Enterprise Architectures: From VPN to ZTNA*. Zscaler, 2023. (CH3, CH4, CH7)

Made in the USA
Monee, IL
14 September 2026

63117174R00252